DATE			

Practical Pharmaceutical Chemistry

Practical Pharmaceutical Chemistry

Third Edition, in Two Parts

by

A. H. BECKETT
Ph.D., D.Sc., F.P.S., F.R.I.C.
Professor of Pharmaceutical Chemistry
Chelsea College of Science and Technology
University of London

and

J. B. STENLAKE
Ph.D., D.Sc., F.P.S., F.R.I.C., F.R.S.E.
Professor of Pharmacy and Pharmaceutical Chemistry
University of Strathclyde, Glasgow

Part One

THE ATHLONE PRESS *of the University of London* 1975

Published by
THE ATHLONE PRESS
UNIVERSITY OF LONDON
4 *Gower Street London WC* 1

Distributed by Tiptree Book Services Ltd
Tiptree, Essex

U.S.A. and Canada
Humanities Press Inc
New Jersey

© *A. H. Beckett and J. B. Stenlake* 1975

ISBN 0 485 11156 x

Set in 10/11 *pt Linotron Times*
at The Universities Press, Belfast,
and printed by photolithography
in Great Britain by J. W. Arrowsmith Ltd, Bristol

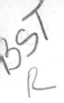

Preface to the Second Edition

The rapid development of chemical, physico-chemical and instrumentation techniques since the publication of the first edition, and their widespread application in pharmaceutical analysis, together with the introduction of many new drugs, has necessitated a complete revision of the book. The selection of new material for inclusion has not been easy, and in order to contain the book within a reasonable compass, and in an attempt to meet the differing needs of students at elementary and advanced levels, the 2nd Edition of Practical Pharmaceutical Chemistry has been produced in two parts.

Part 1, which is concerned solely with general analytical methods, provides an introduction to Pharmaceutical Analysis in a completely revised chapter on the general principles of quality control. Part 1 is also intended to provide the basis for more elementary instructional laboratory classes in pharmaceutical analysis. The selection of examples for inclusion has been based on the current editions of the British Pharmacopoeia, and its Addenda, and whilst coverage is, intentionally, not comprehensive, a wide selection of examples has been achieved.

As in the first edition, a number of accepted pharmaceutical conventions and certain pharmacopoeial practices have been adopted to avoid confusing the student should he wish to refer to the Pharmacopoeia or other texts. In particular, substances which are the subject of official monographs in the Pharmacopoeia are indicated by the use of initial capital letters, e.g. Sodium Chloride. For similar reasons, we have continued to express volumes as litres and millilitres, and to ignore the recent I.S.O. ruling on dm^3 as the unit of volume. The widespread use of demineralised water as an alternative to distilled water has also been recognised and the text amended appropriately by the adoption of the term Water (spelt with an initial capital letter) to indicate either, in distinction to tap water.

Part 2 is intended to cater for the needs of the more advanced students. It covers the physical and instructional methods of analysis included in the first edition, extensively revised and expanded, and in addition includes a separate treatment of physical criteria of pharmaceutical chemicals, particle size analysis, chromatography, infrared spectrometry, nuclear magnetic resonance spectrometry, mass spectrometry and radiochemistry. We are greatly indebted to our colleagues for these expert contributions, which are separately acknowledged at the head of each chapter. The extent to which the treatment of these subjects has been linked to examples of the British Pharmacopoeia has been deliberately limited, as it is hoped there may well be other users of Part 2 for whom this aspect is unimportant. Differences in the emphasis on pharmacopoeial requirements also reflect the extent of progress in the application of the various techniques to pharmaceutical analysis at the time of publication. It is felt, however, that the potential use for such techniques as

nuclear magnetic resonance spectroscopy, and mass spectrometry, warrants
the inclusion of chapters at this stage.

A. H. B.
J. B. S.

Preface to the Third Edition

Since the publication of the second edition in 1968, there has been a number of
important developments affecting the practice of pharmaceutical analysis in
the United Kingdom. These have included the passing into Law of the
Medicines Act, 1968, the entry of the United Kingdom into the European
Common Market, the publication of the European Pharmacopoeia in a series
of volumes and supplements commencing in 1969, and the publication of the
British Pharmacopoeia.

The Third edition endeavours to reflect the repercussion of these developments. Accordingly, the requirements for drug and product assessment imposed by the Medicines Act, 1968, and the implications for the analyst and
compiler of product applications form the subject of an entirely new Chapter.
For similar reasons, we have also greatly extended our treatment of general
analytical procedures applicable to pharmaceutical dosage forms in another
entirely new Chapter. This covers all the important product types including
pressurised aerosols, tablets, capsules, injections, ointments, creams,
lozenges, mixtures, solutions, suspensions, suppositories, and relevant analytical developments relating to particle size control, dissolution testing of
bioavailability and the examination of sustained release preparations.

In other respects, the character of the book has been largely retained,
though with a number of deletions, minor amendments, and additions which
reflect trends and developments in pharmaceutical analysis over the last few
years.

A. H. B.
J. B. S.

Part 1

General Pharmaceutical Chemistry

Revised by
J. B. STENLAKE
D.Sc., Ph.D., F.P.S., F.R.I.C., F.R.S.E.

with contributions by
W. D. WILLIAMS
B.Pharm., Ph.D., F.P.S., A.R.I.C.

Acknowledgements

To the National Physical Laboratory for permission to use Fig. 9 and Table 13 which are taken from their publication *Notes on Applied Science*, No. 7, H.M.S.O., 1954; to the British Pharmacopoeia Commission for permission to reproduce Fig. 35; and to the staff of the Medicines Division, Department of Health and Social Security for help and advice in the preparation of Chapter 2. In particular, our most grateful thanks are due to Sylvia Cohen for a wealth of assistance in the preparation and correction of the manuscript, in proof reading, and in the most demanding and boring job of all, the preparation of the index; without her help, so generously given, the completion of this edition would just not have been possible.

Contents

1 Chemical Purity and its Control

Chemical purity implies freedom from foreign matter. A state of absolute purity is virtually unobtainable, but may be approached as closely as desired, provided sufficient care is taken during the manufacturing process. However, the high costs attendant upon the attainment of the highest standards of purity may render the process economically unsound, so that in practice it is often necessary to strike a balance in order to obtain a product, at reasonable cost, which is sufficiently pure for all pharmaceutical purposes. Thus it would be impracticable to lay down standards for drugs and medicinal substances which would provide for the complete absence of even any one impurity.

The standards demanded of chemicals for pharmaceutical use are determined by a number of factors, which take account of the impurities likely to arise as a result of all known methods of manufacture. The essential criterion is safety in use; particular attention is directed towards impurities which might be toxic or which are capable of causing chemical interference when the substance is formulated (i.e. compounded to give a medicinal preparation in a form suitable for administration). The general stability of the substance is also important; thus if it is chemically unstable, hygroscopic or, alternatively, should it effloresce, then standards must be so adjusted that materials stored with reasonable care and for reasonable periods of time will still comply with the imposed requirements. Standards must guard against the possibility of either accidental contamination or intentional adulteration of pharmaceutical chemicals.

THE SOURCE OF IMPURITIES IN PHARMACEUTICAL CHEMICALS

A knowledge of those impurities which occur in pharmaceutical substances in general use is readily available from actual batch analyses. Experience in the manufacture of any one particular substance often shows that not all the expected impurities are present in practice. But for a substance newly available it is important that one should be able to deduce the impurities with which it is likely to be contaminated. A list of the possible impurities can be readily compiled from a knowledge of the raw materials used, the method of manufacture and the stability of the product.

Raw Materials

Impurities known to be associated with these chemicals may be carried through the manufacturing process and contaminate the final compound. For example rock salt contains small amounts of calcium sulphate and magnesium chloride, so that sodium chloride prepared from this source will almost certainly contain traces of calcium and magnesium compounds.

The Method of Manufacture

This may introduce new impurities due to contamination by reagents and solvents at various stages of the process as described below:

(a) *Reagents employed in the process.* 'Soluble alkali' in Calcium Carbonate arises from sodium carbonate used in the process. Calcium Carbonate is prepared by interaction of a soluble calcium salt and a soluble carbonate, and the product is therefore liable to contain traces of soluble alkali, which the washing process has failed to remove.

(b) *Reagents added to remove other impurities.* Potassium Bromide is liable to contain traces of barium, which is added in the course of the manufacturing process to remove excess sulphate which in turn arises from the bromine used.

(c) *Solvents.* Water is the cheapest solvent available and is used whenever possible, especially in the manufacture of inorganic chemicals. The use of tap water on a large scale can give rise to trace impurities of Na^+, Ca^{2+}, Mg^{2+}, CO_3^{2-}, Cl^- and SO_4^{2-}. Softened water, although less liable to produce contamination, with Ca^{2+} and Mg^{2+}, is more liable to produce Na^+ and Cl^- as impurities, as a result of the usual chemical water-softening processes. These difficulties do not arise with the use of distilled or demineralised water (Purified Water B.P.—denoted as Water in this text), though economic factors often preclude their use on a large scale.

(d) *The reaction vessels.* These may be metallic, such as copper, iron, galvanised iron (which gives rise to Zn^{2+} impurity), lead, but nowadays more usually stainless steel. Solvent action may cause corrosion so that traces of metal ions tend to pass into solution and contaminate the product. Glass vessels similarly may give up traces of alkali to the solvents, though this is unlikely if the vessels are of hard glass.

Atmospheric Contaminants

Widespread atmospheric pollution causes considerable risk of contamination by dust (aluminium oxide, silica, sulphur, soot, etc.), sulphur dioxide, hydrogen sulphide, and arsenic. Carbon dioxide and water vapour are possible atmospheric contaminants of products which are susceptible to their action, and which have been manufactured or stored (see below) under less than ideal conditions.

Manufacturing Hazards

Even in a well-run manufacturing house, certain hazards exist which can lead to product contamination. A well-established system of checks on manufacture and insistence on adequate analytical control of the product are usually sufficient to ensure that standards are well-maintained. Control analysts will be aware of the more likely hazards, but even so, specifications for drugs and formulated products should be designed to exclude contamination ranging from the accidental inclusion of particulate matter to microbial contamination, or even labelling errors.

(a) *Particulate contamination.* The presence of unwanted particulate matter can arise in a number of ways, such as accidental inclusion of dirt, or glass, porcelain, metallic and plastic fragments from sieves, granulating, tabletting and filling machines, or even from product containers. This type of contamination may stem from either bulk materials used in the formulation or from the use of improperly cleaned equipment or containers, but is more likely to be the result of particles being shed through wear and tear of equipment. Clarity of solutions for injection is particularly important.

An unusual example of such contamination is that of metal particles which have been found in eye ointments packed in metal tubes. In this case, the method of tube manufacture gives rise to metal splinters which cannot be detached from the inside of the tube by any of the usual washing and cleaning procedures. Some of the particles, however, are extruded with the ointment, the number of the particles so extruded increasing with the viscosity of the ointment. Both tin and aluminium tubes yield metal particles, though particles from the former tend to be larger than those from the latter.

(b) *Process errors.* Gross errors arising from incomplete solution of a solute in a liquid preparation should be detected readily by the normal analytical control procedures. Minor errors, however, could escape notice, if the manufacturing tolerance for the quantity of active ingredient in the product is wide, and analysis indicates that the lower limit has only just been reached. Whilst errors of this sort are undesirable, they are probably only of serious concern in the case of solutions of potent medicaments. The preparation of such solutions, therefore, calls for special precautions, such as filtration, to avoid the danger of undissolved solute contaminating part of the batch. Uneven distribution of suspended matter during manufacture can, similarly, become the source of batch variation, or even variation within a batch.

Failure to acknowledge the limits of mechanical efficiency of mixing, filling, tabletting, sterilising and other equipment can lead to minor variation, and very occasionally gross error, in the content of both active ingredient and formulatory adjuncts in compounded products. The mixing of powders of significantly different particle size or density calls for special care, and still further care in transfer to avoid separation in later stages of manufacture. Special care is essential to avoid mixing and filling errors in the preparation of low dosage (5 mg and less) forms, such as tablets and capsules of highly potent medicaments, and analytical standards and control procedures for such preparations should be appropriately designed to exclude all but minor variation in the individual dosage units.

Decomposition, and loss of potency, through insufficient attention to chemical stability of pharmaceutical chemicals or additives under the conditions of manufacture represent a further source of impurity. Careful attention to process design, and to the handling of compounds liable to decompose, for example, by hydrolysis, oxidation, or in the case of optically-active drugs such as Ergometrine Maleate and Adrenaline Acid Tartrate, by racemisation, is essential if products are to conform to the highest standards. The use of ionising radiations for sterilising pharmaceutical products may also lead to decomposition with consequent loss of the medicament and possible formation of toxic breakdown products.

(c) *Cross-contamination.* The handling of powders, granules and tablets in large bulk frequently creates a considerable amount of air-borne dust, which, if not controlled, can lead to cross-contamination of products. The danger is well-known to experienced manufacturers, particularly of steroidal and other synthetic hormones. Precautions, such as the use of face-masks and special extraction equipment, already in use to protect operators from undesirable effects of certain drugs of this type, are also suitable for more general use to limit cross-contamination. Manufacturers of penicillin preparations in the United States are required by the Food and Drugs Administration (FDA) to institute adequate control of the manufacture, handling and storage of drugs and their preparations to limit cross-contamination of one by another. The application of special limit tests places a check on contamination by penicillin of other products manufactured on the same premises.

(*d*) *Microbial contamination.* The pharmacopoeial requirement of sterility tests for all products intended for parenteral administration and ophthalmic preparations, irrespective of whether they are prepared by end-sterilisation processes or produced under aseptic conditions, provides an adequate level of control for such preparations. Many other products, especially liquid preparations and creams for topical application to broken skin, or mucous membranes, are liable to bacterial mould and fungal contamination from the atmosphere (or, less frequently, from contaminated equipment) during manufacture. A few materials are self-sterilising, but many products capable of supporting microbial growth require the addition of suitable antibacterial or anti-fungal agents if microbiological spoilage of the product is to be completely avoided. Certain materials, which are particularly prone to microbial contamination, may constitute a health hazard unless they are carefully controlled. These are mainly substances of natural origin, which are known to be liable to contamination, usually with specific organisms. The most satisfactory control, therefore, is one which sets a requirement for freedom from specified microbial contaminants (Table 1).

Table 1. Recommended control of microbial contamination in natural pharmaceutical substances

---

Freedom from Salmonellae
    Acacia; Senna; Tragacanth
Freedom from Salmonellae and Escherichia coli
    Cochineal; Digitalis; Gelatin; Pancreatin; Starch; Thyroid
Freedom from Pseudomonads
    Aluminium Hydroxide Gel; Dried Aluminium Hydroxide Gel; Aluminium Phosphate Gel

---

(*e*) *Packing errors.* Products of similar appearance, such as tablets of the same size, colour and shape, packed in similar containers can constitute a potential source of danger through mis-labelling of either or both. The handling of two such products in proximity should be avoided, and control procedures should guard against the possibility of such mishaps.

Inadequate checks on the issue of labels, on the filling of labelling machines, on the setting of ampoule or other printing machines, and the destruction or return to stock of unused labels also constitutes a major packaging hazard. Such mis-adventures can only be avoided by care in manufacture with particular attention to detail and cross checks in the matter of stock records, process dockets, and batch marking of both raw materials and finished products.

Inadequate Storage

(a) *Filth*. Stored products may become contaminated with dust, the bodies of insects, and even animal and insect excreta, unless adequate precautions are taken. Modern packaging is usually capable of excluding all such contamination from finished pharmaceutical products, but bulk storage of raw materials, and especially that of vegetable drugs, is far more likely to lead to such contamination, and an appropriate test for *filth* is desirable for all materials so stored.

(b) *Chemical instability*. Impurity can also arise during storage as a result of chemical instability, and a number of pharmaceutically-important substances are known to undergo chemical decomposition when stored under non-ideal conditions. The nature of the decomposition, which is often catalysed by light, traces of acid or alkali, air oxidation, water vapour, carbon dioxide and traces of metallic ions can frequently be predicted from a knowledge of the chemical properties of the substance. The accent is on limitation or, if possible, total avoidance of all such decomposition, by adopting suitable storage procedures and conditions. Thus, light-sensitive materials should be stored in darkened glass vessels or metal containers to inhibit photochemical decomposition. An interesting example of special precautions to exclude light is the use of opaque capsule shells to protect Chlordiazepoxide from decomposition by light. Material liable to oxidation or to attack by moisture or carbon dioxide may, if it is especially sensitive, require displacement of air from the container by nitrogen; for less sensitive products, it is usually sufficient to prescribe storage in a *sealed container*.

Oxidation is also preventable by the inclusion of appropriate antioxidants, including such phenols as Butylated Hydroxyanisole (BHA), Butylated Hydroxytoluene (BHT) and Thymol, which are capable of undergoing free radical oxidation at the expense of the material (e.g. Liquid Paraffin) they are being used to protect. Sodium metabisulphite, another antioxidant, is suitable for use in aqueous solution (e.g. Procaine and Adrenaline Injection), depending for its action on the oxidation which, as sodium bisulphite, it undergoes to sodium bisulphate. Exceptionally, suitable additives may be used to neutralise the toxic products of oxidation as in the stabilisation of Chloroform. When pure it readily forms traces of phosgene by a reaction which is light catalysed. In practice, therefore, a stable and non-toxic product is obtained by the addition of 1 or 2 per cent of ethanol, which suppresses the decomposition reaction, and at the same time converts any traces of phosgene to the harmless ethyl carbonate.

(c) *Reaction with container materials*. The possibility of reaction between container and contents constitutes a hazard which cannot be ignored. Creams and ointments liable to react with metal surfaces, e.g. Salicylic Acid Ointment, must not be packed in metal tubes, unless they have been lacquered internally to inhibit reaction. Solutions of alkali-sensitive materials, particularly if subjected to any form of heat treatment during preparation, e.g. Atropine Sulphate Injection, which is sterilised by autoclaving, must be packed in glass ampoules, which comply with the test for *hydrolytic resistance*, European Pharmacopoeia (E.P.). This is the resistance offered by the glass to release of soluble mineral substances into freshly distilled water in the container, and is determined by titration. Glass containers are recognised in three grades, I, II and III. Type I is neutral glass, with high hydrolytic resistance arising from the composition of the glass. Type II glass also has high hydrolytic resistance, but this is due to surface treatment of the glass. It can be distinguished from Type I in a crushed-glass test. Type III glass has somewhat more limited hydrolytic resistance than glass of Types I and II.

All glass containers for injectable preparations must comply with the test for hydrolytic resistance. Aqueous injectable solutions must be packed in containers consisting of glass either of Type I or II. Type III glass may be used only for non-aqueous solutions and injection solids which are stable in this standard of glass.

Plastic containers and closures require careful evaluation, because of the tendency to yield undesirable additives, such as plasticisers, a tendency which increases markedly in the presence of non-aqueous solvents. The plastics materials, mainly polyethylene, polypropylene, polystyrene and polyvinyl chloride, are used in conjunction with appropriate additives, which may consist of antioxidants, colours, antistatics, plasticisers, lubricants, impact modifiers, and mould release agents. Not all these additives are used in the manufacture of any one type of container, but plastic containers for pharmaceutical use irrespective of their composition must be free from undesirable properties which affect the safety and efficacy of the medicament.

Plastic containers for injections should be sufficiently translucent to permit visual inspection of the contents, and if greater than 500 ml capacity, must also comply with tests which limit animal *Toxicity* in the cat, for *Ether-Soluble Extractive*, and for *Metal Additives* with special reference to barium and to the heavy metals tin, cadmium and lead.

Rubber closures, widely used in the packing of multidose injections, are, on the other hand, more prone to absorb medicaments, and also antioxidants and bactericides from solution, unless suitably pretreated by immersion in solutions of the compounds concerned. They should comply in general with the British Standard 3263:1960 for Rubber Closures for Injectable Products, as modified by the requirements of the British Pharmacopoeia.

(*d*) *Physical changes.* The occurrence of change in physical form of the drug during storage is not unknown. Changes in crystal size and form, agglomeration and even caking of suspended particles, which are not always preventable, may lead to marked changes in the efficiency of the product. Thus particle size, and consequently surface area, may be a critical factor in determining the rate of absorption, and hence blood levels, of a drug of low solubility, such as Griseofulvin. Multidose suspensions, which are inefficient through rapid settling or claying, likewise constitute a safety hazard, giving rise first to the possibility of underdosage and later to overdosage, as successive doses are withdrawn from the container. Injectable emulsions in which the globule size has increased in storage may similarly be dangerous in that they could be the cause of fat embolism.

(*e*) *Temperature effects*. The rate at which chemical and also physical change occurs in stored products is conditioned by temperature, and labile products may have temperature storage requirements assigned to them to guard against unnecessary decomposition. The use of subjective instructions such as *store in a cool place* are capable of ambiguous interpretation on the part of the operator, and more precise interpretation are generally placed on these phrases (Table 2).

Table 2. Recommended storage conditions

Instruction	Interpretation
Store in a Refrigerator (U.S.N.F.)	Temperature between 2 and 8°
Store in a Cold Place (U.S.N.F.)	Temperature not more than 8°
Store in a Cool Place (B.P.)	Temperature between 10 and 15°
Store in a Cool Place (U.S.N.F.)	Temperature between 8 and 15°
Protect from Heat	Temperature not more than 30°
Store in a Dry Place	Relative Humidity less than 5 per cent

Products required for tropical use, must, if prepared under temperate conditions, be capable of withstanding the effects of change to tropical temperatures. Thus, pastes and ointments must be so formulated as to retain their consistency at higher temperatures, and appropriate standards laid down for them. The high humidity often encountered in tropical countries also constitutes yet another hazard in that it is conducive to mould contamination of the product, and suitable precautions may be necessary in formulation.

STANDARDISATION OF PHARMACEUTICAL CHEMICALS AND FORMULATED PRODUCTS

Official standards for pharmaceutical chemicals and formulated products are designed primarily to set permissive limits of tolerance for the product at the time it reaches the patient. They do not necessarily equate with manufacturing specifications, since they must also take into account possible degradation of the product during its shelf-life up to the time it is prescribed and ultimately used. Official standards must also encompass all known methods of manufacture, and safeguard against varying standards of purity and impurity patterns, and varying degrees of stability arising from more than one method of manufacture. They are, therefore, a compromise and do not necessarily represent the highest standard attainable by a single efficient producer at the time the product is manufactured.

The prime consideration undoubtedly is that the product should be satisfactory clinically. Variation in biological response is such that it is doubtful whether the difference between one preparation containing 85 per cent of the required amount of active ingredient and another containing 115 per cent

(±15 per cent limit) is detectable in practice, unless a very carefully controlled trial involving large numbers of such preparations and patients were undertaken. The standards, however, are essential to ensure that reasonably reproducible products can be prepared in different laboratories by different operators, and by the same operator on different occasions, and also to ensure that the product retains an acceptable level of potency and freedom from toxicity during storage before use.

Official standards, therefore, do not necessarily equate with manufacturing standards, a frequent misconception, since official standards must be framed to take account not only of the nature and purity of the raw materials and the methods and hazards of manufacture, but also the manner and hazards of storage, and the conditions under which the product is likely to be used.

Manufacturing Variations

The principal hazard is loss of active ingredient, and all official standards for drugs and formulated products must allow for unavoidable decomposition or loss in manufacture, and storage under reasonably adequate conditions for a reasonable period of time.

The limits for compounded preparations take account of such hazards. This is particularly important for small-scale and extemporaneous preparations, which unlike large-scale manufacture are not necessarily subject to rigorous analytical control, and hence do not allow for adjustment of the product. Errors in weighing small quantities on dispensing balances and in the measurement of small volumes of liquid are significant, and limits of error for dispensed mixtures are usually at least of the order of ±5 per cent.

Storage and Dating of Products

Appreciable manufacturing losses or anticipated decomposition on storage may require the inclusion of a definite overage, so that the product will still comply with the official standards after a reasonable period of storage.

This method, which is practised extensively with certain unstable vitamin preparations such as those containing Retinol (Vitamin A) and its esters, is only permissible provided it does not interfere with any precise dosage requirement, and provided the breakdown products are not toxic, and do not further catalyse decomposition. The method is thus undesirable on clinical grounds for use with products containing Calciferol, where consistent overdosage can lead to a condition of hypervitaminosis. Likewise, it is unsuitable for Benzylpenicillin products, in which the primary product of hydrolytic breakdown, penicilloic acid, is capable of accelerating the further decomposition of the parent antibiotic by functioning as an acid catalyst.

The alternative method of stipulating an expiry date is not normally used in official standards. Opinions are very much divided on the use of expiry dates and on the value of the alternative, a declaration of date of manufacture, on the label of products. The calculation of expiry dates is circumscribed by a number of factors, not the least of which is the precise control of storage conditions, particularly temperature and humidity, for sensitive products. Failure to comply with expressly-stated storage conditions, could completely invalidate a carefully calculated expiry date. A declaration of the date of manufacture, on the other hand, merely indicates the age of the product, and gives no guide in itself to the possible life of the product.

Conditions of Use

Official standards must also have regard to the conditions under which the product is iikely to be used. Thus, the geographical distribution for sale is important; semi-solid preparations such as pastes and ointments, prepared under temperate conditions, but required for tropical use, must comply with standards which preclude liquefaction or cracking in tropical climes; similarly, the viscosity requirements of suspensions and emulsions for tropical use must be such that the product retains stability, despite the decrease in viscosity with increased temperature.

Dosage Form

Consideration of dosage form and packs in relation to the intended use of the product is also important in relation to the setting of official standards. This is especially relevant to the preservation and prevention from contamination of such widely diverse products as multidose injections, eye-drops, and products for external application to broken skin surfaces, which may be subject to spoilage with repetitive use.

The presence of toxic impurities or toxic decomposition products consti-tutes a hazard, no less important than loss of active ingredient, which has long been appreciated. Official standards legislate for and seek to control such impurities by limit tests based on qualitative and semi-quantitative proce-dures, which have become all the more sensitive with the refinements intro-duced by such techniques as gas and thin-layer chromatography.

OFFICIAL METHODS OF CONTROL

In practice, official monographs for pharmaceutical chemicals and formulated products are both descriptive and informative in addition to prescribing standards for the product, and conditions for its storage. Official monographs for pharmaceutical chemicals generally embrace:

1. A description of the drug or product.
2. Tests for identity.
3. Physical constants.
4. Quantitative assay of the pure chemical entity in the case of phar-maceutical chemicals, or of the principal active constituents in the case of a formulated product.
5. Limit tests to exclude excessive contamination.
6. Storage conditions.

Tests for Identity

Identification is usually achieved by a combination of simple chemical tests and the measurement of appropriate physical constants. Frequently, however, there is considerable overlap between identity tests, on the one hand, and limit tests which are designed to limit undesirable amounts of impurity. Identity tests, no matter whether physical or chemical, can, if they are sufficiently specific, be used as the basis of a quantitative assay or in the design of specific limit tests, and in practice a single chemical test or physical constant may contribute to both the identification and standardisation of the drug.

Chemical tests, which are used for identification, are included to establish as far as possible the presence of the required active groupings. They may often be criticised on the grounds that they are far too general in character, but they are usually considered sufficiently specific to meet the need when taken in conjunction with the other requirements of the monograph. For inorganic substances, identification tests are usually based on those in use in general qualitative inorganic analysis. Organic substances are identified by the characteristic reactions of one or more of the functional groups present in the molecule.

Measurement of Physical Constants

Physical constants, such as melting point, boiling point, refractive index, and solubility in water and in organic solvents are characteristic properties, useful for both identification and in the maintenance of standards of purity.

Melting Point

The limitations of melting point data as a criterion of purity are now generally recognised, and, even when such factors as sample size, capillary dimensions, temperature of insertion into the heating bath or block, and rate of temperature rise are standardised, it is difficult to ensure reproducibility from one laboratory to another. Melting range is, therefore, a more practical criterion of identity and purity than melting point. Most organic substances have low melting points (i.e. below about 300°), and these melting points are more often than not depressed in the presence of an organic contaminant. In some cases, even traces of impurity are sufficient to cause an appreciable melting point depression, and this fact is occasionally utilised to dispense with the assay process as the principal means of exercising control of purity. Such melting points are however often preferably stated as a melting range as, for example, in Isoniazid (m.p. 170–173°) and Phenobarbitone (m.p. 174–177°). This provides a certain tolerance in manufacture and allows for the effects of storage over a reasonable length of time.

Solubility

Precise solubility requirements are seldom used as official standards for pharmaceutical chemicals. Statements included in official monographs of the British Pharmacopoeia under the heading *Solubility* are only approximate, and are intended solely for information (see British Pharmacopoeia General Notices). Statements under qualified headings, such as *Solubility in alcohol*, however, do express an exact requirement, and hence form part of the official standards for the substance. All are measured at 20°.

Weight per Millilitre and Refractive Index

These constants are widely used as standards for liquids, including fixed oils (e.g. Olive Oil), synthetic chemicals (e.g. Glycerol) and solutions (e.g. Syrup). Because of Excise requirements, however, *Specific gravity* (20°/20°) and not *Weight per ml* continues to be used in the United Kingdom in the control of Alcohol and preparations containing alcohol (e.g. Tinctures and Extracts). The term *Relative Density* (20°/20°), which is synonymous with *Specific Gravity*, is used in the European Pharmacopoeia. Measurements are usually made at 20°, but occasionally higher temperatures must be used as in the case of Theobroma Oil which is a waxy solid melting at about 33°; the refractive index is measured at 40°.

Light Absorption

Measurement of light absorption in the visible and ultraviolet range is now widely used as a means of identification (e.g. Chlorine Theophyllinate, λ_{max} 275 nm in *0.01N* NaOH). It is especially useful when the observed spectrum shows more than one characteristic maximum (e.g. Nitrofurantoin, λ_{max} at 266 and 367 nm in buffer) or alternatively where the intensity and position of the maxima are influenced by pH (e.g. Triprolidine Hydrochloride, λ_{max} 290 nm in *0.1N* H_2SO_4, and λ_{max} 230 nm and 276 nm in water). In some cases, the extinctions at specified wavelengths (not necessarily maxima) are used as the basis of limit tests to exclude undesirable impurities (see Tricyclamol Chloride). The ratio of extinctions at two maxima also provides a useful criterion of identity (Procyclidine Hydrochloride) or purity (Viprynium Embonate).

Infrared Absorption

Comparison of the infrared absorption spectrum in the region $4000-667$ cm^{-1} with that of an authentic specimen is becoming increasingly important in the identification of drugs of widely differing character, as for example with Amitriptyline Hydrochloride, Nialamide, Iothalamic Acid and Sulphadimethoxine. Authentic specimens required for comparison purposes can be obtained from the Offices of the British Pharmacopoeia Commission or the European Pharmacopoeia Commission.

Optical Rotation

The optical rotation of chemically inhomogeneous liquids (e.g. Castor Oil) and, more particularly, measurement of *specific rotation* of compounds provides a valuable means of controlling the optical purity of pharmaceutical chemicals, in which pharmacological or physiological activity is highly correlated with molecular configuration, as for example with Adrenaline Acid Tartrate, Prednisolone Sodium Phosphate and Dextropropoxyphene Hydrochloride. In Levodopa, for which the observed rotation at the D line of sodium is inconveniently small, measurement of the rotation of a suitable complex at a specific wavelength in the ultraviolet region provides a more effective means of ensuring that only the correct isomer is present. Circular dichroism spectrometry has also been shown to provide a more satisfactory means than simple optical rotation measurements for isomer control in Phenthicillin (Part 2).

The optical rotation may have a special significance in the control of Camphor where the use of either natural $\{[\alpha]_D^{20} + 40°$ to $+43°$ $(c, 10, 95$ per cent ethanol)$\}$ or synthetic camphor $([\alpha]_D^{20}\ 0°)$ but not mixtures of the two is permitted. This latter point, although not specified in these terms, is definitely implied in the limits of specific rotation for synthetic camphor, which are $-1.5°$ to $+1.5°$.

Rotation measurements are also used as an assay to control the content of Dextrose in Dextrose Injection, and also of dextrans in various Dextran Injections.

Viscosity

Viscosity measurements are used as a means of distinguishing various grades of liquid and solid paraffins. They are also used to control the molecular size of dextrans in Dextran Injection, the composition of Iron-Sorbitol Injections, the extent of nitration of Pyroxylin, and in the standardisation of Methylcellulose.

Polymorphism and Particle Size

The physical state of insoluble drugs, which are necessarily administered in solid form, can markedly influence the rate of both solution and absorption of the drug. Crystal form and particle size are the important factors controlling surface area, the property which determines the activity of such drugs as Griseofulvin, Phenothiazine, Spironolactone, Digoxin and Phenindione. Thus the official monograph for Griseofulvin provides for standardisation of the powder such that the bulk of the material consists of particles not more than 5 μm in maximum dimension, with the occasional particle up to 30 μm.

Control of *Sediment volume*, as measured by the depth of sediment after standing for a period of hours in a vessel of specified dimension, provides a simple alternative means of controlling particle size in Tetracosactrin Zinc Suspension.

Dihydrotachysterol exists in two polymorphic forms. The α-form melts at about 113°, and resolidifies to the β-form which melts at about 129°. The β-form can be reconverted to the α-form by recrystallisation from methanol (95 per cent). Solubility differences exist between the two forms.

Chloramphenicol Palmitate exists in polymorphic forms, of which one is biologically inactive. Infrared spectroscopy is used to compare samples of Chloramphenicol Palmitate recovered from Chloramphenicol Palmitate Mixture with an artificially prepared mixture of authentic material containing 10 per cent of the inactive polymorph.

Paper Chromatography

Paper chromatography, although much less widely used than formerly, still provides a useful test of homogeneity for certain peptides such as Polymyxin B Sulphate and Colistin Sulphate. Paper chromatography of the amino acids, produced by acid hydrolysis of peptides, is also used to distinguish Coliston from Polymyxin, the former containing leucine and threonine, and the latter phenylalanine and serine, all of which can be identified by direct comparison chromatographically with authentic samples.

Thin-layer Chromatography

The very much more rapid and more sensitive technique of thin-layer chromatography forms the basis of a number of important identity tests, as for example for propranolol in Propranolol Injection and Tablets, and for Hydrochlorthiazide, Hydroflumethazide, Bendrofluazide and related substances. Thin-layer chromatography is also used with Framycetin Sulphate in order to distinguish it from Neomycin Sulphate.

The technique is still more widely used as the basis of a number of important limit tests (q.v.).

The apparatus and methods used in the measurement of these physical properties are described elsewhere in this text.

Assay

Non-specific assay methods are frequently used. This applies particularly in the case of acid-alkali titrations for bases and acids. Many inorganic salts, also, are determined simply on the content of one of the ions present; thus Sodium Sulphate is assayed on its sulphate content by precipitation as barium sulphate. Even when modern physical methods are used, as for example the measurement of ultraviolet absorption, this may be by no means specific. For example, most simple aromatic substances show an absorption, which can form the basis of assay, in the region 260–300 nm; this absorption is characteristic of the aromatic ring, and while differences exist in the spectra of different aromatic substances, these are seldom completely characteristic. The assays are therefore non-specific.

Despite elements of non-specificity, however, the assay process is, nevertheless, regarded as being sufficiently specific when taken in conjunction with the other requirements of the monograph, as discussed above. Details of the methods used for assay are described elsewhere in this text.

Assay Tolerances

Other factors of considerable importance in fixing standards for pharmaceutical preparations are the assay tolerances. These include the limits of error of the actual assay process for the active ingredients, manufacturing tolerances for the particular dosage form, and sampling errors. The limits of error of the assay method itself depend upon the method adopted. Volumetric and gravimetric methods usually have quite narrow limits of error, e.g. Sodium Chloride ($\not< 99.5$ per cent calculated with reference to the dried substance). Non-aqueous titrations are usually less precise, and titres are often marginally high. Typical limits are therefore somewhat wider, as in Mepyramine Maleate (99–101 per cent) and Atropine Sulphate (98.5–100.5 per cent). Spectrophotometric assays and assays involving extraction or other complex manipulations are usually subject to greater variation than volumetric and gravimetric methods. Thus, the limits applied to the ultraviolet absorption assay of Testosterone are 99–103 per cent, whilst those for Phenobarbitone Sodium, which involves an extraction with ether, are 98–100.5 per cent.

Assay methods based on the development of specific colours by treatment with a suitable reagent are even less precise, as for example in the control of Cortisone Acetate (96–104 per cent).

The effect of allowances for manufacturing tolerances is seen in the standards proposed for many injections and tablets. This is particularly evident in injection solutions, which may well be relatively dilute, and in tablets containing high dilutions of medicaments in inactive excipients. Even with Sodium Citrate, a relatively innocuous substance administered in doses up to 10 g daily, the assay tolerances permitted for 125 mg Tablets are 95–105 per cent, compared with 99–101 per cent for Sodium Citrate itself by the same volumetric method of assay. The somewhat wider tolerances necessary for a formulated product assayed by aqueous extraction, derivatisation and weighing, is seen in the 300 mg Tablets of Piperazine Adipate, which despite their much higher active ingredient contents, have limits of only 92.5–107.5 per cent compared with 98.5–100.5 per cent for the parent substance. Similarly, wide tolerances are adopted for Morphine Sulphate Tablets (20 mg) which are required to conform to a content of 92.5–107.5 per cent of the labelled strength. Likewise, many injection solutions controlled by spectrophotometric assay, e.g. Cortisone Injection and Testosterone Propionate Injection, have assay limits as wide as 90–110 per cent of the declared content.

Sampling Procedures and Errors

Errors, due to sampling, arise in the selection of material for analysis where the material selected is not truly representative of the batch as a whole. The problem does not arise with homogeneous liquids or solutions unless these are distributed in separate containers which may contain material from different batches. Heterogeneous liquids, such as emulsions and suspensions, and solid samples are subject to mixing variations and to separation, and the problem of obtaining a representative sample for analysis then becomes important.

One of two methods is generally used for solid samples, depending upon whether or not the bulk material is reasonably homogeneous or not. Random sampling is used for products which are not subject to gross variation; the bulk is divided into real or imaginary units and a sample proportionate to the size of each unit is collected, the selection of individual samples being carried out in a random manner. For heterogeneous materials, such as crude drugs comprising various parts of the plant, representative samples of each type of material present are selected in a random manner in proportion to the amount of that material present in the whole.

The difference between the mean value of the active ingredient in the samples and the true value is known as the sampling error. The error due to sampling is inversely proportional to the square root of the number of samples taken. This error is additional to those introduced by the analytical method and by the fact that, in a formulated product, the active ingredient is itself subject to a permitted variation.

Where the material is packed in a number of separate containers, representative samples for analysis must be selected from a proportion of the containers taken at random. Emulsions, suspensions, pastes and ointments should be shaken or stirred to ensure the best possible mixing before sampling; fluid samples of emulsions and suspensions should be taken as far as possible from different parts of the container and/or from a suitable selection of containers. Pastes and ointments should be sampled with an auger to give cored samples from various parts of the container(s).

Unit-dose Variations

Products such as injections, prepared by dissolving the contents of a sealed container before use (pro-injections), powders, capsules and tablets, which are supplied in unit-dose forms, are as a rule subject to greater variation than comparable preparations supplied in multi-dose forms. Standards are, therefore, applied to capsules, tablets and pro-injections to ensure *Uniformity of Weight* and to single-dose injections for *Uniformity of Volume*.

Uniformity of weight. This as applied to capsules (120 mg or less in weight), demands that the contents of each of eighteen out of twenty capsules shall not differ in weight from the average weight by more than ±10 per cent; the remaining two capsules may differ from the average by from ±10 to ±20 per cent. Somewhat stricter tolerances (±7.5 per cent) are demanded for capsules over 120 mg in weight. Similar provisions are made for the Uniformity of Weight of uncoated tablets, and for ampoules containing pro-injections. The permitted percentage deviation decreases with increasing size of tablet or injection weight as shown in Table 3.

Table 3. Uniformity of weight of uncoated tablets

Average weight of tablet (g)	Permitted deviation (%) for ≮18 of 20 tablets weighed	Permitted deviation (%) for ≯2 of 20 tablets weighed
80 mg or less	±10	±15
80–250 mg	±7.5	±12.5
250 mg or more	±5	±10

The weights of at least eighteen of the twenty tablets examined must fall within the permitted deviation; the weights of the remaining two tablets must not deviate from the average by more than the percentage deviation indicated in the Table. It is permissible, if twenty tablets are not available, to examine only ten; not less than nine must be within the permitted deviation from the average weight, and one may deviate by not more than the permitted deviation indicated. The same range of percentage deviations from the stated weight for pro-injections is allowed as for tablets.

Uniformity of diameter. Tablets are also required to conform to standards for *Uniformity of Diameter* to ensure that tablets containing the same active ingredient and of the same strength produced by different manufacturers still have similar appearance. The object of these requirements is to avoid confusion of the patient which might arise if the same product were supplied in varying sizes and shapes on succeeding occasions.

Injection overages and uniformity of volume. Standards laid down by the British Pharmacopoeia for injections supplied in single-dose containers stipulate that a specified volume in excess of the nominal volume of the injection be supplied. A table of excess volumes, which vary for mobile and viscous injections, is given. Different tolerances for *Uniformity of Volume* are stipulated for nominal volumes up to 20 ml and for those in excess of 20 ml. For the former, the average volume in ten single-dose containers must not deviate by more than 5 per cent of the requirements for total volume (i.e. nominal volume plus excess volume) and no single container must deviate from the total volume by more than 10 per cent. Where the nominal volume is greater than 2 ml, the total volume in each container must not be less than the nominal volume and not more than 5 per cent greater than the total volume.

Content. Standards are also fixed for the content of active ingredients in pro-injections, capsules, and tablets which incorporate the unit weight variations. For capsules and tablets, the assay is carried out on a portion of the mixed contents of twenty capsules or on a portion of the mixed powder from twenty tablets. The average amount of active ingredient calculated from the assay, and from the average weight per capsule or tablet must lie within the permitted limits. A similar procedure is adopted for pro-injections, except that the amount of active ingredient in one container of the ten examined, may be up to twice the degree of tolerance permitted by the monograph.

Uniformity of content. Tablets containing highly potent medicaments present in milligram or microgram doses may be subject to large inter-tablet variation, as a result of failure to achieve a homogeneous mix of active ingredient and excipients during manufacture. Tests for *Uniformity of Content* are applied to individual Digoxin Tablets and tablets of other highly potent medicaments in addition to the assay which controls the average content of the mix. The content of each of ten individual tablets is required to be between 80–120 per cent of the average of the ten, except that one tablet is permitted the wider limits of 75–125 per cent of the average.

Bio-availability

Manufacturers of potent pharmaceutical products are becoming increasingly conscious of the need to control the availability of active ingredients from particular dosage forms, notably tablets. Two types of product are of particular concern. One consists of tablets containing potent active ingredients of low solubility and hence low release rates. The other consists of products specially designed to slow the release rate, and so extend the period during which high blood levels are attained.

All tablets, capsules and suppositories are subject to *Disintegration Tests* to ensure that they break down readily in the gastro-intestinal tract. In a limited number of cases, the release of insoluble material from the relatively coarse particles formed on disintegration also requires control to ensure an adequate rate of uptake into solution. This control is achieved by including a *Dissolution Test* in the specification (Chapter 14). A commonly used form of dissolution test is one in which a number of tablets (usually six) are rotated inside a small cylindrical wire mesh basket at a fixed speed (usually between 80 and 150 rpm) in an appropriate solution, often water, with or without adjustment of pH to simulate the effect of gastric secretions on the active ingredient. The solution is sampled at fixed time intervals, and the samples assayed to determine the rate of solution. Reasonable correlations are usually obtained with rates of uptake into the blood stream actually measured from the tablets in human volunteers. Dissolution tests should be conducted under what are described as *sink conditions*, with the volume of solvent such that the dissolving active ingredient only reaches concentrations which are a fraction of its actual solubility. This technique is, therefore, a reasonable simulation of the dynamic nature of absorption processes in the gastro-intestinal tract.

Solution rate tests are also applied to preparations such as Slow Lithium Carbonate Tablets and Slow Orphenadrine Citrate Tablets to ensure a reduced but steady rate of release. A small number of tablets is tested using an extraction thimble as container in the standard disintegration test apparatus, with aqueous hydrochloric acid as the solvent. The solution is sampled at appropriate intervals, and the release rate is required to conform to specified patterns appropriate to the products concerned.

LIMIT TESTS

Limit tests are quantitative or semi-quantitative tests designed to identify and control small quantities of impurity which are likely to be present in the substance. The quantity of any one impurity in an official substance is often small, and consequently the visible reaction response to any test for that impurity is also small. The design of individual tests is therefore important if errors are to be avoided in the hands of different operators. This is accomplished by giving attention to a number of factors, which are discussed below.

Specificity of the Tests

Any test used as a limit test must, of necessity, give some form of selective reaction with the trace impurity. Many tests used for the detection of inorganic impurities in official inorganic chemicals are based upon the separations involved in inorganic qualitative analysis. A test may be demanded which will exclude one specific impurity, but highly specific tests are not always the best; a less specific test which limits several likely impurities at once is obviously advantageous, and in fact can often be accomplished. An example of such a test is the heavy metals test applied to Alum, which not only limits contamination by lead, but also other heavy metal contaminants precipitated by thioacetamide as sulphide at pH 3.5.

Sensitivity

The degree of sensitivity required in a limit test varies enormously according to the standard of purity demanded by the monograph. The sensitivity of most tests is dependent upon a number of variable factors all capable of strict definition, and all favourable towards the production of reproducible results. Thus the precipitation of an insoluble substance from solution is governed by such factors as concentration of the solute and of the precipitating reagent, duration of the reaction, reaction temperature, and the nature and concentration of other substances unavoidably present in solution. As a general rule, cold dilute solutions give light precipitates, whereas more granular ones are obtained from hot concentrated solutions. Many of the limit tests, however, are concerned with very dilute solutions which are often slow to react, and here sensitivity of the reaction can often be increased by extending the duration of the reaction or by raising the reaction temperature. Similar considerations apply in the design of colour and other tests employed as limit tests. With suitable control of the factors described the same degree of reproducibility can be guaranteed in all cases.

Control of Personal Errors

It is essential to exclude all possible sources of ambiguity in the description of a test. Vague terms such as 'slight precipitate', should be avoided as far as possible. The extent of the visible reaction to be expected under the specified test conditions should be clearly and precisely defined. This is usually accomplished in one of three ways.

(a) **Tests in which there is no visible reaction.** A definite statement is incorporated in the wording of the test, which states that there shall be no colour, opalescence or precipitate, whichever is appropriate to the particular test. One example of this type of requirement is the test for *barium* and *calcium* in Dilute Hypophosphorous Acid (B.P. Appendix I), where the addition of dilute sulphuric acid under precisely controlled conditions shall produce 'no turbidity, or precipitate' within one hour. The time factor is used here as a means of increasing the sensitivity of the test.

Tests such as these which give negative results do not necessarily imply the complete absence of the impurity, the test as laid down merely indicating the absence of an undesirably large amount of the impurity.

(b) **Comparison methods.** Tests of this type require a standard containing a definite amount of impurity, to be set up at the same time and under the same conditions as the test experiment. In this way the extent of the reaction is readily determined by direct comparison of the test solution with a standard of known concentration. The official limit tests for chlorides, sulphates, iron and heavy metals are based on this principle. The limit tests for lead and arsenic are, in practice, also comparison methods. They are, however, so designed that they can be readily applied as quantitative determinations.

(c) **Quantitative determinations.** Quantitative determination of impurities is only applied in special circumstances, usually in those cases where the limit is not readily susceptible to simple and more direct chemical determination. The method is used in the following different types of test:

   (i) Limits of insoluble matter
  (ii) Limits of soluble matter
 (iii) Limits of moisture, volatile matter, and residual solvents
 (iv) Limits of non-volatile matter
  (v) Limits of residue on ignition
 (vi) Loss on ignition
(vii) Ash values
(viii) Precipitation methods.

Limits of Insoluble Matter

An example of this type of control is the requirement for *alcohol-insoluble substances* in Boric Acid, which is required to be almost completely soluble in alcohol (95 per cent). Monographs on the soluble barbiturates such as Barbitone Sodium include tests for *neutral and basic substances*, which are extractable with ether, primarily to exclude traces of the parent barbiturate. *Phenylbarbituric acid* is specifically limited in Phenobarbitone by the requirement that 1 g shall be completely soluble in 5 ml of boiling ethanol (90 per cent) within three minutes. Sodium chloride impurity in Chloramine is controlled by solubility of the material in cold dehydrated alcohol. Similarly in Crystal Violet a limit of *alcohol-insoluble matter* guards against heavy contamination with inorganic salts, which are often used in the isolation of the dyestuff. A limit test of *water-insoluble matter* for Lead Acetate controls contamination with basic acetates.

Tests for *clarity of solution*, similarly constitute a means of limiting insoluble parent drugs in their more highly water-soluble derivatives as for example phenytoin in Phenytoin Sodium. The test may also be combined with a limit on *colour* (Choline Theophyllinate, Nicoumalone and Sulphamethoxydiazine), or even *alkalinity* (Methohexitone Injection) or *acidity* (Oxytetracycline Injection).

Clarity of solutions for injection is particularly important to ensure reasonable freedom from *particulate matter*. Solutions for injection must appear to be free from insoluble matter when viewed by eye against a black background in upright, horizontal and vertical positions under a screened light source of specified intensity. Solutions should also be viewed in plane-polarised light to detect cellulose fibres. Large volume intravenous injections, such as Dextrose Injection, are further subject to a *limit test for particulate matter* using an instrument capable of counting particles within specific size ranges. This test requires the mean count of particles per ml from five containers to be $\not> 1000$ greater than 2 μm, and $\not> 100$ greater than 5 μm in equivalent sphere diameters.

Colour of solution is also frequently subject to control by the methods of the European Pharmacopoeia. A solution may be described as *colourless* if it has the appearance of water when examined under specified conditions. Varying degrees of colour (*degree of colouration of liquids*, E.P.) are permissible, and controlled by direct comparison of solutions with specified standard colour solutions.

Limits of Soluble Matter

These are applied to limit soluble impurities in official substances which are themselves completely insoluble in a particular solvent, the object usually being to detect some specific impurity. The control of *water-soluble barium salts*, which are highly toxic and therefore stringently excluded from Barium Sulphate required for X-ray work, is an excellent example of the value of this type of test. A general limit of matter soluble in dilute acid is applied to Light Kaolin by refluxing with 0.2N HCl, filtering and evaporating the filtrate. Limits of *water-soluble* and *acid-soluble matter* are applied to Talc to control trace impurities of water-soluble salts on the one hand, and of oxides, hydroxides and carbonates on the other. *Water-soluble acids* in Undecanoic Acid are controlled by shaking a sample with warm water, filtering and titrating the filtrate with standard sodium hydroxide solution. *Matter soluble in light petroleum* is similarly controlled in Griseofulvin, while *free cyclobarbitone* in Cyclobarbitone Calcium is limited by extraction with benzene.

Limits of Moisture, Volatile Matter and Residual Solvents

Many substances absorb moisture on storage. Deterioration of this nature is readily limited by a requirement for the loss in weight (*Loss on drying*) when the substance is dried under specified conditions. The amount of heat to which the substance is submitted varies considerably according to the nature of the substance. The temperature must be sufficiently high to produce the required result within a reasonable time, but not so high as to cause decomposition. If the substance is stable this is usually applied by drying to constant weight at 105°, as for example with the steroid hormones, Ethisterone and Ethinyloestradiol; also with Sodium Benzoate, and Sodium Bromide. Special modifications are adopted for thermolabile substances: in the case of Hyoscine Hydrobromide the sample is first dried at room temperature *in vacuo* for one hour, and then heated to constant weight at 105°, presumably because it is unstable only in the presence of moisture. Vacuum drying at various suitable temperatures is used for the hydrated salt, Isoprenaline Sulphate, and also for Cycloserine, Dequalinium Chloride and a number of other organic compounds, whilst Phenylmercuric Nitrate, Menadione, Maleic Acid and Chlorphenesin are dried *in vacuo* over phosphorus pentoxide.

Inorganic salt hydrates (Sodium Sulphate, Na_2SO_4, $10H_2O$) lose all or part of their water of crystallisation on drying. The loss in weight is often considerable and in fact forms a secondary assay process. Such limits for loss in weight when a salt is dried are often necessarily wide (Sodium Sulphate loses between 52 and 57 per cent) and may allow for either efflorescence or deliquescence. Where an official moisture limit is specified, this is usually taken into account in calculating the results of the assay process (Cyclizine Hydrochloride is required to contain not less than 98.0 per cent $C_{18}H_{22}N_2$, HCl calculated with reference to the substance dried to constant weight at 130°). This obviates the necessity of using dried samples for assay, which is advantageous because (*a*) the assay and loss by drying procedure can be carried out simultaneously, and (*b*) dried (or ignited) substances, because of their hygroscopic character, are more difficult to weigh than air-dried samples.

In certain other substances, where the content of water is critical or where it is important to distinguish between moisture and other volatile matter, water may be determined titrimetrically with Karl Fischer Reagent. This method, which is described in detail on p. 260, is used to determine *water* in Procaine Penicillin, Erythromycin, Dipipanone Hydrochloride and a number of other substances.

The introduction of gas-liquid chromatography provides a useful alternative method for limiting specific volatile substances. The method is used for the determination of *methanol* in Orciprenaline Sulphate, from which it cannot be removed quantitatively by drying at 105°. Gas-liquid chromatography is also used to control *residual solvent* in Novobiocin Calcium and Novobiocin Sodium, using propanol as an internal standard.

Determination of the Loss on Drying of Sodium Chloride

Method. Heat a clean, dry, shallow, stoppered weighing-bottle in an air-oven at 130°, for about 20 minutes. Transfer to a desiccator, cool for at least 20 minutes with the stopper on its side in the neck of the bottle, insert the stopper and reweigh. Reheat and reweigh until constant weight is attained. Introduce about 1 g of the sample into the bottle, insert the stopper and reweigh. Place in the drying oven with the stopper on its side in the mouth of the bottle and heat in the oven at 130° for 2 hours, cool in a desiccator for at least 20 minutes, insert the stopper and weigh. Repeat the drying and cooling until constant weight is attained. Calculate the percentage loss in weight.

The following example illustrates the calculations involved in utilising the figures for moisture content in conjunction with the assay figures.

Calculation. A sample of Sodium Chloride was found to contain 98.90 per cent NaCl (from the assay figures) and upon drying lost 0.760 per cent of its weight. Calculate the percentage of NaCl with reference to the dried sample.

$$\therefore \quad 100 \text{ g sample (undried)} \quad \text{contains} \quad 98.9 \text{ g NaCl}$$

$$\text{But} \quad 100 \text{ g sample (undried)} \equiv 100.00-0.76 \equiv 99.24 \text{ g dried sample}$$

$$\therefore \quad 99.24 \text{ g sample (dried)} \quad \equiv 98.9 \text{ g NaCl}$$

$$\therefore \quad 100 \text{ g sample (dried)} \quad \equiv \frac{98.9 \times 100}{99.24}$$

$$\equiv 99.65 \text{ g NaCl}$$

$\therefore$ *The sample of Sodium Chloride was found to contain 99.65 per cent NaCl calculated with reference to the substance dried to constant weight at 130°.*

Limits of Non-volatile Matter

These limits are applied to both inorganic and organic substances which are readily volatile. The Pharmacopoeia draws a distinction between substances which are readily volatile (as on a boiling water bath) and those which are volatile only when ignited strongly. Limits of *non-volatile matter* are applied to the former group to control contamination by unspecified inorganic matter such as other salts and dirt. Ammonia Solution, Hydrogen Peroxide Solution, Water for Injections, and a number of readily volatile organic substances including Alcohol (95 per cent), Isopropyl Alcohol, Chloroform, Halothane, Methanol, Solvent and Anaesthetic Ether, Vinyl Ether, Chlorocresol, and Trichloroethylene are examined in this way. The determination of non-volatile matter in Hydrous Wool Fat by heating to constant weight on a boiling water bath provides an approximate determination of the amount of water present in the sample.

Limits of Residue on Ignition

These are applied to two classes of substance; those which are completely volatile when ignited and those which undergo a major decomposition leaving a residue of definite composition. The former include certain inorganic substances such as Mercury, and the test limits contamination by unspecified inorganic matter. Calamine, a basic zinc carbonate, decomposes when ignited to give carbon dioxide and water, leaving the oxide as residue. The process is therefore a non-specific assay. Ignition is also used as a method of assay to control *zinc oxide* in Zinc Cream.

Loss on Ignition

This type of test is applied to stable substances which are liable to contain thermolabile impurities. Contamination of Light Magnesium Oxide by carbonate is controlled in this way, and the loss in weight is of the order of 5 per cent. *Carbonate and moisture in Lead Monoxide are also controlled in the same way.*

Determination of the Loss on Ignition of Zinc Oxide

Method. Heat a clean crucible supported on a pipeclay triangle in a hot Bunsen flame for 10 minutes. Cool in a desiccator for about 20 minutes and weigh. Repeat until constant weight is attained. Introduce about 1 g of sample into the crucible, reweigh and then heat strongly in a hot Bunsen flame for about 1 hour. Cool in a desiccator and reweigh. Repeat the heating for a further period of 30 minutes, cool and reweigh. Repeat the heating and cooling until constant weight is attained.

Ash Values

The ash content of a crude drug is the inorganic residue remaining after incineration. It represents not only the inorganic salts such as calcium oxalate occurring naturally in the drug, but also inorganic matter from external sources. The official ash values are chiefly of use in examining powdered drugs. Thus they may have one or more of the following applications:

(a) To ensure the absence of an undue proportion of extraneous mineral matter introduced accidentally or by design at the time of collection or in subsequent treatment, e.g. earth, sand, and floor sweepings.

(b) To ensure absence of other parts of the plant, e.g. in Cardamom Fruit B.P.

(c) To detect adulteration with exhausted drug e.g. in Ginger B.P.

(d) To detect adulteration with material containing stone cells or starch which would modify the ash values.

The four types of ash value used in the British Pharmacopoeia are *ash*, *acid-insoluble ash*, *water-soluble ash* and *sulphated ash*.

Ash

A figure for the total *ash* content is valuable for drugs in which little calcium oxalate is present, e.g. limed ginger or nutmeg. If much calcium oxalate is present then the value for the acid-insoluble ash is a better criterion of purity. *Ash* in Sterilizable Maize Starch limits the amount of magnesium oxide which it is permitted to contain.

Determination of Ash in Ginger

Method. Ignite a silica dish to constant weight at a dull red heat over an Argand or other suitable burner (Note 1). Spread 2 to 3 g of the powdered drug evenly over the bottom of the dish and weigh. Ignite at low temperature until vapours have almost ceased to be evolved and then increase the temperature slightly, not exceeding 450°, to burn off the carbon (Note 2). Cool in a desiccator and weigh. Repeat the heating and cooling until constant weight is attained.

If a carbon-free ash is not obtained in this way, extract the residue with hot water, filter through an ashless filter paper, incinerate the residue and filter paper in the silica dish until all carbon is removed, using a higher temperature than that used previously if necessary (Note 3). Add the filtrate to the dish, evaporate to dryness and ignite to constant weight using not more than a dull red heat.

Calculate the percentage of ash with reference to the air-dried drug.

Note 1. An Argand burner gives a low uniform temperature over a wide area.

Note 2. Too high a temperature would cause a loss of such substances as alkali chlorides which are volatile at high temperatures.

Note 3. Sometimes heating at a low temperature does not give a carbon-free ash because carbon particles are trapped in fused alkali carbonates or phosphates. Treatment of the ash with hot water dissolves the salts. The insoluble residue can then be ignited rapidly at a higher temperature to remove carbon without loss of inorganic material; the filtrate is added and subsequently evaporated to dryness when the whole residue can be ignited at the lower temperatures, at which alkali chlorides are not volatile.

Acid-insoluble Ash

Crude drugs containing calcium oxalate can give variable results upon ashing depending upon the conditions of ignition. Treatment of the ash with acid leaves virtually only silica. Hence *acid-insoluble ash* forms a better test to detect and limit excess of soil in the drug than does the total ash.

Determination of Acid-insoluble Ash in Rhubarb

Method. Wash the total ash (obtained as above) from the silica dish into a 100 ml beaker using 25 ml of dilute hydrochloric acid. Boil for 5 minutes. Transfer the insoluble residue to a previously prepared, ignited and weighed Gooch crucible, wash well with hot water, allow suction to remove most of the water and then ignite at a temperature not exceeding 450° inside an ordinary crucible used as a 'jacket crucible'. Cool in a desiccator and weigh. Calculate the percentage of acid insoluble ash with reference to the air-dried drug.

Water–soluble Ash and Water–soluble Extractive

These determinations are only specified in the case of one official drug, Ginger B.P., where it is helpful in detecting samples which have been extracted with water.

Determination of Water-soluble Ash in Ginger

Method. As for acid-insoluble ash, but use 25 ml of water in place of the 25 ml of acid; subtract the weight of the insoluble residue from the weight of the total ash; the difference in weight represents the water-soluble ash. Calculate the percentage of water-soluble ash with reference to the air-dried drug.

Sulphated Ash

The determination of *sulphated ash* is used in the case of unorganised drugs such as Colophony, Podophyllum Resin, Wool Alcohols and Wool Fat. It is also applied very widely to control the extent of contamination by inorganic impurities in organic substances and as an assay for Calcium Acetate, and Anhydrous Sodium Acetate. The substance is ignited with concentrated sulphuric acid, which decomposes and oxidises organic matter, leaving a residue of inorganic sulphates. The test is used, for example, to control traces of alkali metals in Chlorbutol arising as a result of the method of preparation (i.e. heating acetone and chloroform in the presence of potassium hydroxide). It is also used to exclude contamination with calcium in Citric Acid, which is isolated via its calcium salt. Non-volatile inorganic impurities in Aspirin, Colchicine, Lignocaine Hydrochloride, Phenobarbitone, Benzyl Benzoate and numerous other organic compounds are also controlled in this way. Reproducible results are more readily obtained in this determination than in a total ash determination because, in general, metal sulphates are stable unless heated very strongly.

Determination of Sulphated Ash in Aspirin

Method. Ignite a large silica crucible to constant weight. Add approximately 5 g of sample, reweigh, moisten with sulphuric acid and heat gently at first and then more strongly as the volatile matter is removed. Ignite more strongly to remove the carbon, *cool*, remoisten with sulphuric acid and then re-ignite to constant weight . Calculate the percentage of sulphated ash.

Note. This determination must be carried out in a fume-cupboard.

Precipitation Methods

These are used when moderate amounts of impurities such as *iron, aluminium and silica* are permitted as for example in Potassium Hydroxide. A sample (5.0 g) is boiled with excess dilute hydrochloric acid, and then made alkaline with ammonia. Any precipitate of metallic hydroxide or silica is collected, dried and after ignition should weigh $\not> 5$ mg (limits of Fe, Al, and matter insoluble in hydrochloric acid).

Similar tests are used to limit *iron, aluminium and phosphate* in Calcium Hydroxide and Chalk.

LIMIT TESTS FOR METALLIC IMPURITIES

The Pharmacopoeia places the greatest emphasis on the control of physiologically harmful impurities. Contamination by arsenic and lead is widespread, largely as a result of atmospheric pollution, and quantitative limit tests of wide general applicability are specified for them. A general limit test is also specified for iron.

The Limit Test for Lead

The test is based on the conversion of traces of lead salts to lead sulphide, which is obtained in colloidal form by the addition of sodium sulphide to a slightly alkaline solution buffered by a fairly high concentration of ammonium acetate. The brown colour, due to colloidal lead sulphide in the test solution, is compared with that obtained from a known amount of lead.

The colour in the two solutions is compared by examination in special vessels known as Nessler Glasses (Fig. 1). An identical pair of Nessler Glasses (British Standard Specification No. 612; 1966) should be used, i.e. a pair made of the same glass and also of the same diameter. The graduation mark must also be at the same height from the base in both vessels. Comparison is made by viewing down through the solution against a light background.

Fig. 1.
Nessler Glass

The test is complicated by the fact that many substances submitted to the test affect the colour of the colloidal lead sulphide produced. These effects are qualitative and the difficulty can be overcome by comparison of two solutions of the substance, known respectively as the primary and auxiliary of which the former contains more of the sample under test than does the latter. It has been found that the influence of substances in solution on the colour of the lead sulphide reaches a maximum when their concentration is about 4 per cent. The auxiliary solution, in addition, contains a known amount of Dilute Lead Solution PbT. The stringency of the test is varied considerably from substance to substance and is fixed by the weight of sample used in the preparation of the two solutions and by the volume of Dilute Lead Solution PbT added to the auxiliary. Contamination by lead varies greatly with the substance, depending upon the source and the method of manufacture. The stringency of the tests is fixed with these factors in mind, but owing to the toxicity of lead salts is also determined by the dosage in which the substance is usually employed, and the usual duration of administration.

Determination of Lead

Dilute Lead Solution PbT is a solution of lead nitrate (0.016 g) containing nitric acid (0.5 ml) per litre. Each ml contains the equivalent of 0.01 mg of lead.

Method. The weights of the sample required for the preparation of the primary and auxiliary solutions may be found by reference to the appropriate appendix of the British Pharmacopoeia. The two solutions are prepared as follows:

Primary Solution	*Auxiliary Solution*
(a) Dissolve x g of sample in hot Water containing the specified volume of acetic acid.	Dissolve y g of sample in hot Water containing the specified volume of acetic acid.
(b) Boil, if necessary, to remove carbon dioxide which may have been generated.	Boil, if necessary, to remove carbon dioxide which may have been generated.
(c) —	Add the specified volume of Dilute Lead Solution PbT.

(d) Make each solution alkaline by the addition of Ammonia Solution PbT, and add Potassium Cyanide Solution PbT (10 per cent; 1 ml). The latter prevents the precipitation of iron and copper as sulphides at a later stage in the test by converting them to complex cyanides which are unaffected by sodium sulphide.

(e) Filter the solutions, if turbid.

(f) At this stage the colour of the two solutions may be different, due to the different amounts of sample used for each. If necessary, match the two solutions by the addition of a few drops of a burnt sugar solution.

(g) Dilute the solutions to 50 ml with Water, and add 2 drops of Sodium Sulphide Solution PbT to each.

(h) Stir well and compare the colours of the two solutions. The colour of the primary solution should not be darker than that of the auxiliary.

As usually applied, the test records the fact that the sample merely complies, or fails to comply, with the official requirements for lead. The test can be made quantitative by observing the amount of Dilute Lead Solution PbT which must be added to the auxiliary solution, in order that, after dilution to 50 ml, equal colours will be produced when the test is completed. If more than 15 ml of Dilute Lead Solution PbT is required, a smaller quantity of substance must be taken.

Calculation. Acetic Acid is required to contain not more than 1 part per million of Pb. The solutions are prepared as follows:

Primary Solution	*Auxiliary Solution*
12 g of sample	2 g of sample.
	Add 1 ml of Dilute Lead Solution
	PbT ($\equiv$ 0.00001 g of Pb).

Thus the comparison is between 10 g (i.e. 12 − 2) of sample and 0.00001 g of Pb. If the colours are equal, then 10 g of sample contains 0.00001 g of Pb.

∴ 10/0.00001 g of sample contains 1 g of Pb.

∴ 1 000 000 g of sample contains 1 g of Pb. (i.e. 1 part per million)

Modifications of the General Procedure for the Quantitative Test for Lead

A number of official substances require treatment by a modified process. Some of the more important modifications are discussed below.

Omission of acetic acid. If the substance is acidic, then addition of acetic acid is unnecessary and can be omitted. This applies to Acetic Acid, Citric Acid, Lactic Acid, Nitric Acid, Sodium Acid Phosphate, and Tartaric Acid.

A number of other substances, which would be insoluble in acetic acid, are dissolved by the addition of Ammonia Solution PbT. These include Benzoic Acid, Boric Acid, Methylthiouracil, Nicotinic Acid, Propylthiouracil, Saccharin, Saccharin Sodium and Sodium Salicylate.

Substances precipitated by acetic acid. Phenytoin Sodium would give a precipitate of the parent hydantoin upon acidification with acetic acid. Therefore the parent compound is removed by dissolving the sample in water and precipitating with dilute hydrochloric acid. The filtrate is used for the test.

Addition of acetic acid to Antimony Sodium Tartrate would precipitate sodium acid tartrate, which has a low solubility. Acetic acid is therefore omitted. The aqueous solution of the sample is made alkaline with sodium hydroxide instead of ammonia which is usually used in the lead limit test. Both these reagents precipitate antimony trioxide, which would interfere with the test, but the oxide is soluble in excess sodium hydroxide to give a solution of sodium antimonite. Potassium cyanide solution is added and the test may then be completed in the usual way.

Sulpha drugs are relatively insoluble in both acetic acid and ammonia. With sodium hydroxide, they form water-soluble sodium salts and hence this reagent is used. The sulphonamide derivatives, Probenecid and Chloro-thiazide, are treated similarly.

Substances precipitated by ammonia. Substances in this category are mainly oxides, hydroxides and carbonates of magnesium and calcium. Calcium Hydroxide is dissolved in a mixture of dilute hydrochloric and nitric acids and boiled to remove carbon dioxide. Addition of ferric chloride and ammonia gives a precipitate of ferric hydroxide on which lead hydroxide is adsorbed. The precipitate is washed with water and dissolved in hot dilute hydrochloric acid. The solution is treated with ammonium thiocyanate and extracted with a mixture of amyl alcohol and ether to remove iron. The aqueous solution is further acidified with citric acid, made alkaline with ammonia and tested with sodium sulphide. The addition of iron is omitted in lead tests on Chalk as iron is present as impurity in sufficient quantity for the test to work.

The Limit Test for Arsenic

The test is a modification of the Gutzeit test in which all arsenic is converted into arsine (AsH_3) by reduction with zinc and hydrochloric acid. Reaction of the issuing gases with mercuric chloride paper produces a yellow stain, which can be compared with that produced from a known amount of arsenic. The apparatus used is of the type shown in Fig. 2. The capacity of the bottle should be about 120 ml, and the dimensions of the tube are required to comply with certain definite specifications (length 200 mm, internal diameter 6.5 mm). The tube is open at both ends and preferably should have a ground surface at the upper end. A small hole in the side of the tube at the lower end is necessary to prevent condensed liquid from being forced up the tube by the pressure of hydrogen developed in the bottle thus blocking the tube. The tube is packed with cotton wool previously impregnated with lead acetate solution and dried. This serves to remove traces of hydrogen sulphide from the liberated gases, which would otherwise interfere with the test. A small extension tube of the same internal diameter and similarly flanged at one end is used to fix the mercuric chloride paper in position in such a way that all the arsine will pass through a circle of paper 6.5 mm in diameter. The two tubes with the mercuric chloride paper in place are held together by a clip or elastic band.

Hydrogen gas is generated in the solution by the action

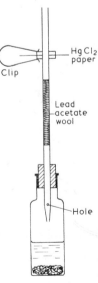

Fig. 2. Arsenic limit test apparatus

of stannated hydrochloric acid on arsenic-free granulated zinc. The presence of a small quantity of stannous chloride in the hydrochloric acid ensures a rapid reaction between the acid and the zinc and a steady evolution of hydrogen. Pure zinc is not very reactive toward hydrochloric acid, and the presence of a small quantity of tin salt increases the reaction rate by the formation of localised spots of Sn/Zn electrolytic couples. Stannous chloride also acts as a reducing agent so that any pentavalent arsenic is reduced to the trivalent state. The hydrogen formed reduces any arsenic present to arsine, AsH_3, which is carried through the tube by the stream of hydrogen and out through the mercuric chloride paper. A reaction occurs between arsine and mercuric chloride which may be represented:

$$2AsH_3 + HgCl_2 \longrightarrow Hg\begin{smallmatrix} \diagup AsH_2 \\ \diagdown AsH_2 \end{smallmatrix} + 2HCl$$

Other products such as $AsH(HgCl)_2$, $As(HgCl)_3$ and As_2Hg_3 may also be formed. These reactions result in the formation of a yellow or brown stain on the mercuric chloride paper. Provided the diameter of the paper exposed to the issuing gases is constant (6.5 mm; 5 mm in the European Pharmacopoeia), the depth of colour produced is proportional to the amount of arsenic present. Mercuric chloride paper becomes discoloured on exposure to light and should, therefore, be stored in the dark. Discoloured papers must not be used in the test. The test is comparative in that two solutions, one containing the sample under test and the other containing a known amount of arsenic, are submitted to the test at the same time. The stains are then compared by daylight.

The stringency with which the test is applied depends upon a number of factors, of which the dose of the substance under examination is probably the most important. Arsenic is a cumulative poison and substances likely to be administered in large doses, or repeatedly in small doses over a long period must be strictly controlled with respect to arsenic impurity. Other factors such as the origin of the medicament and the processes used in its manufacture will also have been taken into account in fixing the permitted limit for arsenic. In the test itself, stringency is determined by two factors, the weight of sample used in the preparation of the solution to be examined, and the volume of Dilute Arsenic Solution AsT used in the preparation of the standard stain. In practice the latter is usually kept constant, comparison being made with what is known as a 1 ml Standard Stain. The stains fade on keeping and must be freshly prepared.

Determination of Arsenic in Calcium Lactate

Dilute Arsenic Solution AsT. This is a standard solution prepared from arsenic trioxide (0.0132 g), hydrochloric acid (5 ml) and sufficient water to produce 1000 ml. Each ml of this solution contains 0.01 mg of As.

Method. The method which should be used in the preparation of 'the solution to be examined' varies considerably and is specified in the appendix to the British

Pharmacopoeia. The general method is to dissolve the substance in a mixture of water and stannated hydrochloric acid. For the examination of Calcium Lactate the test solution and standard solution are set up as follows:

Test	Standard
(a) Dissolve 5 g of sample in 50 ml of Water and 12 ml of stannated hydrochloric acid, and transfer to the bottle.	Dilute Arsenic Solution AsT (the appropriate amount, which is between 0.2 and 1 ml, as specified). Add 10 ml of Stannated Hydrochloric Acid AsT. Add 50 ml of Water.

(b) Add potassium iodide (1 g) to each solution. This liberates hydriodic acid which assists in the reduction of pentavalent arsenic to the trivalent state from which arsine is then formed by reduction.

(c) Add granulated zinc AsT (10 g) to each solution, and immediately insert the bung with the tube (already assembled with the mercuric chloride paper in position, as in Fig. 2).

(d) Allow the reaction to proceed for 40 minutes at room temperature. Evolution of hydrogen may be slow at first, and if necessary, it is permissible to raise the temperature to 40° to ensure a steady but not too vigorous evolution of gas. Both solutions must be treated in a like manner.

(e) After 40 minutes, compare the test and standard stain. The substance is said to comply with the requirements of the test if the colour of the test stain is not darker than that of the standard stain.

Calculation. Calcium Lactate is required to contain not more than 2 parts per million of As. The colours obtained from 5 g of sample and from 1 ml of Dilute Arsenic Solution AsT are compared. If the colours are equal then:

5 g of sample contains 0.00001 g of As

∴ 500 000 g of sample contains 1 g of As

∴ 1 000 000 g of sample contains 2 g of As (i.e. 2 parts per million).

Modifications of the General Method

Insoluble substances. No special treatment is used for insoluble substances such as Barium Sulphate, Bentonite, Heavy and Light Kaolin and Magnesium Trisilicate. Their insolubility does not interfere with the solution of arsenic and its reduction to arsine. Such substances are therefore merely suspended in water and stannated hydrochloric acid, and the normal test is applied.

Substances which evolve carbon dioxide with hydrochloric acid, or which react vigorously with hydrochloric acid. At this stage, arsenic is usually converted to arsenic trichloride which is volatile and therefore may be carried off with large volumes of carbon dioxide if these are produced at the same time. The use of brominated hydrochloric acid avoids this difficulty, the bromine oxidising arsenic to the pentavalent state, in which it is non-volatile. The following substances are treated in this way: Chalk, Calcium Hydroxide, Magnesium Carbonates, Light Magnesium Oxide, Potassium Hydroxide and Sodium Hydroxide. Ammonia Solution is reduced to low bulk by evaporation before the addition of brominated hydrochloric acid.

Sparingly-soluble organic acids and their salts. Special treatment is necessary to prevent frothing of the liquid in the bottle, which would otherwise occur, and to overcome the difficulty that some of the arsenic might remain occluded in the solid particles.

Free acids are converted to their sodium salts with sodium carbonate and arsenic oxidised to the pentavalent state by evaporating the solution with bromine. The residue is gently ignited until carbonised to destroy organic matter, arsenic being retained as non-volatile sodium arsenate. The residue is dissolved in brominated hydrochloric acid and the test completed in the usual way. Aspirin and Saccharin are treated in this way. Alkali metal salts of organic acids including Sodium Aminosalicylate, Sodium Salicylate and Saccharin Sodium are also examined by a similar method.

Sodium metabisulphite. Acidification of this substance would give a precipitate of sulphur. To prevent this occurring it is oxidised to sodium sulphate by heating with potassium chlorate and hydrochloric acid. Most of the excess chlorine is removed by heating gently, and the final traces with stannous chloride.

Precipitated sulphur. This is converted to yellow ammonium sulphide by digestion with ammonia under reflux. All the arsenic goes into solution as ammonium thioarsenite or thioarsenate, and residual sulphur can be removed by filtration. The filtrate is evaporated to dryness and the residue of sulphur and arsenic pentasulphide boiled with sodium carbonate solution to give a solution of sodium thioarsenate together with some sodium polysulphides. Sulphides are oxidised to sulphate by the addition of bromine, and excess bromine is removed by boiling the solution. The last traces of bromine are removed by addition of stannous chloride, which at the same time reduces the arsenic to the trivalent state, before application of the general test procedure.

Potassium nitrate. This substance produces an oxidising solution with hydrochloric acid and would yield oxides of nitrogen on the addition of zinc. Under these circumstances, no hydrogen would be produced for the reduction of arsenic. The sample is evaporated with concentrated sulphuric acid until the liquid commences to fume. Dilution with water decomposes nitrosylsulphuric acid which may have been formed, the decomposition being completed by a second evaporation until white fumes are again evolved. The usual test procedure may then be applied.

Metals which interfere with the normal reactions involved in the test. Iron will deposit on the surface of zinc, depressing the rate of reaction between the zinc and the acid.

(a) *Ferrous sulphate.* The sample is dissolved in water and stannated hydrochloric acid is added to convert all arsenic to the trivalent state as arsenic trichloride. Arsenic trichloride is volatile and may then be separated by distillation from the other metallic salts present, and the distillate examined in the usual way.

(b) *Ferric ammonium citrate.* Arsenic is fixed as non-volatile sodium arsenate by evaporation with sodium carbonate and bromine solution before the sample is ignited to destroy organic matter. The residue is leached out with brominated hydrochloric acid, arsenic reduced to the volatile trivalent state with stannous chloride and separated by distillation. The distillate is submitted to the normal test procedure.

(c) *Antimony potassium tartrate; antimony sodium tartrate.* Antimony compounds are reduced by zinc and hydrochloric acid to stibine (SbH_3) which reacts with the mercuric chloride paper to give a stain. The sample is distilled with hydrochloric acid to give a distillate containing all of the arsenic, but only a fraction of the antimony, as this is relatively non-volatile. A second distillation removes the last traces of antimony.

Medicinal dyestuffs (Brilliant Green, Methylene Blue, Indigo Carmine and Crystal Violet). The organic matter is destroyed by heating with nitric and sulphuric acids. Nitrous fumes are removed by a double evaporation with sulphuric acid. Stannated hydrochloric acid is added, the solution distilled and the test performed on the distillate.

The Limit Test for Iron

The test is based on the formation of a purple colour by reaction of the iron with thioglycollic acid in a solution buffered with ammonium citrate, and comparison of the colour produced with a standard colour containing a known amount of iron (0.04 mg of Fe). The purple colour is due to the formation of the co-ordination compound, ferrous thioglycollate, ferric iron being reduced to the ferrous state by the reagent:

$$2Fe^{3+} + 2CH_2SH.COOH \longrightarrow 2Fe^{2+} + COOH.CH_2S.S.CH_2.COOH + 2H^+$$

$$Fe^{2+} + 2CH_2SH.COOH \longrightarrow \begin{array}{c} CH_2.SH \\ | \\ CO.O \end{array} \diagdown \underset{\diagup}{Fe} \diagup \begin{array}{c} O.CO \\ | \\ HS.CH_2 \end{array} + 2H^+$$

Determination of Iron in Sodium Chloride

Standard Iron Solution FeT is a solution of ferric ammonium sulphate (0.173 g) containing dilute hydrochloric acid (5 ml) diluted to 1000 ml. Each ml contains 0.02 mg of iron. The suffix, FeT, applied to other reagents used in the Iron Limit Test, is used to denote a substance of usual reagent standard, but of such quality that it complies with additional tests to ensure the absence of iron.

Method. The solutions are prepared in a pair of Nessler glasses as follows:

Test	*Standard*
(a) Dissolve the sample (weight specified in monograph) in Water (35 ml) or treat as otherwise specified in the monograph.	Dilute 2 ml of Standard Iron Solution FeT with 40 ml of Water.
(b) Add 2 ml of 20 per cent w/v aqueous citric acid FeT.	Add 2 ml of 20 per cent w/v aqueous citric acid FeT.
(c) Add 2 drops of thioglycollic acid.	Add 2 drops of thioglycollic acid.
(d) Make alkaline with ammonia solution FeT.	Make alkaline with ammonia solution FeT.
(e) Dilute to 50 ml with Water.	Dilute to 50 ml with Water.

Compare the colour of the two solutions after five minutes. The colour of the test solution should be not more intense than that of the standard.

Qualitative Tests for Metallic Impurities

Other metallic impurities are normally detected by the usual precipitation and colour reactions of inorganic qualitative analysis. A *no-reaction* method is usually adopted in limit tests of this type, i.e. the weight of sample and volume of reagents specified should give no visible reaction.

Heavy Metals

All metals other than alkali metals and alkaline earths are tested for in the simplest possible way. The two reagents which are used are ammonia, and either hydrogen sulphide or sodium sulphide. A combination of the two reagents is also employed.

Table 4 indicates the heavy metal impurities which are precipitated by ammonia under various conditions.

Certain specific conclusions may for example be drawn from the absence of a precipitate or colour on the addition of excess of ammonia to a solution of the sample. Absence of a precipitate or opalescence, would limit Pb, Bi, Al and Fe^{3+} (and the less likely impurities Hg, Sn, Sb). It should be noted that addition of ammonia to a solution of a soluble salt such as sodium chloride, would produce some ammonium chloride (in solution), so that absence of a precipitate would not then be a reliable indication of the absence of Co, Mn or Mg. Absence of colour, would act as a suitable limit test for copper or nickel.

Similarly, Table 5 indicates the heavy metal impurities precipitated as sulphides (with H_2S or Na_2S) under the conditions specified. The following examples serve to illustrate the application of such limit tests to inorganic substances.

Since heavy metals themselves react, the above tests are unsuitable for the detection of other heavy metal impurities in salts of heavy metals, and they are either modified, or if this is not possible, replaced by a more specific test. The following are typical examples:

Limit of zinc and lead *in Copper sulphate.* Copper itself would not be precipitated by H_2S in the presence of potassium cyanide. The absence of opalescence indicates a limit of zinc, and the requirement that there shall be

Table 4

Insoluble in excess NH_4OH	Soluble in excess NH_4OH	Precipitation prevented by adding ammonium salts	Precipitated in the presence of phosphate
Hg^+, Hg^{2+}, Pb, Bi, Sn, Al, Fe (Co) (Mg) (Mn)	Cu ⎫ Ni ⎭ Blue solution Ag ⎫ Cd ⎬ Colourless Zn ⎭ solution Cr Pale violet solution	Co, Ni, Mn, Zn, Mg, Fe^{2+}	Mg, Ca, Sr, Ba

Table 5

Precipitated in the presence of mineral acid	Precipitated in the presence of a weak acid (Acetic acid)	Precipitated by ammonia and hydrogen sulphide	
		KCN absent	KCN present
Ag ⎫ Hg ⎪ Pb ⎪ Bi ⎪ Qualitative Cu ⎬ analytical As ⎪ Groups I Sb ⎪ and II Sn ⎪ (Cd)⎭	Zn, Ni, Co	All the metals in the preceding columns, except As, Sb, (Sn) *Note*: Al and Cr will also be precipitated as hydroxides	A large group of metals form complex salts with potassium cyanide, from which they are *not* precipitated as sulphides by hydrogen sulphide. The following give precipitates in the presence of potassium cyanide: Pb (black), Zn (white), Cd (yellow)

not more than a slight darkening controls the amount of lead, which may be present.

Heavy Metal Impurity in Organic Compounds

Similar tests are also used to control heavy metal contamination in organic medicinal substances. *Copper and iron* are limited in Tartaric Acid by addition of an excess of ammonia and sodium sulphide solution when not more than a slight colour should be produced. *Mercuric salts and heavy metals* in Mersalyl Acid and Phenylmercuric Nitrate are limited by specifying absence of an immediate colour on addition of sodium sulphide in the absence of ammonia. The test for *mercuric salts* in Thiomersal with ammonium sulphide is more stringent, in that the white precipitate which is formed must not darken within thirty minutes. The test for *heavy metals* in Chlorpropamide and Tolbutamide is carried out on the sulphated ash, which is dissolved in hydrochloric acid, and then treated with sodium sulphide and ammonia.

Zinc salts are used extensively in the dyestuffs industry to assist precipitation of organic dyes in a crystalline condition, and although zinc salts are not so used in the production of dyes for medicinal purposes, a test is always included to guard against accidental contamination in this way. The colour of the dye would mask any precipitation test, so the organic molecule is first destroyed by ignition with concentrated sulphuric acid. The residue which contains any zinc as sulphate is dissolved in dilute acid, the solution made alkaline with ammonia and then treated with ammonium sulphide. Any zinc will appear as an opalescence or as a white precipitate. Brilliant Green, Crystal Violet and Methylene Blue are examined in this way.

Acetomenaphthone, which is prepared by the reduction of 2-methyl-1,4-napthaquinone with zinc and acetic acid, must also be examined for contamination by *zinc*. The latter is obtained in solution by boiling with dilute hydrochloric acid in which Acetomenaphthone is insoluble, and then precipitated by adding potassium ferrocyanide solution.

Wool Alcohols may contain traces of copper arising from the use of copper catalysts during manufacture. *Copper* is controlled in the residue remaining on ignition by the specific reaction with sodium diethyldithiocarbamate.

The *heavy metals* test of the European Pharmacopoeia, which is often used in place of a specific lead limit test, makes use of colour development (brown) with thioacetamide using a standard lead solution as control.

Palladium, Nickel and Zinc

Traces of *palladium* in Carbenicillin Sodium, nickel in certain other compounds, and *zinc* in various Insulins are controlled by the use of atomic absorption spectroscopy.

Alkaline Earth Metals

Of these, barium is a likely impurity, since barium salts are often used for the removal of large amounts of sulphate ions. Barium sulphate is insoluble in water and dilute acids, and the metal is invariably tested for by the addition of dilute sulphuric acid. Quinine Hydrochloride and Quinine Dihydrochloride, which are prepared from the corresponding sulphate derivatives by treatment with barium chloride, may be contaminated by *barium*, and this impurity is controlled by the addition of dilute sulphuric acid when no precipitate should be obtained.

Calcium sulphate is very much more soluble than barium sulphate and does not precipitate simultaneously with barium unless calcium impurity is present in very high concentration. Strontium is an unlikely impurity, but would be precipitated inder the conditions of the test for barium. The test for *calcium* in Sodium Sulphate is precipitation as its insoluble oxalate by the action of ammonium oxalate in the presence of ethanol. This test is unsuitable for insoluble substances such as Magnesium Carbonate and Oxide and the sample is dissolved in dilute acetic acid, and ethanol is added. Whilst magnesium oxalate is readily soluble in dilute ethanol, calcium oxalate is quite insoluble and is, therefore, precipitated. The precipitate is compared with a standard calcium oxalate precipitate. In Sodium Acid Phosphate, *calcium* is controlled by precipitation with magnesium as phosphate in the presence of ammonia. In other substances, such as Sodium Sulphate, *magnesium* is controlled separately from calcium, colorimetrically.

Alkali Metals

No specific tests are applied, but tests for these elements are included in various composite tests, as in the following examples. For insoluble substances such as Magnesium Carbonate, determination of *soluble matter* includes alkali metals. The sample is boiled with water, the solution filtered and the filtrate evaporated; the weight of residue is limited.

Sodium in Liothyronine Sodium is controlled within specified limits by gentle ignition with sulphuric acid to form sodium sulphate.

Ammonium Salts

Ammonia is used in the preparation of a number of official organic substances, and these may therefore be contaminated by ammonium salts. The usual test is treatment with sodium hydroxide solution, when there should be no odour of ammonia. A test for *ammonium compounds* in Saccharin involves warming with magnesium oxide in water, a procedure which would liberate ammonia from ammonium salts, but not from the imido group of saccharin.

LIMIT TESTS FOR ACID RADICAL IMPURITIES

The most common acid radical impurities are chloride and sulphate which generally arise from the use of tap water in manufacturing processes. Because of widespread contamination by these two impurities, the British Pharmacopoeia specifies general limit tests for them which are applicable with minor modifications to a larger number of medicinal substances.

The Limit Test for Chlorides

This test, which is mainly used to control chloride impurity in inorganic substances, depends upon the precipitation of the chloride with silver nitrate in the presence of dilute nitric acid, and comparison of the opalescent solution so obtained with a standard opalescence containing a known amount of chloride ions. The opalescence in the two solutions is compared by examination in special vessels known as Nessler Glasses (see p. 25), by viewing them transversely through the solution against a dark background. Borderline cases may often be decided by comparison over the printed page of a book, when the definition of the print is readily discernible and can be used as the criterion of the depth of opalescence.

Determination of Chloride in Sodium Sulphate

Method. The two solutions used in the test are prepared as follows:

Test	Standard
(a) Dissolve the sample (weight specified in monograph) in Water, or treat as otherwise directed in the monograph.	1 ml of *0.01N* HCl
(b) Add 10 ml of dilute nitric acid (omit if used in the preparation of the above solution).	Add 10 ml of dilute nitric acid.
(c) Dilute to 50 ml with Water.	Dilute to 50 ml with Water.
(d) Add 1 ml of 5 per cent silver nitrate solution.	Add 1 ml of 5 per cent silver nitrate solution.
(e) Stir immediately.	Stir immediately.

Compare the opalescence of the two solutions after five minutes. The opalescence of the test solution should not be greater than that of the standard.

The stringency of the test depends on the amount of material which is used. The greater the weight of sample specified, the more stringent does the test become. Apart from such variations as the weight of material examined, the test is always carried out as far as possible under the same conditions. The acidity of the solution is kept equivalent to 10 ml of dilute nitric acid. Additional acid is therefore used in sufficient amount to neutralise alkali metal hydroxides and carbonates, and leave the required 10 ml excess. Insoluble substances, such as the Magnesium Oxides and Carbonates, Calcium Hydroxide and Calcium Lactate, which react with nitric acid, or are dissolved by it, are treated similarly. Additional nitric acid is also used with salts of organic acids such as Potassium Citrate to counteract the buffering action of the acid radical. Nitric acid itself is partly neutralised by the addition of ammonia before the test is applied.

Insoluble substances such as Light Kaolin and Magnesium Trisilicate are boiled with a mixture of water and dilute nitric acid, the solution filtered, and the test applied to the filtrate. Insoluble organic substances such as Promethazine Theoclate and Propyl Gallate are merely shaken with cold water, filtered, and the test applied to the filtrate. A similar modification is applied to metallic salts of aromatic acids such as Sodium Salicylate, Sodium Benzoate, and also Sodium Calciumedetate, after acidification with nitric acid. Chlorbutol ($CCl_3.C(CH_3)_2OH$) is examined as a 5 per cent solution in ethanol (95 per cent), to suppress hydrolysis of the substance itself, which would yield chloride ions in aqueous solution.

Coloured substances, which would interfere, are specially treated before applying the usual test, e.g. Potassium Permanganate is decolorised by reduction with ethanol, filtered from the precipitated manganese dioxide and the test applied to the filtrate.

Special Limit Test for Chloride in Bromides

Chloride in Potassium Bromide, E.P., is limited to not more than 1 per cent, and determined in an extension of the assay. The latter consists of of oxidation with acidified potassium permanganate, which removes the bromide as bromine. The remaining solution which contains only chloride is determined volumetrically by a Volhard titration.

The method used for the control of *chloride* in Sodium Bromide, B.P. similarly depends on oxidation of bromide and its removal as bromine. Total halide is first determined by the Volhard method (p. 224), and the result then corrected to give bromide.

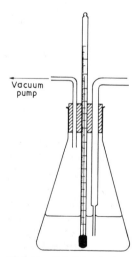

Vacuum pump

Fig. 3. Chloride limit test in bromides apparatus

Method. Boil the sample (1 g) in Water (75 ml) with nitric acid (25 ml) in a 500 ml flask fiitted with a bung carrying a thermometer, a short air outlet tube and a tapered air inlet tube (Fig. 3). Adjust the bung so that the inlet tube is above the level of the liquid. Connect the outlet tube to a source of vacuum (water pump) and allow a gentle stream of air to pass over the surface of the liquid. Heat the solution to boiling, lower the tube into the liquid by inserting the bung into the neck of the flask (as in Fig. 3) and continue to boil the solution for 1 minute. The temperature must be maintained at 105° to 106° throughout this period of boiling, being controlled with the aid of a thermometer immersed in the liquid (see diagram). Allow to cool for 10 minutes with a brisk stream of air passing through the solution. Oxidation occurs as indicated by the equations:

$$KBr + HNO_3 \rightarrow KNO_3 + HBr$$
$$2HBr + O \rightarrow Br_2 + H_2O$$

Add *0.1N* $AgNO_3$ (5 ml pipette) and 5 drops of dibutyl phthalate, and shake. The latter forms a film over the particles of silver chloride so that excess silver nitrate can be titrated without the need for filtration (see p. 221). Back-titrate the excess silver nitrate with *0.1N* NH_4SCN using ferric ammonium sulphate solution as indicator. Not less than 3.7 ml of *0.1N* NH_4SCN should be required.

A limit test for chloride in alkali metal iodides is not necessary since the assay process (KIO_3 titration) is specific for iodides.

Special Limit Test for Chloride (and Bromide) in Iodine

The necessity for this test arises because free iodine is soluble in water sufficiently to interfere in the general test. The difficulty is overcome by reducing the free iodine to iodide by the addition of zinc.

Method. Triturate the sample of Iodine (3.5 g) with Water (35 ml) and filter. Decolorise the filtrate by the addition of a small quantity of zinc powder. Take 25 ml of the solution, add dilute solution of ammonia, and then silver nitrate solution (5 ml) gradually. Silver iodide, which is insoluble in ammonium hydroxide is precipitated, whilst silver chloride dissolves. Filter, dilute the filtrate with Water to 50 ml and remove the ammonium hydroxide by acidification with nitric acid (4 ml). Any opalescence is due to chloride and bromide and should not be greater than the standard opalescence in the general limit test (above).

Special Limit Test for Chloride in Purified Water

A standard which closely approaches complete freedom from chloride impurity is required for Purified Water. A sample (10 ml) is tested with silver nitrate solution (1 ml), without acidification, and should remain clear and colourless for five minutes. This test will also limit other acid radicals precipitated by silver nitrate in neutral solution.

Special Limit Tests for Chloride Contamination in Organic Compounds

The general limit test for chlorides is only directly applicable to those water-soluble organic compounds, which are not precipitated under the conditions of the test. In a few cases, a quantitative test depending upon a similar reaction with silver nitrate in the presence of nitric acid is used. Such a test is applied to Neostigmine Methylsulphate, Sodium Stibogluconate and Sulphobromophthalein Sodium, using smaller quantities of sample and reagents than in the standard test, presumably on grounds of economy. A similar test is also carried out for *chloride* in Glycerol using an exceptionally large sample to increase the stringency of the test. Special semi-quantitative tests are also used for substances which are heavily contaminated with chloride. Liothyronine Sodium and Thyroxine Sodium, which precipitate in acid solution, are first dissolved in alkali, and the solution then acidified with nitric acid and titrated potentiometrically with silver nitrate solution. *Chloride* in Sodium Lauryl Sulphate is controlled by a Mohr titration with silver nitrate after neutralisation with nitric acid.

The general test is not applicable to water-immiscible liquids such as Methoxyflurane, Halothane, Trichloroethylene and Tetrachloroethylene. *Chloride* is extracted by shaking the sample (20 ml) with freshly boiled and cooled Water (20 ml), separating, and testing the aqueous layer with silver nitrate in the presence of dilute nitric acid. The test is stringent and requires that there shall be no opalescence. *Chloride* in Wool Fat is controlled by testing a solution in dilute ethanol with ethanolic silver nitrate and nitric acid using a comparison method.

Limits of Iodide

Iodide in Liothyronine Sodium is controlled by titration with silver nitrate in dilute nitric acid. Colorimetric methods involving oxidation of iodide to iodine, and extraction of the latter into chloroform are applied to Sodium Diatrizoate, Iodipamide Meglumine Injection and Acetrizoic Acid.

The Limit Test for Sulphates

This test is designed for the control of sulphate impurity primarily in inorganic substances. It depends upon the precipitation of the sulphate with barium chloride in the presence of hydrochloric acid and traces of barium sulphate. The latter assists rapid and complete precipitation by seeding. The opalescent solution so obtained is compared with a standard turbidity containing a known amount of sulphate ion.

Determination of Sulphate in Lactic Acid

Barium Sulphate Reagent. Dilute *0.5M* barium chloride solution (15 ml) and sulphate-free alcohol (95 per cent; 20 ml) in a 100 ml graduated flask with Water (55 ml). Add potassium sulphate solution (0.0181 per cent w/v; 5 ml), dilute to 100 ml with Water and shake to mix.

Method. Prepare two solutions in a pair of 50 ml Nessler glasses as follows:

Test	Standard
(*a*) Dissolve the sample (weight specified in monograph) in Water, or treat as otherwise described in the monograph.	Pipette 1.25 ml of *0.01N* H_2SO_4 into the glass.
(*b*) Add dilute hydrochloric acid (2 ml) (omit if used in the preparation of the above solution).	Add dilute hydrochloric acid (2 ml).
(*c*) Dilute to 45 ml with Water (a preliminary graduation of the Nessler glass is necessary).	Dilute to 45 ml with Water.
(*d*) Add 5 ml of *freshly prepared* barium sulphate reagent.	Add 5 ml of *freshly prepared* barium sulphate reagent.
(*e*) Stir immediately.	Stir immediately.

Compare the turbidities of the two solutions after five minutes by viewing the solutions transversely.

The stringency of the tests is controlled in much the same way as in the limit test for chlorides. The acidity of the solution is similarly controlled, being kept equivalent to 2 ml of dilute hydrochloric acid. The solubilities of barium sulphate precipitates are very much affected by the acid concentration.

Other Acid Radicals

Specific tests are also applied for a number of other acid radical contaminants.

Arsenate

Acetarsol is prepared by a process in which aromatic intermediates (phenol, aniline) are condensed with arsenic acid (H_3AsO_4). *Inorganic arsenates* are extremely toxic and careful control is maintained by the addition of magnesium ammonio-sulphate solution to an aqueous solution of the sample. This reagent gives a white precipitate with inorganic arsenates.

Inorganic arsenic in Melarsoprol Injection, which is insoluble in aqueous ammonia, is controlled by dissolving in dilute hydrochloric acid, and treating with hydrogen sulphide.

Carbonate

Carbonate impurity in Amylobarbitone Sodium may arise from contamination with atmospheric carbon dioxide, which becomes fixed as sodium carbonate and at the same time releases free amylobarbitone. Contamination of the substance in this way reduces its solubility and renders it unsuitable for the preparation of Amylobarbitone Injection. Samples are therefore required to give a clear solution in water at a specified concentration, since amylobarbitone is insoluble.

Cyanide

Cyanide in Disodium Edetate is determined by titration with silver nitrate in neutral solution using an adsorption indicator.

Nitrate

Nitrate in Trichloroacetic Acid is controlled by the use of indigo-carmine, a blue dyestuff which is decolorised when oxidised by nitric acid. The sample is dissolved in water and treated with nitrogen-free sulphuric acid and indigo-carmine for one minute, when the blue colour should not be discharged.

Oxalate

Oxalate is a common impurity in organic acids and their salts, due to the use of oxalic acid to remove calcium during the manufacturing processes. Oxalic acid is also used in the isolation and purification of organic bases such as ephedrine, which form well-defined crystalline oxalates. The standard reagent for the detection of these impurities is calcium chloride, with either ammonia or acetic acid, though colorimetric methods, such as that used for example with Sodium Citrate, are more sensitive.

Phosphate

A limit on *total phosphate* in Sodium Phosphate (^{32}P) Injection is imposed by means of a yellow colour reaction with ammonium vanadate and ammonium molybdate in the presence of perchloric acid. The exact composition of the molybdovanadophosphoric acid complex is uncertain. A similar test is used to control phosphate in Tetracosactrin Zinc Injection.

Silicate

Silicates and silica are insoluble in dilute hydrochloric acid, and are usually limited by reference to solubility in this acid. They may also be limited along with aluminium and iron by a precipitation test with ammonia. For example, Chalk is boiled with excess hydrochloric acid, ammonia added, the solution filtered, and the residue ignited and weighed.

LIMIT TESTS FOR NON-METALLIC IMPURITIES

Boron

Boron, which is capable of causing unwanted skin reactions, and is toxic to nervous tissue and muscle, is rigorously limited in compounds such as Salbutamol, where it arises as a result of the use of sodium borohydride in manufacture. Boron is converted to borate and organic matter destroyed by fusion with sodium carbonate. The boron is then determined colorimetrically.

Free Halogen

Organic iodo-compounds such as Di-iodohydroxyquinoline are liable to contain traces of *free iodine*. This is limited by shaking the sample with aqueous potassium iodide and chloroform. Iodine, if present, gives a violet colour in the chloroform layer. Iodipamide Meglumine Injection is examined using starch. *Iodine* in Iophendylate is controlled by measuring the extinction of an aqueous solution at 485 nm due to the absorption of the element. *Chlorine* in Chloroform and in Tetrachloroethylene is limited by shaking with an aqueous solution of cadmium iodide. Iodine is displaced in the presence of chlorine and gives a positive reaction with starch mucilage.

Selenium

Selenium dioxide is used as a reagent in the preparation of the steriod Methandienone. Selenium is toxic and contamination is controlled by an absorptiometric method after destruction of the organic compound with fuming nitric acid. The latter leaves the selenium as selenous acid, which when treated with

3,4-diaminophenylpiazselenol

3,3'-diaminobenzidine under controlled conditions, gives the highly-coloured piazselenol. The latter is extracted with toluene after making the aqueous solution alkaline, and the colour compared with a standard prepared similarly from a known amount of selenium.

THE CONTROL OF ORGANIC IMPURITY IN ORGANIC MEDICINAL SUBSTANCES

Physical Methods

Contamination of organic medicinal substances occurs in exactly the same way as for inorganic materials. The wide range of chemical types, and the even more varied nature of the contaminating impurities covered by official substances, makes the design of general tests for organic impurities difficult. Considerable reliance is now placed on the use of physical separation methods, particularly thin-layer chromatography, and on the measurement of appropriate physical characteristics, in combination with specific chemical tests designed to limit particular likely impurities.

Thin-Layer Chromatography

The speed and ease with which thin-layer chromatographic separations can be made, and the high sensitivity of the method, make this technique a particularly attractive one as the basis of a rapidly-growing number of important specific limit tests. It is used to control *decomposition products* in Chlordiazepoxide Hydrochloride, *free prednisolone* in Prednisolone Sodium Phosphate, and *related foreign steroids* in Fluocinolone Acetonide. Other important applications of thin-layer chromatography include the control of 3-t-*butyl-4-methoxyphenol* in Butylated Hydroxyanisole, and 4-*chlorophenol* in Dichlorophen. In all these examples, the authentic samples of the impurity, which is limited, are available for direct comparison. Such tests, however, suffer from the disadvantage that other impurities arising from alternative methods of manufacture or unexpected contamination may be separated by the solvent system used or may be much less sensitive to the visualising agent employed. Unfortunately, there is no fool-proof answer to this problem.

Other problems arise where for one reason or another reference samples of the expected impurity are not available. In such cases, the usual device is to use suitable dilutions (often 1 or 2 per cent) of the drug substance under examination. This method, however, has the disadvantage that impurity spots, which are referred in terms of intensity to those of the dilutions, may not only have different sensitivities to the visualising agent, but also may well appear at considerable distances from the reference spot, increasing the difficulties of comparing their relative intensities.

Thin-layer chromatography is sufficiently sensitive to provide a means of controlling impurity arising from epimers in optically active compounds. Thus, it is used to control the inactive 4-epi-*tetracycline*, as well as the closely related *anhydrotetracycline* in Tetracycline Hydrochloride. It has also proved valuable in providing a satisfactory method for limiting closely related, but unwanted derivatives in Pethidine Hydrochloride, which otherwise were undetected, even when present in fairly substantial amounts.

Paper chromatography still finds some use in particular limit tests such as that for the control of retinol in Vitamin A Ester Concentrate, *related biguanides* in Phenformin Hydrochloride, and for controlling the composition of Capreomycin Sulphate (≮ 90 per cent Capreomycin I).

Gas-liquid Chromatography

Gas-liquid chromatography is now widely used in the control analysis of pharmaceutical chemicals, wherever its greater sensitivity and efficiency are demanded. A typical example of its use arises in the control of *foreign substances* in Fenfluramine Hydrochloride. The free fenfluramine base is isolated by basification and extraction with chloroform, and chromatographed on a suitable column using authentic Fenfluramine Hydrochloride for comparison against N,N-diethylaniline as internal standard. Similar glc procedures are used both as assay and to control *foreign substances* in Fenfluramine Tablets.

Even when the particular impurity requiring control is identified, as for example *formamide* in Allopurinol, the nature of the impurity, and column characteristics suitable for the separation of the parent compound and impurity may demand that some appropriate, but more suitable alternative substance be used as internal standard. In this case, dimethylformamide, which is readily available, is used. The latter, however, gives a much greater response than formamide to the flame ionisation detector which is used. For this reason, the internal standard is used at a concentration one-eighth that of the limit for formamide.

Gas-liquid chromatography provides an even more sensitive method than thin-layer chromatography for separating and distinguishing between closely related materials, as for example the *cis-* and *trans*-isomers of Tranylcypromine Sulphate.

Ion-exchange Chromatography

A column exchanger, using an appropriate ion-exchange resin in the hydroxyl form, is used to separate *kanamycin B* from *kanamycin A* in Kanamycin Sulphate, and limit the former to not more than 3 per cent of the latter. A similar procedure is used to separate neamine (*neomycin A*), *neomycin B* and *neomycin C*, and to limit the latter to not more than 3 per cent of the content of the main component *neomycin B* in Neomycin Sulphate. The Pharmacopoeia permits these tests to be run as an automated procedure.

Electrophoresis

Both paper and gel electrophoresis are used in certain limit tests. Paper electrophoresis in a formic acid- glacial acetic acid- acetone system is used to control *related substances* in Cephaloridine in a test which uses crystal violet to visualise the mobility of the test and control spots, and cyanocobalamin to indicate the position of the true baseline. The behaviour of the sample under test is compared with that of authentic cephaloridine at 1 per cent and 0.1 per cent levels. A similar test is applied for *related substances* in Cephalexin. Gel electrophoresis is used in a test to limit *benzylpenicillin sodium* in Carbenicillin Sodium.

Absorptiometric Methods

A number of important tests limiting organic impurity in organic medicinal substances, are based on the development of specific colours and comparing them photoelectrically with standard solutions. The Beer-Lambert Law, which applies to such measurements, gives the following relationships:

$$E_{test} = \log_{10} \frac{I_0}{I_{test}} = kc_1t_1$$

$$E_{standard} = \log_{10} \frac{I_0}{I_{standard}} = kc_2t_2$$

so that

$$\frac{E_{test}}{E_{standard}} = \frac{c_1t_1}{c_2t_2}$$

When visual colorimeters are used, for example, the Klett or Dubosqu, the thickness t is varied until $E_{test} = E_{standard}$. At this point

$$c_1t_1 = c_2t_2$$

and

$$\frac{c_1}{c_2} = \frac{t_2}{t_1}$$

As c_2 is the standard, and therefore known, the concentration c_1 can be calculated.

In photoelectric instruments, the thickness t is constant and therefore

$$\frac{c_1}{c_2} = \frac{E_{test}}{E_{standard}}$$

where the extinctions are measured with reference to a blank of reagents. For routine work, a calibration curve can be used, so that even though Beer's Law may not be obeyed, a particular method can still be of use.

The method used for the development of specific colours varies in both method and detail of procedure from compound to compound. A few selected examples are considered.

Acetone in *Methanol.* The detection and determination of *acetone* is based upon the formation of indigo in the presence of *o*-nitrobenzaldehyde and sodium hydroxide.

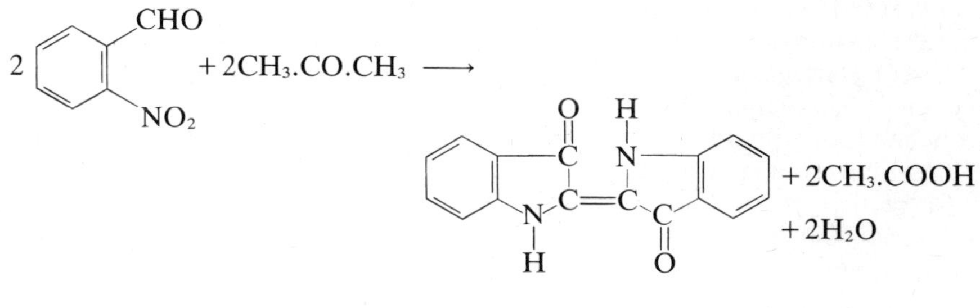

The reaction is carried out in Nessler glasses and the limiting colour corresponds to 0.05 per cent v/v acetone.

3-Aminophenol in *Sodium Aminosalicylate*. 3-*Aminophenol* is detected by treatment with *N*,*N*-diethylphenylenediamine sulphate solution in the presence of dilute ammonia solution. The reaction is given by the following equation:

$$Et_2N-\!\!\!\bigcirc\!\!\!-NH_2 + \overset{H_2N}{\underset{}{\bigcirc}}\!\!\!-OH \longrightarrow Et_2N-\!\!\!\bigcirc\!\!\!-N=\overset{H_2N}{\underset{}{\bigcirc}}=O + 2H_2O$$

The oxidation is effected by potassium ferricyanide. The product is soluble in toluene, and hence is easily freed from any coloured product that sodium aminosalicylate may form by washing with ammonia and water. The limiting colour corresponds to 0.03 per cent of 3-aminophenol.

4-Aminophenol in *Paracetamol*. 4-*Aminophenol* is detected by the blue colour which it gives with alkaline sodium nitroprusside solution. The test conditions are unusual in that 4-aminophenol-free paracetamol is included in the solution used for comparison. The limit corresponds to 0.005 per cent.

3-Amino-2,4,6-tri-iodobenzoic acid in *Iodipamide Meglumine Injection*. 3-*Amino*-2,4,6-*tri*-*iodobenzoic acid* is detected by diazotisation and coupling with 1-naphthol in the presence of sodium hydroxide. The extinction is measured at 361 nm. $E_{1cm}^{1\%}$ for coloured impurities is 207 at 361 nm. The limit is not more than 4.0 per cent.

Chloroaniline in *Chlorhexidine Hydrochloride*. *Chloroaniline*, an intermediate used in the preparation of chlorhexidine, is limited to not more than 0.01 per cent in the solution. The primary amine is diazotised and coupled with *N*-(1-naphthyl) ethylenediamine hydrochloride, excess sodium nitrite being removed with ammonium sulphamate. Any magenta colour produced is compared with a standard obtained by treating 4-chloroaniline in the same way. *Chloroaniline* in Proguanil Hydrochloride is controlled similarly.

Chlorophenol in *Chlorphenesin*. 4-*Chlorophenol* impurity is determined by the action of the phenolic group on sodium molybdophosphotungstate under alkaline conditions. The blue colour due to 'molybdenum blue' is compared with a standard which limits the impurity to not more than 0.055 per cent.

Coloured impurities in *Cyanocobalamin*. *Coloured impurities* are separated from cyanocobalamin by paper chromatography (see Part 2) and, after removal of the cyancobalamin band, are concentrated at one side of the paper. Elution with water gives a solution which is diluted further and filtered. The extinction is measured at 361 nm. $E_{1cm}^{1\%}$ for coloured impurities is 207 at 361 nm. The limit is not more than 4.0 per cent.

Coloured impurities in *Hydroxocobalamin* and *Hydroxocobalamin Injection* *Hydroxocobalamin*. Column chromatography (Part 2) is used to detect other cobalamins which are limited to not more than 3.0 per cent. $E_{1cm}^{1\%}$ at 361 nm is 207 for the other cobalamins. Two columns are used, the first being the weakly basic diethylaminocellulose and the second the weakly acidic carboxymethylcellulose. The weakly basic column retains acidic impurities for which a limit of 3.0 per cent is set. They are removed by elution with 1 per cent solution of sodium chloride and determined by measuring the extinction of the eluate at the maximum between 351 and 361 nm in a 1 cm cell. $E_{1cm}^{1\%}$ is 190 for the acidic impurities. The coloured eluate from the methylcellulose column is examined at 361 nm.

Hydroxocobalamin Injection. After conversion to cyanocobalamin by treatment with hydrocyanic acid, all coloured matter is extracted by small quantities of phenol/chloroform mixture. Phenol is removed by repeated washing of the extracts with anaesthetic ether in a centrifuge tube and the residue is used for the test for coloured impurities described under Cyanocobalamin.

Formaldehyde in *Sterilizable Maize Starch. Formaldehyde* is determined by means of the red colour produced when traces of formaldehyde are treated with phenylhydrazine followed by potassium ferricyanide. The limit corresponds to 0.01 per cent w/w of HCHO.

Formaldehyde in *Oxidised Cellulose.* Traces of formaldehyde are detected in an aqueous extract by the very sensitive reagent chromotropic acid. The colour is compared at 570 nm with that of a standard which corresponds to a limit of 0.001 per cent HCHO. Formaldehyde in Absorbable Gelatin Sponge is controlled similarly.

Free Aminoacid in *Acetarsol.* Incomplete acetylation of 4-hydroxy-3-aminophenylarsonic acid during the manufacture of acetarsol, or hydrolysis during storage, will be indicated by the presence of free amino acid. Detection of this impurity is based upon diazotisation of the primary aromatic amino group and coupling with a suitable phenol to give a colour. Diazotisation is carried out at a low temperature (5°) to avoid decomposition of the diazo compound, and excess of nitrous acid is removed by means of sulphamic acid.

$$HNO_2 + H.SO_3.NH_2 = H_2SO_4 + N_2 + H_2O$$

β-Naphthol is used as the coupling phenol, and a standard colour is produced from a known quantity of amino acid which has been obtained by acid hydrolysis of acetarsol. The limit is quite small, being equivalent to about 0.015 per cent of free amino acid.

Free Sulphathiazole in *Phthalylsulphathiazole. Free sulphathiazole* which would result from hydrolysis on storage is limited to not more than 5 per cent. The test is the formation of a coloured Schiff's base with *p*-dimethylaminobenzaldehyde in an acid buffer solution.

The compound has an absorption maximum at 455 nm, and comparison with a standard of sulphathiazole treated in the same way is carried out at that wavelength.

Free Sulphathiazole in *Succinylsulphathiazole. Free sulphathiazole* in this compound is determined by diazotisation and coupling with N-(1-naphthyl) ethylenediamine hydrochloride under acid conditions. Sulphamic acid is used to remove excess of nitrite ion in the diazotisation procedure. The standard colour prepared in the same way from sulphathiazole corresponds to about 0.75 per cent of free sulphathiazole in the sample.

Noradrenaline in *Adrenaline* and *Adrenaline Acid Tartrate.* The Pharmacopeia directs that adrenaline which has been prepared from natural sources must be substantially free from noradrenaline. The test used to distinguish and limit the noradrenaline is the reaction with sodium β-naphthoquinone-4-sulphonate in alkaline media. Under these conditions, a deeply-coloured *p*-quinonoid condensation product is formed by the reaction of the two amine hydrogens of noradrenaline with the sulphonic acid group of the reagent.

The coloured complex is extracted with toluene, in the presence of benzalkonium chloride. The colour of the organic phase is compared with that obtained by treating a known amount of noradrenaline-free adrenaline acid tartrate in the same way. The excess reagent remains in the aqueous phase and does not interfere.

Other Digitoxosides in *Digitoxin Tablets. Other digitoxosides*, left on the column in the assay process, are eluted with chloroform. After removal of the solvent the residue is treated with glacial acetic acid containing sulphuric acid and a trace of ferric chloride. The colour is compared at 590 nm with that developed with a solution of the reference preparation of digitoxoside treated with the reagent.

Salicylic Acid in *Aspirin. Salicylic acid*, which is formed from aspirin by hydrolysis, is controlled by comparing the violet colour formed with ferric chloride and the phenol with a standard corresponding to a limit of 0.05 per cent salicylic acid in the sample.

Suitable modifications of this test employing preliminary solvent extractions are used to control the quality of aspirin in Aspirin Tablets, Soluble Aspirin Tablets, Codeine Tablets, and Aspirin, Phenacetin and Codeine Tablets.

Ultraviolet Light Absorption Method

The identification of a substance by its absorption characteristics cannot be accepted as conclusive evidence of identity and other tests, chemical and physical, must also be used. The absorption, however, does offer an easy method for the routine control of organic substances and its use can be extended to improve the value of the test. Thus for Cyanocobalamin, Procyclidine Hydrochloride and Phytomenadione the extinctions at several wavelengths are measured and the calculated ratios are limited to certain values. For example, the extinction of the solution used in the assay of Cyanocobalamin is measured at 278, 361 and 550 nm. The ratios of the extinctions at 278 and 550 nm to that at 361 nm should be about 0.57 and about 0.3 respectively. This method is particularly useful in the detection of irrelevant absorption, i.e. absorption caused by impurities.

In other cases, when the medicinal compound is transparent, impurities which absorb light are readily detected. This method is used in the Pharmacopoeia to detect or limit the amount of noradrenalone in Noradrenaline Acid Tartrate. The latter, although absorbing strongly at 279 nm shows no absorption at 310 nm at which wavelength noradrenalone does absorb.

Ultraviolet absorption data for important pharmaceutical chemicals in appropriate solvents are classified by wavelength of peak absorptions in Part 2 of this book.

Infrared Spectroscopy

Infrared spectroscopy is used as a limit test for the biologically inactive, *polymorph A*, in Chloramphenicol Palmitate Mixture, B.P.C. (p. 334). It is also used to control the ratio of *cis-* and *trans*-isomers of *clomiphene* base in Clomiphene Citrate, which have characteristic maxima in carbon disulphide solution at 13.51 and 13.16 μm. The mixture is required to be within the range 30–50 per cent *trans*-isomer.

Chemical Methods

Specific organic impurities are often controlled in much the same way in different organic medicinal substances. Some of the more widely used tests are grouped together in the following sections.

Contamination by Organic Halogen Compounds

Combined halogen is usually converted to halide ion before examination in the usual way. *Halides* in Dimercaprol would represent contamination by 1,2-dibromopropanol, an intermediate in the synthesis of Dimercaprol. The bromine is obtained in an ionised form by refluxing the sample with ethanolic potassium hydroxide. Oxidation of the Dimercaprol to the corresponding disulphonic acid with hydrogen peroxide prevents any interference in the subsequent Volhard titration for bromide. *Chlorinated compounds* (dichlorodiethyl ether) in Vinyl Ether are converted to chloride by refluxing with sodium in amyl alcohol, and determined by Volhard's method. *Chlorinated compounds* in Sodium Benzoate, which arise as a result of the method of preparation from toluene, are controlled by ignition and titration. *Halogen-containing substances* (1,3-dibromopropane and 1-bromo-3-chloropropane) in Cyclopropane, arising from the processes of manufacture, are decomposed by oxidation over platinised quartz, and then examined by a method similar to the general limit test for chloride using a standard solution prepared from potassium bromide.

Contamination by Aldehydes, Ketones, Sugars and Unspecified Reducing Substances (Oxidisable Matter)

A considerable variety of test is used to control contamination by aldehydes. Tests range from the use of sensitive reagents such as alkaline potassium mercuri-iodide solution which gives a yellow precipitate with aldehydes, and is applied to the anaesthetics Chloroform and Ether, to the use of sodium hydroxide in ethanol. The latter reagent brings about polymerisation of aldehydes containing the grouping >CH.CHO, to give a resinous product which is yellow or brown in colour. *Aldehydic substances* in Acetic Acid will reduce sodium metabisulphite quantitatively. *Aldehydes and ketones* in Iso-propyl Alcohol, and *acetaldehyde* in Paraldehyde are limited by adding hy-droxyammonium chloride ($NH_2OH.HCl$) and titrating the acid released (see p. 131). Fehling's Solution gives a red precipitate (Cu_2O) in the presence of *reducing sugars* and is used to control contamination by such substances (presumably lactose) in Lactic Acid.

Several monographs make miscellaneous provisions to guard against the presence of 'reducing substances' and 'oxidisable matter'. Decolourisation of potassium permanganate is a suitable test similar to that used for *oxidisable matter* in Purified Water, provided that the substance to which the test is applied is stable to the reagent, as for example Nikethamide. The same test is also used to limit *reducing substances* in Nitrous Oxide. Where a less sensitive test is required, potassium dichromate and sulphuric acid followed by addition of potassium iodide is used as in the control of *formic acid and oxidisable impurities* in Acetic Acid. In these tests, a limited amount of potassium dichromate is added and the test solution allowed to stand for a specified time. Iodine liberated from unused potassium dichromate by the addition of potassium iodide is titrated with standard sodium thiosulphate solution. Glycerol provides an example of an official substance unstable to both potassium permanganate and potassium dichromate. *Certain reducing substances* are therefore controlled in Glycerol by the much milder oxidising agent, ammoniacal silver nitrate, to which glycerol is stable.

Contamination by Unsaturated Substances

Potassium permanganate is also used as a reagent to test for *cinnamylcocaine* in Cocaine and Cocaine Hydrochloride. Cinnamylcocaine is oxidised at the carbon-carbon double bond by the reagent giving first benzaldehyde and then benzoic acid. Cocaine itself is stable towards potassium permanganate.

A much more delicate test is necessary in the control of *unsaturated substances* in Cyclopropane, which is used as a general anaesthetic. The gas is passed through a specified volume of a solution of iodine monochloride (in glacial acetic acid) and then through a solution of potassium iodide. The two solutions are mixed with the result that unreacted iodine monochloride gives free iodine which can be titrated with $0.1N$ sodium thiosulphate. Iodine monochloride adds readily to organic substances which contain a carbon-carbon double bond in the molecule.

$$\text{C=C} + \text{ICl} \longrightarrow \underset{\text{I} \quad \text{Cl}}{\text{C}-\text{C}}$$

Acetylenic compounds in Trichloroethylene and Tetrachloroethylene are limited by shaking the sample with an ammoniacal solution of copper nitrate and hydroxylamine. The latter acts as a reducing agent to give cuprous salts, which would then precipitate acetylene in the form of red cuprous acetylide.

Contamination by Peroxides

Peroxides may be present in both Solvent and Anaesthetic Ether as a result of light-catalysed air oxidation. Peroxides are toxic and moreover can give rise to mixtures which are explosive when distilled. This danger can be largely removed by distillation in the presence of a small quantity of ferrous sulphate. Removal of peroxide impurity during preparation does not, however, guarantee freedom from peroxides after a period of storage and careful tests are essential to guard against the harmful effects of these impurities. The test is applied by shaking solvent ether with twice its volume of potassium iodide in a stoppered tube. Air must be completely excluded from the tube by filling it to the brim. After standing for 30 minutes in the dark the aqueous phase must not be more yellow, due to liberation of iodine from the potassium iodide (by peroxides), than the same volume of aqueous potassium iodide containing 0.5 ml of *0.001N* iodine. A more sensitive test is obtained for Anaesthetic Ether by shaking under similar conditions with aqueous potassium iodide containing starch mucilage. Iodine liberated must not impart a brown colour to the solution.

Ethyl Oleate, which is used as a solvent for injections of certain steroid hormones, may contain toxic *peroxides* formed by air oxidation at the double bond of the oleate radical. The reaction with potassium iodide must be carried out in a mixed solvent of chloroform and glacial acetic acid, since Ethyl Oleate is insoluble in water. Under these conditions, peroxides liberate iodine from the potassium iodide, which after dilution with water, can be titrated with standard sodium thiosulphate solution. Similar tests for *peroxides* are also applied to Vitamin A Ester Concentrate and Paraldehyde.

Contamination of Alcohol and Ether by Methanol

Such contamination is likely to arise in practice through the use of Industrial Methylated Spirits in place of ethanol. The test, which is used to limit *methyl alcohol* in Ether, depends upon a selective oxidation of methanol to formaldehyde by potassium permanganate in the presence of phosphoric acid. Excess potassium permanganate is destroyed by the addition of a mixture of oxalic acid and sulphuric acid. Formaldehyde so produced can be detected by the addition of Schiff's reagent to give a pink colour. Thirty minutes is allowed for development of the colour with this reagent, which normally gives a rapid reaction, except in very dilute solutions. The absence of colour after this time has elapsed therefore gives a strict control over possible contamination. Ether is not miscible with potassium permanganate solution, and is shaken in a separator with 10 per cent ethanol to transfer any methanol to the aqueous phase.

In the determination of alcohol in galenicals, the distillate obtained must comply not only with the necessary limits for refractive index and specific gravity, but also with a glc test for methanol.

Contamination of Steroids by Related Foreign Steroids

Comparison with authentic samples by thin layer chromatography is used to limit *related foreign steroids* in the adrenocortical steroids (Cortisone Acetate, Fludrocortisone Acetate, Methylprednisolone, Dexamethasone, Hydrocortisone, Prednisolone, and Prednisone). The chromatograms are developed with alkaline triphenyltetrazolium chloride solution, when red spots corresponding to the sample under test and the authentic material appear; there should be no secondary spots.

Contamination of Barbiturates by Neutral and Basic Substances

Neutral and basic substances in Amylobarbitone, and a number of other barbiturates are controlled by dissolving in aqueous sodium hydroxide, extracting with ether, and limiting the weight of residue obtained from the ether on evaporation.

Primary Aromatic Amines in Neutral and Acidic Substances

Considerable use is made of the diazo reaction of primary aromatic amines to limit their presence in related neutral and acidic drugs, which do not react in the test. *Chloroaniline* in Proguanil Hydrochloride and *sulphathiazole* in Succinylsulphathiazole are controlled in this way. A solution in dilute hydrochloric acid is diazotised with sodium nitrite and coupled with N-(1-napthyl)ethylenediamine hydrochloride. The colour produced should not be more than that of a comparison standard. 3-*Amino*-2,4,6-*tri-iodobenzoic acid* in Iodipamide Meglumine Injection is limited by a similar reaction using 1-napthol as coupling agent.

Free Bases in Neutral Compounds

Pyridine in Cetylpyridinium Chloride and *ammonia* in Benzalkonium Chloride Solution, which are instantly recognised by odour, are controlled by an olfactory test, after treating the drugs with sodium hydroxide solution. *Free amines* in Crotamiton are limited by dissolving in ether, extracting with hydrochloric acid and evaporating the acid extract. The weight of residue is limited.

Contamination of Alkaloids and their Salts by other Alkaloids

Plants usually produce not a single alkaloid, but rather a whole group of chemically related substances. The relationship is often quite close e.g. quinine and the related cinchona alkaloids; morphine and certain opium alkaloids; consequently separation may be a tedious and lengthy process. Contamination by other alkaloids is therefore something which is extremely difficult to exclude. Nevertheless it must be kept under control and most monographs contain some provision for the exclusion (or limitation) of some such impurity. Some of these tests, with a brief explanation are summarised in Table 6 (p. 52–3).

Contamination of Synthetic Substances by Chemical Intermediates

Synthetic organic medicinal substances are liable to be contaminated by traces of the intermediates used in their production, and many official monographs make provision for tests to exclude definite impurities which arise in this way. Some of the more important of these tests are set out in Table 7 (p. 54–5).

Control of Fixed Oils, Fats and Waxes

The close similarity between the physical constants (Table 8) for various fixed oils lays the more expensive ones open to sophistication by admixture with cheaper oils. Adulteration of this kind is not always readily detected by chemical determination of acid value (p. 136), saponification value (p. 151), iodine value (p. 204) etc., and some special tests are therefore incorporated in the various official monographs. General tests for arachis oil, cottonseed oil and sesame oil in other oils are described.

Test for the Absence of Arachis Oil in Other Oils

This test depends upon the very much lower solubility and higher melting point of arachidic acid, the product of hydrolysis of arachis oil, than the acids obtained from the oil under examination (Almond, Maize and Olive Oils). The conditions of experiment must be adhered to rigidly. Boil 1 ml of the oil with *1.5N* potassium hydroxide under reflux for 10 minutes. Add 70 per cent ethanol (50 ml) and concentrated hydrochloric acid (0.8 ml). With a thermometer immersed in the liquid, cool slowly with continuous stirring (rate of cooling 1° per minute). No turbidity should appear above 4° for Almond Oil, above 11° for Maize Oil or above 9° for Olive Oil. Should a precipitate be obtained, further detailed tests may be applied, involving multiple recrystallisation of the precipitate, which should not melt above a stated temperature.

Test for the Absence of Cottonseed Oil in Other Oils

Mix equal volumes of the oil, amyl alcohol, and carbon disulphide containing 1 per cent of precipitated sulphur. Place in a glass-stoppered boiling tube and tie in the stopper. Immerse in a boiling water-bath to one third of its depth and heat for 30 minutes. There should be no crimson colour. It is best to carry out the test on an authentic specimen at the same time. The test is not sensitive to much less than 10 per cent of cottonseed oil in the sample. It is applied to Almond, Arachis, Maize, and Olive Oils.

Test for the Absence of Sesame Oil in Other Oils

Mix 2 ml of oil with 1 ml of concentrated hydrochloric acid containing 1 per cent w/v of sucrose, and set aside for 5 minutes. The acid layer should not be coloured pink. The test is sensitive only for concentrations greater than about 5 per cent, and is applied to Almond, Arachis, Maize and Olive Oils.

Tests Specifically Applied to Individual Oils

A number of other less general tests are also applied in the same way, similarly to guard against adulteration, and these are dealt with under the individual oils.

Almond Oil. Tests are included to guard against adulteration with cheap oils such as *apricot-kernel and peach-kernel oils*. Shake 5 ml of oil vigorously with a mixture (freshly prepared) of equal parts by weight of sulphuric acid,

Table 6

Official substance	Other alkaloid	Test	Remarks
Apomorphine Hydrochloride	Morphine	Addition of solution of potassium mercuri-iodide to a solution in dilute hydrochloric acid does not give more than a slight opalescence	1. Apomorphine is prepared synthetically from morphine 2. Amorphine gives no reaction with Mayer's reagent; morphine gives a gelatinous precipitate
Codeine Phosphate	Morphine	A solution in dilute hydrochloric acid treated with sodium nitrite followed by ammonia should give a yellow colour not deeper than that similarly obtained from a stipulated amount of morphine	Codeine does not give a yellow colour under these conditions
Diamorphine Hydrochloride	Phenolic substances	A solution in dilute hydrochloric acid treated with sodium nitrite followed by ammonia should give a yellow colour not deeper than that similarly obtained from a stipulated amount of morphine	This colour reaction is only given by phenolic alkaloids in the morphine series
Emetine Hydrochloride	Cephaëline and other alkaloids	Thin-layer chromatography using authentic samples of cephaëline, isoemetine and o-methylpsychotrine for comparison	
Homatropine Hydrobromide	Atropine, hyoscyamine and hyoscine	A small quantity treated with fuming nitric acid and evaporated to dryness on a water bath does not give a violet colour on further treatment with acetone and methanolic potassium hydroxide	The test is a specific colour reaction for atropine and a few closely related compounds, and homatropine does not interfere. A positive reaction is associated with the tropic acid portion of the molecule

Substance	Impurity	Test	Remarks
Hyoscine Hydrobromide	Foreign alkaloids	An aqueous solution gives no turbidity on the addition of ammonia	Hyoscine is sufficiently soluble in water not to be precipitated under the conditions of the test, whilst the other alkaloids are insoluble
Morphine Sulphate	Other alkaloids	A solution in sodium hydroxide yields only a limited residue to chloroform	Morphine, which is a phenolic alkaloid, gives a water-soluble sodium derivative; other non-phenolic alkaloids (e.g. codeine) are extractable by chloroform
Papaverine Hydrochloride	Codeine and morphine	A solution in dilute hydrochloric acid treated with vanillin, does not give a violet colour (iodine) when shaken with carbon tetrachloride	The reaction is a specific test for codeine and morphine. Papaverine does not interfere
Pilocarpine Nitrate	Foreign alkaloids	An aqueous solution gives no turbidity either with ammonia or solution of potassium dichromate	Pilocarpine is not precipitated from aqueous solution at the concentration used in the test, whereas 'other alkaloids' are insoluble
Quinine Sulphate	Other cinchona alkaloids	Thin-layer chromatography in comparison with authentic cinchonine	
†Digitoxin	Digitonin	A solution in ethanol (95 per cent) does not give a precipitate with an ethanolic solution of cholesterol	Digitonin forms an insoluble complex with $3\text{-}\beta\text{-}$hydroxysteroids such as cholesterol
†Digoxin	Gitoxin	A solution in chloroform-methanol (1:1) treated with hydrochloric acid and glycerol shows an absorption maximum at 352 nm not greater than that due to a specified amount of gitoxin	

† Note, these are glycosides, not alkaloids.

Table 7

Official substance	Contaminant	Test	Remarks
Aspirin	Salicylic acid	A solution in diluted ethanol gives a violet colour with ferric chloride, which is not greater than that produced by a standard of salicylic acid	The violet colour is characteristic of the phenolic hydroxyl group, which is acetylated in acetylsalicylic acid (and therefore non-reactive). The test serves to control hydrolysis which may occur when Aspirin is stored under unsuitable conditions
Benzocaine	Acidity	A solution in diluted ethanol requires only a limited volume of 0.1N sodium hydroxide for neutralisation to phenolphthalein	The test limits unesterified p-aminobenzoic acid
Benzoic Acid (and Benzyl Alcohol)	Chlorinated compounds	Ignition with sodium carbonate, acidification and treatment with silver nitrate	The chloro compounds are nuclear-substituted aromatic chloro compounds formed as by-products in the production of benzoic acid by chlorination and oxidation of toluene
Carbimazole	Methimazole (2-Mercapto-1-methylglyoxaline)	Thin-layer chromatography in comparison with authentic methimazole	
Mepacrine Hydrochloride	3-Chloro-7-methoxyacridone	A sample extracted with anaesthetic ether shows only a limited fluorescence	3-Chloro-7-methoxyacridone may be produced by decomposition of mepacrine. The former is soluble in ether whereas the latter, being present as a salt, is insoluble
Methoin	5-Ethyl-5-phenyl hydantoin	Treat with ammonia and extract with chloroform. The ammoniacal solution on acidification, and extraction with chloroform should yield no residue (in the chloroform)	The imide-nitrogen of the impurity leads to solubility in ammonia not shown by the parent compound

Substance	Impurity	Test	Remarks
Methylthiouracil (Propylthiouracil)	Thiourea	A solution (saturated) treated with sodium acetate and silver nitrate gives only a limited colour	Thiourea is used in the synthesis of methylthiouracil. Sulphur present in thiourea is labile under the conditions of the test and gives a brown precipitate of silver sulphide
Neostigmine Bromide	3-Hydroxy-NNN-trimethyl-anilinium bromide	Soluble in chloroform to give a clear solution	The test is based on the difference in solubilities in chloroform
Phenacetin	p-Phenetidin	An aqueous ethanolic solution treated with one drop of $0.1N$ iodine and boiled does not acquire a red tint	
Phenobarbitone	Phenylbarbituric acid	The sample gives a clear solution when boiled with ethanol	
Riboflavin	Lumiflavin	A solution of specified strength in chloroform is not deeper in colour than that of a comparison standard	Lumiflavin is one of the decomposition products of riboflavin which are formed on exposure to light. Lumiflavin, unlike riboflavin, is soluble in chloroform
Saccharin	Foreign substances	Thin-layer chromatography in comparison with authentic 4-sulphamoylbenzoic acid and toluene-2-sulphonamide	
Stilboestrol	3,4-Di-(4-methoxy phenyl) hex-3-ene	Extinction in dehydrated ethanol ≯0.5 at 325 nm	
Theophylline	Caffeine and theobromine	A clear solution is obtained in either potassium hydroxide solution or dilute ammonia solution	Theophylline is soluble in ammonia; caffeine and theobromine are not
Urethane	Urea	A strong aqueous solution treated with nitric acid gives no precipitate	Urea nitrate is obtained as a crystalline precipitate under the conditions of the test

Table 8

Oil, fat or wax	Acid value	Saponification value	Iodine value	Ester value	Ratio number	Unsap. matter	Wt./ml.	Refractive index	Other characteristics
Almond Oil	≯2	188–196	95–102	—	—	—	0.910–0.915	1.470–1.473	Freezing point ≯ –18°
Arachis Oil	≯0.5	188–196	85–105	—	—	—	0.911–0.915	1.468–1.472	
White Beeswax	18–24*	—	—	70–80	3.3–4.2	—	—	—	Mp 62–64°
Castor Oil	≯2	177–187	82–90	—	—	—	0.953–0.964	1.477–1.481	Optical rotation ≮ +3.5°; Acetyl value 140
Cetostearyl Alcohol	†	≯1	≯3	—	—	—	—	—	Determination of alcohols: Solidifying point 45–53°
Cod-liver Oil	≯1.2	180–190	150–180	—	—	≯1.5%	0.917–0.924	1.478–1.482	
Emulsifying Wax	†	≯1	≯3	—	—	≮88	—	—	Determination of alcohols
Ethyl Oleate	≯0.5	—	75–84	—	—	—	0.869–0.874	—	
Halibut-liver Oil	≯2	≯180	≮112 (PyBr₂)‡	—	—	≮7%	0.915–0.925	—	Iodine value glycerides 112–150 (PyBr₂)‡
Oleic Acid	195–202	—	85–92	—	—	—	0.889–0.895	—	
Olive Oil	≯2	190–195	79–88	—	—	—	0.910–0.913	1.468–1.471	
Soft Soap, fatty acids	≯205	—	≮83	—	—	—	—	—	Solidifying point ≯31; Limit test for resin
Theobroma Oil	≯4	188–196	35–40	—	—	—	—	1.456–1.458 at 40°	Mp 31–35°
Undecanoic Acid	—	—	135–140	—	—	—	—	1.448–1.450 at 25°	Freezing point 21–24°
Wool Alcohols	≯2	≯12	—	—	—	—	—	—	Mp ≮58°
Wool Fat	≯1	90–105	18–32*	—	—	—	—	—	Mp 36–42°

* Special conditions   † Limit test for acidity   ‡ Pyridine Bromide

fuming nitric acid and water, keeping cool. The whitish mixture should show no pink colour after fifteen minutes.

Arachis Oil. A test for *other vegetable oils* is included which is based on the general test for the absence of arachis oil in other oils (p. 51). Boil 1 ml of the oil under reflux with *1.5N* ethanolic potassium hydroxide for five minutes. Add 1.5 ml of acetic acid and 70 per cent ethanol (50 ml), warm until the solution is clear and then cool slowly with a thermometer in the liquid. The solution should commence to become turbid at a temperature not lower than 37°.

Cetostearyl Alcohol. *Hydrocarbons* are limited by chromatographing a solution in light petroleum (b.p. 40–60°) on a column of alumina under specified conditions and weighing the residue obtained from the first portion of eluate.

Cod-Liver Oil. The presence of *stearin* is controlled by requiring that the oil shall remain clear and bright when cooled to 0° and maintained at that temperature for 3 hr. The value of the oil depends largely on its content of vitamins A and D. Vitamin A is determined by a spectrophotometric method, and Vitamin D by biological assay.

Oleic Acid. Tests are included to exclude adulteration by *mineral acids, neutral fats* and *mineral oils*.

Mineral acids. On shaking with water, the aqueous phase, after filtration, is not acid to methyl orange.

Neutral fats and mineral oils. Boil 1 ml with N Na_2CO_3 and water (25 ml); oleic acid is soluble. The solution, while hot, is clear or, at the most, opalescent. Neutral fats and mineral oils are insoluble in dilute aqueous sodium carbonate.

Congealing point. Dry a sample by heating at 110°, with constant stirring. Cool a small sample in a tube (20 mm in diameter), immerse in a suitable water-bath at 15° and cool steadily at a rate of 2° per minute, stirring continuously. It does not become cloudy above 10°, and congeals to a white solid mass at about 4°.

Undecanoic Acid. A limit of *neutral fats* and *mineral oils* is imposed by the requirement that a sample boiled with aqueous sodium carbonate should give a clear solution whilst hot.

Vitamin A Ester Concentrate. An Acid Value is imposed to guard against decomposition of the vitamin. *Retinol* is controlled by the application of paper chromatography (Part 2).

Wool Fat. *Paraffins* are limited by chromatography in light petroleum on an alumina column, and elution of paraffin hydrocarbons in the same solvent.

RADIOPHARMACEUTICALS

Radionuclide Purity

Standards for most radiopharmaceuticals incorporate a test for *radionuclide purity*. This in effect is the proportion of the total redioactivity present in the form of the named radionuclide, including that due to daughter radionuclides, which by convention are not considered as impurities. Ideally, the activity of every radionuclide present should be measured. This, however, is seldom practical, and the usual method is to examine for certain specific radionuclides by a defined method. This is usually gamma scintillation spectrometry, either directly or, if long-life impurities are present in short-life radionuclides, by observation of time-dependent spectrum changes. The usual level of radionuclide purity demanded is not less than 99 per cent within the life of the product specified by the expiry date.

Some short-lived radionuclides present special problems. Sodium Pertechnetate (^{99m}Tc) Injection is usually prepared in a generator from molybdenum-99. The resulting technecium-99m has a half-life of only 6 hr, and the standard for the Injection is such that up to the date and hour at which the product may be used, not more than 0.01 per cent of the total radionuclides must be due to radionuclides other than technecium-99m or its decay product technecium-99. The detection of precursor molybdenum-99 can only be properly achieved by waiting until the product technecium-99m has decayed to a sufficiently low level. Such a procedure, however, is precluded by the short shelf-life of the Injection. A much more approximate test is, therefore, used in which the gamma-ray spectrum of the injection solution is examined through a lead shield (6 mm). The latter absorbs the lower energy (0.140 MeV) technecium-99m radiation, so that any response recorded by the detector is due to the more energetic (0.780 MeV) molybdenum-99.

A further problem arises with certain radiopharmaceuticals, as for example, Chlormerodrin (^{197}Hg) Injection, and similar short-life injections, which are sterilised by *filtration*. The half-life of mercury-197 is only 64 hr, so that the *test for sterility* required of an injection prepared by filtration, cannot be completed until after the injection has been used. The Pharmacopoeia, however, nonetheless requires a sterility test to be completed, so that the general procedures used in preparation are subject to examination, albeit in retrospect. Thus, whilst the control of each preparation is negative, there is a positive on-going supervision of the general preparation process.

Radiochemical Purity

The *radiochemical purity* of a radiopharmaceutical product is a statement of the proportion of radionuclide in the designated chemical form. Thus, not less than 95 per cent of the iodine-131 in Sodium Iodohippurate (^{131}I) Injection is in the form of sodium *o*-iodohippurate. The determination of radiochemical purity usually involves some form of chromatography which is specific for the particular drug substance. Thus, a dilution of the Sodium Iodohippurate (^{131}I) Injection is diluted to contain about 1mCi per ml. Paper chromatography of this dilution under standardised conditions should yield a chromatogram with the main spot characteristic of *o*-iodohippuric acid, visible in u.v. light at 366 nm, in which not less than 95 per cent of the radioactivity is concentrated.

In some cases, non-radioactive parent compound is used as a carrier to aid detection of the spots on the chromatogram. In the special case of L-Selenomethionine (^{75}Se) Injection, L-methionine rather than L-selenomethionine is used as the carrier in order to limit the extensive oxidation of sample selenomethionine which would otherwise occur.

2 Registration and Assessment of Medicines

Legislation

Under the Medicines Act, 1968, *product licences* are required for the manufacture of all medicines for sale and supply in the United Kingdom, for importation and in some cases for export. Similar provisions relate to the issue of *clinical trial certificates* for the trial of medicinal products by doctors and dentists in human patients, and to the issue of *animal test certificates* for comparable trials of veterinary products.

All licences and certificates are issued by the appropriate Licensing Authority which is the Department of Health and Social Security, Medicines Division, for the issue of product licences for human medicines and clinical trial certificates, and the Ministry of Agriculture, Fisheries and Food for veterinary product licences and animal test certificates. In practice, *product licences* for human medicines and *clinical trial certificates* are issued by the Licensing Authority normally on the recommendations of the *Committee on Safety of Medicines*, whose task it is to advise on the safety, quality and efficacy of the products concerned. Additionally, the Committee on Safety of Medicines is also charged with the collection, investigation and dissemination of information on *adverse reactions* of drugs and medicines, so that effective advice on matters of safety may be given.

Under the Act, the Licensing Authority before refusing to issue a product licence or issue a clinical trial certificate must consult the Committee, and before issuing a licence or certificate may require amendments to be made to the application. If the Committee have in mind a refusal, however, the applicant has the right to be heard and/or to make representations in writing to the Committee. He may appeal to the Medicines Commission against any adverse recommendation of the Committee on Safety of Medicines. Additionally, even when licences and clinical trial certificates are in force, the Licensing Authority has the power to suspend, revoke or vary them for various reasons including matters of safety or failure to maintain agreed standards or conditions.

Clinical Trial Certificates and Product Licences

Applicants for product licences and clinical trial certificates are required to present individual applications in respect of each product or trial giving appropriate data on the lines set out in *notes for guidance* issued by the Licensing Authority. In general, the information required on chemistry, pharmacy, standards and animal studies for the issue of a clinical trial certificate is much the same as for a product licence to cover the manufacture, sale or supply of a product, the clinical effectiveness of which has already been supplied. In both instances, the product is for administration to human subjects, and safety is all important. There can be no question of double standards, as is sometimes argued on the grounds that fewer patients are at risk in a clinical trial than when the product is generally available on licence. If anything, risk to the individual is greater at the clinical trial stage, when the drug is still untried and its safety still not fully evaluated. Adequate safeguards in matters of safety are, therefore, every bit as important at the stage of clinical trial if not more so than at the time of issue of a product licence. Quality, too, must be safeguarded for the purposes of a clinical trial just as much as for the manufacture and sale of a product of proven efficacy, since, in the absence of adequate safeguards, the trial may well be rendered uninformative, or in an extreme case be completely vitiated if a product of doubtful composition or stability were used.

Some relaxation on the complete characterisation of potential impurities or in the matter of complete stability studies on tentative formulations of dosage forms, which may well require modification in the course of a clinical trial, may be conceded provided there are reasonable safeguards. Thus, the method by which a particular compound is synthesised for the purposes of a clinical trial may not necessarily be the method ultimately used for the production of much larger batches for sale when a product licence is ultimately issued. In these circumstances, the impurity patterns may be completely different, and the need for an extensive investigation and characterisation of all significant impurities in compounds for clinical trial might on occasion be obviated by the presentation of more limited data, provided this is adequately supported by sufficient additional information showing the product to be reasonably free from toxicity under the maximum dosage regimens proposed for the trial. Nevertheless, specifications must be such as to ensure that the essential characteristics of both the drug substance and its dosage form remain reasonably constant throughout the trial, and are reproducible in any product proposed in subsequent marketing applications. Otherwise, the proposed trial ceases to be a meaningful exercise as a guide to the safety, quality and efficacy of the final marketable product.

On similar grounds, stability testing of the drug substance, which may on occasion be related to its particular impurity pattern, and stability studies of the provisional dosage form may in certain circumstances reasonably be less extensive than those required for the issue of a product licence. The essential criterion is that they should relate appropriately to the method of handling and storage proposed and to the detailed arrangements to control the life of the product and the disposal of unused remainders during the course of the trial. Apart from any such provisos limiting the stringency of requirements for products on clinical trial, requirements for submissions both for clinical trial certificates and for product licences are essentially the same in respect of basic information on active ingredients (drug substances) and other components of medicines (dosage forms). Product licence submissions must, however, provide appropriate evidence of clinical efficacy, and safety in use in patients.

Codes of Good Manufacturing Practice

The quality of medicinal products is dependent just as much on the conditions under which they are manufactured, as on the requirements of the product licences granting authority for their manufacture. The development within the industry over many years of practices designed to ensure reliability and safety in manufacture have resulted in the establishment of guidelines to good manufacturing practice which all manufacturers of repute would wish to emulate.

These now widely accepted practices, which stem from the high ideals of dedicated production teams within the industry, have been codified and issued in written form by such international authorities as the World Health Organisation and various other national authorities, as for example in the *Guide to Good Pharmaceutical Manufacturing Practice* published by H.M.S.O. (1971) on behalf of the Department of Health and Social Security. The latter provides a guide to good practice in such matters as:

(a) the location, construction and adaptation of buildings,
(b) the design, construction, location and maintenance of equipment,
(c) cleanliness, and general and personal hygiene,
(d) production procedures and the keeping of records,
(e) quality control,
(f) transportation and storage, and
(g) supervision.

Special provisions for safeguarding the manufacture of sterile products apply not only to all injectable products, but also to certain other products specially liable to microbial contamination, such as eye preparations and solutions for internal irrigation (Report on the Prevention of Microbial Contamination of Medicinal Products, H.M.S.O. 1973).

Quality control should be the responsibility of a specifically designated *Quality Controller* appointed by the management. His independence and authority are critical to the success of any policy of good manufacturing practice. He must be completely free of all responsibility for actual production processes, so that objective criticism of production methods can be made should this be necessary. The Quality Controller should also be equally independent of other divisions of the firm, so that all decisions on matters of quality and safety may be reached on their merit without pressure or the threat of being overruled on grounds of commercial expediency.

Satisfactory laboratory facilities must be provided under the responsibility and supervision of the Quality Controller to permit such testing that he may deem necessary to decide on the acceptance or rejection of raw materials, materials in-process, finished products and packaging materials. Raw materials may only be released for use in manufacture when the batch has been passed as satisfactory by the Quality Controller. Where in-process control is in effect, material at each stage must have the approval of the Quality Controller before being passed to the next stage of the process. Similarly, the release or rejection of each batch of finished product for packaging and for distribution must have the agreement of the Quality Controller.

The duties of the Quality Controller must also include the compilation and approval of Specifications for all active ingredients and excipients used in the process, for in-process controls, and for finished products. Duties also extend to examination and evaluation of the stability of each product, the retention of representative samples from production batches and the determination of shelf-life and expiry dates. In order to ensure true independence of control, samples for analysis should be taken by the quality control staff using approved sampling methods. The only allowable exception to this rule is in respect of samples for in-process control, which may be taken by the production staff, provided laid-down sampling procedures are followed.

The preservation of analytical reports relevant to each batch, and of the signed and dated authorisations for each batch of product released for sale, is imperative.

PRODUCT LICENCE APPLICATIONS

Applicants for product licences are required to submit particulars of the product according to the official notes for guidance under the following headings:

    (a) Summary of particulars
    (b) Chemistry and Pharmacy
         (i) Active Ingredients (Drug Substances)
         (ii) Dosage Form
    (c) Reports of Experimental and Biological Studies
    (d) Reports of Clinical Trials

The division of information is largely one of convenience for assembly, presentation and submission of data for assessment, but it cannot be emphasised too strongly that it is the sum total of all information presented which must be considered by the Licensing Authority in terms of safety, quality and efficacy, as required by the Medicines Act. The work of the pharmaceutical analyst is concerned primarily with quality control, but the extent and effectiveness with which controls are applied also depends very much on the need for the product to meet requirements for safety and efficacy. Accordingly, whilst the emphasis in the sequel rests primarily on the requirements in respect of the Chemistry and Pharmacy section of submissions, its relationship to information presented under the other sections must always be kept continuously in mind. Applicants should, therefore, take care to ensure that a reasonable and meaningful correlation exists between data recorded in different laboratories and presented in separate sections of the submission. This can only be achieved by careful appraisal and editing of all the information by one person, who should be an experienced scientist with a broad overall view of the submission's requirements.

Whilst general guidelines are issued, the precise information to be supplied is a matter of judgement on the part of the applicant. Clearly, it varies according to the type of product, depending upon whether the application is for an entirely new drug, a new presentation of an existing drug, a new combination of existing drugs, or merely a re-formulation for the purpose of updating and improving the product, or meeting some change in source or supply of a particular excipient. The precise format is, therefore, a matter for decision on the part of the applicant, but whatever its form and content, it should be, if nothing else, accurate, scientifically sound, strictly relevant, concise, and free from major inconsistencies.

Summary of Particulars

General Information

The purpose of this section of a submission is to give certain general information relating to the applicant, the licencee and the product. Thus, it should clearly indicate whether or not the product is to be manufactured in the United Kingdom, and in the case of importation or assembly of the final product, the precise origin of the imported material. Constituents of medicines or complete products, manufactured overseas to national pharmacopoeial or other foreign standards, must be shown to conform either to the standards in force in the United Kingdom for use of such substances in medicine (i.e. British Pharmacopoeia or, where appropriate, European Pharmacopoeia standards), or if they do not precisely meet the appropriate standards, to deviate from them only in trivial respects which do not affect their safety or stability. In this connection, it cannot be assumed that the Licensing Authority and its advisers are *au fait* with standards published in languages which are unfamiliar and little used in the United Kingdom. The onus is, therefore, on the applicant to ensure that his application is complete in such details, including authenticated translations of relevant monographs, if these are appropriate.

Description and Name of Product

The Summary of Particulars should also provide a brief description of the pharmaceutical form of the medicine (i.e. tablets, capsules, suspension, cream, injection, etc.), including size, shape, colour and markings of solid dosage forms (tablets and capsules) a statement of the active ingredients, and an indication of the manner in which these will be stated on the label of the medicine and in any associated descriptive material. British Approved Name(s) should be used, but in the absence of an Approved Name(s), the International Non-Proprietary Name, or failing that, the Trade Name(s) will do.

Physical Properties

The physical characteristics of the active constituents should be reported. These may range from such basic characteristics as colour, odour, taste, melting point, specific gravity, refractive index, viscosity and light absorption, which taken together provide a proper basis for identification, to more specialised characters, such as crystal habit, polymorphism, particle size, specific surface area, and bulk density. The latter are important criteria in determining the batch to batch consistency of both drug substances (active ingredients) and finished products, and the availability in terms of release rate, blood levels and clinical efficacy of active ingredients from the product (Chapter 1).

Information on solubilities relates in some degree to the choice of dosage form and proposed route of administration. For solid dosage forms, and suspensions administered orally or as depôt injections, drug solubility provides a useful criterion by which to assess the need for controlled release rates and uptake of the medicament from the product (i.e. bio-availability).

Solubilities in water and chloroform provide the best indication of likely behaviour in terms of absorption, distribution and excretion in animals and human subjects, since solubility in chloroform is a reasonably good guide to solubility in biolipids. In submitting data on acids, bases and their salts, it is important to bear in mind the effects of pH on dissociation, both in dissolution tests and biological media. The pK_a, and both water and chloroform solubilities of the parent acids and bases as well as solubilities of the salt actually used in compounding the medicine, provide valuable data against which to assess the potential efficacy of a particular dosage form in terms of its bio-availability. For similar reasons, an inappropriate choice of solvent in an *in vitro* dissolution test can, however, give misleading indications. Thus, Chlorpropamide, which has a weakly acidic sulphonamido group, has been known to show good release rates from tablets measured in Tris buffer (triethanolamine), due to the formation of a water-soluble salt, despite the fact that its bio-availability from the same batch of tablets, measured in terms of actual blood levels, may be poor. The latter correlates with its low water-solubility, and the limiting factor in uptake from the acidic milieu of the upper gastro-intestinal tract is, in fact, the specific surface area of the powder.

Recommended Clinical Use and Dosage

Statements on the recommended clinical use, the dosage and route of administration are of material concern in deciding acceptable levels of permitted impurities. Levels of trace elements and other impurities, which would be acceptable in a product for administration at milligram level on a single occasion, may be wholly unacceptable in comparable treatments requiring dosage in grams, or for repeated administration, even at intermediary dose levels over long periods of time. Other things being equal, control of impurity patterns might reasonably be less stringent in a highly potent medicament which is only required for occasional administration than for, say, an antibiotic used in the treatment of a deep-seated infection, such as tuberculosis, or a symptomatic drug for the relief of rheumatism, hay fever, catarrh, or mental disorders where long-term therapy is the order of the day.

In determining the maximum acceptable level of any permitted impurity, it is essential to consider the nature of the hazard involved. Clearly, inactive but relatively non-toxic isomers or by-products of the manufacturing process cause less concern than contaminants which show adverse clinical reactions such as nausea, vomiting, skin reactions, liver or kidney damage, teratogenicity or carcinogenicity. Each drug substance and impurity pattern requires individual consideration in the light of the proposed dosage form and route of administration. Trace elemental contamination is, however, widespread due in part to atmospheric pollution, but mainly as a result of the use of particular catalysts or reagents in organic synthesis. Certain elements are prone to concentrate in particular tissues irrespective of the route by which the contaminated drug is administered, but it is evident, for example, that contaminants capable of causing undesirable skin reactions should be particularly stringently controlled in products for application directly to the skin. Similar contaminants in products administered internally, although equally undesirable, may well be tolerable at higher levels due to more efficient metabolic detoxification and excretion. By the same criteria, contaminants known to cause lung or eye irritation and damage should be more stringently controlled in inhalant aerosols and eye preparations respectively. It is, however, difficult to get quantitative data on the toxicity of particular trace element impurities. Table 9, however, shows some qualitative data on tissue deposition, vulnerable tissues, organs and enzymes, and indicates potential areas of hazard arising from trace elemental contamination.

Place of Manufacture; Sale and Supply

In addition to these aspects, which are of direct relevance to the scientific assessment of the product, the Summary of Particulars is also required to state the place of manufacture to establish that the premises are properly licensed under the Act. They must also indicate whether or not quality control will be exercised over the process and product, the type of container to be used for the product, including details of its size, shape and other distinguishing features or markings. Labelling particulars must be given, and the method proposed for sale and supply, i.e. whether the product is proposed for general sale, supply through registered pharmacies only, either over the counter or on prescription only, or for distribution via alternative outlets, such as hospitals or herbalists.

Active Ingredients (Drug Substances)

This section of the application is intended to provide the Licensing Authority with detailed information on source, manufacture, identity, standards and stability of the active constituents and other ingredients of the medicine which forms the subject of the submission. It should also provide information on the rationale underlying the choice of the proposed dosage form, the method of manufacture, the stability of the product, and control standards and methods.

Nomenclature and Composition

In addition to the Approved, Non-proprietary, or other name of each active ingredient, the submission must give for each substance, which is not the subject of a pharmacopoeial monograph, the systematic chemical name, its molecular and structural formulae and state its molecular weight. In some cases, it may not be possible to meet all these requirements precisely. Thus, it may not be possible to state the precise composition of certain hydrates, solvates, clathrates and inclusion complexes, but merely to indicate the limits within which the composition of the active ingredient will be controlled. Similar criteria also apply to the composition of certain inorganic materials (e.g. Hydrotalcite, an aluminium magnesium hydroxide carbonate hydrate) to organic intermolecular complexes (e.g. Dichloral Phenazone, Tetracycline Phosphate Complex) and, also, to polymeric products (e.g. Polyglactin, Malethamer, Polysorbates) in which only an indication of average molecular weight can be given.

This information is necessary to define the composition of the product, and also for labelling purposes. To the expert, it provides a basis for predicting physical, chemical and biological properties, and hence for critical appraisal

Table 9. Potential toxic hazards arising from trace element contaminants

Element	Specific tissue deposition	General systemic poisons	Vulnerable Tissues and Organs							Potential carcinogens	Vulnerable enzymes
			Skin	Lung (Mucous membranes)	Eye	Nerve and brain	Muscle and heart	Liver	Kidney		
Antimony (organo)			+	+				+			Phosphoglucomutase
Arsenic (organo)	Skin, hair, nails, liver, bone	+	+	+	+					+	Phosphatases and other sulphydryl enzymes
Barium	Bone, lungs			+							
Beryllium	Bone			+			+			+	Alkaline phosphatases
Boron	Brain, nerve		+				+				
Cadmium		+		+		+				+	
Chromium			+	+						+	
Cobalt			+	+						+	
Gallium	Bone / Malignant tissue					+					
Indium	Calcification at injection site							+	+		
Lanthanum	Nucleic acids (complex formation)										
Lead and organo lead	Bone / Brain	+									
Mercury (organo)	Nerve		+	+		+					Catalase
Nickel	Nerve, lungs, heart		+	+				+	+	+	
Palladium	Heart	±									
Platinum			+	+			+	+			
Selenium			+		+						
Thallium		+	+								
Tin (organo)		+									
Tungsten				+							
Vanadium			+	+							

of data supplied on these matters, on stability, and on the nature and properties of the proposed dosage form.

Manufacturing Process

Details of the method of manufacture are essential as evidence of identity and chemical structure. Specifications for starting materials, and purification and control of intermediates, are required to provide a firm basis for the assessment of impurity patterns. The complete synthetic route from appropiate intermediates which are readily available items of commerce must be specified, whether the latter be of synthetic or natural origin. Source materials must be specified in sufficient detail to ensure satisfactory identification by physical, chemical, or sensory characters, adequate quality by assay, and satisfactory control of contaminants which may be deleterious either to the process or the final product. It is frequently argued where several stages are involved in the synthesis of the active ingredient that the quality of the starting materials is immaterial, as impurities will be removed in the course of the process itself. This may well be true in many cases, and adequate evidence of absence or reduction to insignificant levels of unwelcome impurities, whose origins lie in the starting materials, may suffice to allay fears of contamination at an undesirable level. Much, however, depends on the contaminant and the process, since in certain circumstances the latter may serve to concentrate rather than eliminate undesirable impurities arising from impure source materials. Thus, ethyl iodide may contain small amounts of methyl iodide, which can arise from the use of industrial methylated spirits as source material. If used in the quaternisation of tertiary bases such ethyl iodide can give rise to undesirably high levels of the methiodide as a contaminant of the product ethiodide, because of the much higher rate of reaction with methyl iodide compared to that with ethyl iodide. Similarly, steric effects which inhibit the rate of reaction of aromatic compounds in the *ortho* position compared with that in the *para* position can lead to concentration of p-substituted intermediates and products in reactions conceived essentially as an attack on the *ortho* position.

Reagents, catalysts and solvents used in manufacture, which are likely to be retained in the final product, must also be considered as potential impurities, both from the point of view of toxicity and their effects on stability. In an ideal process, these will be eliminated or reduced to innocuous levels, but traces of nickel and palladium hydrogenation catalysts, also boron, arising from the use of boron trifluoride or sodium borohydride, should be particularly carefully controlled on account of their potential for producing toxic hazards (Table 9). Retained solvents are seldom likely to be present in quantities which present serious toxic hazards. Significant levels of methanol have, however, been found in some antibiotics, such as Streptomycin Sulphate, which may well be administered in substantial doses for several weeks or months at a time. Such contamination is clearly undesirable and strict upper limits for solvent contamination should be imposed in the finished product specification for any material administered on this scale. Contamination with benzene and hydrocarbon solvents, which can cause liver damage, should be even more strictly controlled.

In-process Control

Chemical syntheses carried out on a production scale should be subject to in-process checks on intermediate stage products, in which the material is positively identified, checked in some way for quality, and if critical, for freedom from particular contaminants. Some processes such as, for example, the Merrifield type synthesis of peptides, in which the product is built up unit by unit on an ion-exchange resin support without isolation of intermediates, are not amenable to control in this way. In cases such as this, the success of the process depends on rigidly defined and strictly controlled operating conditions, backed by tight specifications designed to limit any inadequacies in the process.

Potential Impurities and Their Detection

Submissions should include details of methods used to detect likely impurities in the final product. Thin-layer chromatography is widely used for this purpose, but gas-liquid chromatography, paper chromatography, and high pressure liquid chromatography provide useful supplementary and alternative techniques for the identification and quantification of impurities which are often closely related chemically to the parent compound. A critical approach to these techniques is essential. A thin-layer method which shows excessive tailing or one in which one of the principal impurities has an R_F value barely distinguishable from that of the parent compound is unlikely to suffice unless it can be quite clearly shown that this is the best that can be attained after several likely alternative solvent systems and supports have been tried. Systems in which likely impurities sit on the baseline or run very close to the solvent front are equally unlikely to be acceptable. A clear indication of the sensitivity of the method including evaluation of the technique used for detection of the spots, together with the presentation of visual evidence in the form of copies of relevant chromatograms, and where appropriate photographic records, are the hall-mark of a good submission.

Similarly, glc systems with grossly asymmetric peaks, inadequate separation of solvent, main and impurity peaks, or with either inordinately long or widely separated retention times are undesirable. But, whatever the technique employed, whether it be for example chromatographic, spectroscopic, enzymatic, it should give a clear distinction on a quantitative or semi-quantitative basis between individual impurities and the parent compound. In particular, the development work must establish that the product is free from harmful amounts of occluded solvents, and trace elements derived from the process, special attention being paid to known potential carcinogens (arsenic, beryllium, cadmium, chromium, cobalt, nickel) and other toxic elements, including boron, mercury, lead, barium and palladium, capable of causing nerve, muscle, liver or skin damage (Table 9).

Evidence of Chemical Structure

Unequivocal evidence for the structural formula of the active ingredients must be presented. This should rely primarily on the synthetic route, but should be supplemented by spectroscopic data derived from ultraviolet, infrared, nmr, mass, optical rotatory and circular dichroic spectrometry, as appropriate to the structure concerned. Sufficient evidence should be provided to demonstrate that the structure is unequivocal. Examples will arise from time to time where for solubility or other reasons none of these techniques is capable of providing the required information. In such an event, the product at the sub-terminal stage or a suitable derivative may prove more amenable to spectroscopic examination, or useful information may be available from thermal analysis or even X-ray crystallography.

Evaluation of evidence relating to the structure of intermolecular complexes should take account of the solvents used in spectroscopic studies. Thus, unequivocal evidence of a complex, formed by two compounds present in stoichiometric proportions, and obtained solely in non-aqueous media, may have little relevance to the physical state of the complex in essentially aqueous biological media. For example, the well-defined hydrogen bonded complexes of chloral hydrate with phenazone and acetylglycinamide, which are physically distinguishable from chloral hydrate by melting point and by the complete masking of the latter's characteristically unpleasant taste, are almost certainly decomposed in aqueous solution and merely provide an alternative means of administering chloral hydrate. Such lack of stability in biological media is relevant and the identity of the pharmacologically effective derivative should be clearly apparent to the reader of submissions relating to complexes of this sort.

Specifications and Batch Analyses

Proposed specifications of the drug substance should take into account all aspects of the manufacturing process. They should provide for the proper identification of the product to avoid mistakes in handling, and for the maintenance of adequate standards for content of the drug substance. Specifications should also give reasonable control of impurities actually found to occur in batches of product at the time of manufacture, or arising subsequently as a result of instability on storage or as a result of mishandling. Standards for content and impurity should take account of the precision of the methods employed, and should be reasonably related to the levels encountered in actual batch analyses. They should provide for the maintenance of reasonable standards whilst allowing latitude for acceptable operator and instrument errors of particular techniques. It would not be reasonable, however, to submit specifications allowing assay tolerances down to a minimum active ingredient content of, say, 94 per cent by non-aqueous perchloric acid titration of a base, a technique which would normally be expected to give results within 99 to 101.5 per cent, when a set of six batch analyses for the material indicate levels of 98.2, 99.5, 100.3, 99.7, 99.5 and 98.6 per cent. Similarly, limits of trace impurities detectable at 0.1 per cent level by tlc, and batch analyses indicating maximum and minimum levels found as 0.3 per cent and 0.1 per cent would not call for a specification limit of 1 per cent without good reasons being given.

The specifications must also take into account any fall in potency on storage during the expected life of the product, and consequent increase in decomposition products. They should, however, be sufficiently strict to exclude products which have been subjected to mishandling with the attendant possibility of an unnecessarily high rate of decomposition. Evidence of excessive decomposition on storage of the bulk drug cannot be made the excuse for a lax specification. Exceptionally, where products with important medical indications can only be produced in a form susceptible to significant decomposition on storage, it may be necessary to have a realistic release specification applicable at the time of manufacture, and a separate *check specification* permitting somewhat lower minimum standards to apply at some later date when the product is incorporated into a particular dosage form.

Stability Studies

Stability data used to determine shelf-life, and co-related specifications, must take full account of the chemistry of the active ingredient and its likely vulnerability to degradation by oxidation, carbon dioxide, moisture, heat and light, and container materials. Product analyses conducted in connection with stability must, therefore, be reasonably comprehensive. Assay figures alone, particularly where the assay process is relatively non-specific and unlikely to distinguish the drug substance from anticipated decomposition products, are not sufficient. Any properly conducted stability study must also include an examination of specific decomposition products by appropriate techniques to establish the identity and relative toxicity of the decomposition products and the concentrations in which they are formed.

Stability studies should not only take account of the physical state in which the compound is likely to be used, but also the immediate biological environment likely to be met on administration. Thus, substances for tabletting, encapsulation, and the preparation of inhalant cartridges or suspensions, should be examined primarily in the solid state. Substances for injection, which must on this account be subjected to some form of sterilisation procedure, must be examined particularly for stability at elevated temperatures, for possible hydrolysis or rearrangement in aqueous media and the effects of exposure to carbon dioxide and light. Similarly, all substances intended for oral administration must be chemically stable to the pH and enzymic conditions likely to be met in the gastrointestinal tract. Exceptions clearly apply where, for example, the drug substance is specifically designed to release the active ingredient by such decomposition, but if this is the case the submission should make this evident, and include those conditions under which the compound is expected to be stable.

To encompass all these requirements, therefore, stability studies must be conducted on the drug substance in the solid state over a range of temperatures, at varying degrees of humidity, in both light and dark, in air and with air excluded. Also, if the product is such that it is likely to be subjected in use to widely varying temperature fluctuations, i.e. a product to be used in multiple dose form in the tropics, which should be stored ideally in cool or refrigerated conditions, then stability tests should include a study of the effects of fluctuating temperatures.

Pharmacology and Toxicity

The pharmacology of the active ingredient must be clearly established and reported, including all effects relevant to the proposed use of the medicine.

All other actions must be reported, including possible drug interactions which could provide a key to undesirable side-effects or otherwise affect its safety in use.

Animal toxicology, covering acute (single dose) and chronic (long-term) toxicity, carcinogenicity, teratogenic and fertility studies in appropriate species, are essential. Reports must be presented in detail giving particulars of the strain, diet, age, sex, weight and number of animals used in each experiment. In all repeat dose studies, the route of administration, method of dosage, and food consumption must be relevant to that proposed for use in man. Reports are expected to include the results of haematological studies, biochemical investigations and urinalysis. Post-mortems, including histopathological studies, must be carried out on all animals dying in the course of experiments and the cause of death established. Where internal effects or significant lesions are found, the extent of reversibility on withdrawal of the drug should be examined.

Dosage Forms

Formulation

The formulation of the finished product or dosage form must be declared so far as is practicable in such a way as to show *either* the amount of each of the active ingredients and excipients per dose *or* if this is not possible, the proportion of each on a percentage basis. The formulation may in some cases contain an *overage* to compensate for loss of potency due to decomposition of active ingredients either during manufacture or on storage. Certain preparations, notably single dose injections may, additionally, contain an *overfill*, in order to permit easy withdrawal of the correct dosage volume.

All constituents, active ingredients and excipients alike, must conform to a fully stated specification, pharmacopoeial or otherwise, appropriate to their use in human medicine. Certain exemptions from licensing of medicines are allowed under the Medicines Act, 1968 in respect of herbal remedies, provided the process of manufacture consists only of *drying*, *crushing* or *comminuting*, and provided the product is sold under a designation which only specifies the plants used and the process by which it is compounded. No other name may be applied nor must there be any written recommendation for use, otherwise the product ceases to be exempt and is classifiable as a medicine.

Any formulation may be proposed for licensing as a medicine, but combinations of active ingredients which are not already recognised for use in medicine may require justification in terms of both safety and efficacy. The possibilities which exist for drug interactions, whether these be in a biological sense leading to modified clinical response or alterations in the pattern of metabolism, or merely physico-chemical interactions affecting absorption, distribution and excretion kinetics, suggest a need for caution in proposing the use of particular unproven drug combinations.

Any formulation proposed for a medicine in a marketing application must either be identical with that used during clinical trials of the active ingredient, or failing that must be capable of being meaningfully equated with the trial formulation(s). Thus, product licence submissions which propose major changes from clinical trial methodology concerning the route of administration, type of product, excipients, physical properties or manufacturing process, should be backed up by adequate data, which clearly demonstrates comparable bioavailability, safety and stability.

The choice of a new route for administration, for example by metered-dose aerosol directly into the lungs, when the same compound has previously been administered orally for absorption from the gastro-intestinal tract, may pose new and important questions. These concern absorption, and ultimate fate, which can perhaps only be answered by fresh pathological studies and clinical pharmacology relevant to local toxicity, carcinogenicity or even teratogenicity. Actual experimental findings are essential, since although drugs administered other than by the oral route may by-pass the liver and hence be subjected to metabolism by non-hepatic pathways, the amount of drug actually reaching the bronchi from a single puff of a metered aerosol suspension is very much dependent on the particle size and other physical characteristics of the active ingredient, and can be as low as 7 per cent of the average dose per puff. Thus, part of the drug may be absorbed from the buccal mucosa, the stomach, the intestine or combination of these, depending on its physico-chemical characteristics.

Excipients

The choice of excipients should be carefully considered, and be capable of rationalisation. Ideally, these should be limited to the minimum necessary to ensure uniformity of dosage and stability throughout the period of the proposed shelf-life of the product. There may be no legal barrier to the marketing of products with multitudinous unnecessary excipients, but there is little scientific or technical justification for products which might be colloquially described as containing 'the kitchen sink'. Ideally, the excipients used in a preparation should be limited to the minimum necessary to achieve correct and uniform dosage and stability throughout the shelf-life of the product. However, some latitude is allowable in the proportions of excipients to provide reasonable manufacturing tolerances, but these should be clearly stated.

It is equally important that there should be no chemical reaction between excipients and the active ingredients, though the manufacturing process may be permitted to incorporate such reactions, as for example the in-process formation of a particular salt of an organic acid or base, provided the true nature and form of the product active ingredient so formed is properly declared on the label. On the other hand, physical interactions may often provide the *raison d'être* for a particular excipient, such as a wetting agent used in tabletting of a relatively water-insoluble drug to aid its solution and more rapid absorption, or the incorporation of a particular solid support such as silicon dioxide to enhance the surface area of a silicone used as an antiflatulent.

The Manufacturing Process

The method of manufacture must be set out in sufficient detail to ensure its reproducibility if placed in the hands of a suitably qualified operator without special knowledge or experience of the process. This is particularly important in respect of processes leading to the production of products which are essentially physical mixtures, whether these be solid (pro-injection solids, tablets and capsules), semi-solid (creams and ointments) or liquid products (emulsions and suspensions), and especially so when the fine detail of the process may well provide the principal means of controlling physico-chemical characteristics. Thus, milling processes and micronisation of active ingredients which are relatively insoluble must be so specified as to ensure reasonable batch to batch reproducibility of particle size or better still specific surface area, and to ensure that any changes in solid state characteristics which occur in the process are consistent batchwise. Similarly, solid products and suspensions incorporating wetting and dispersing agents can show marked differences in the rate and extent of bio-availability of insoluble components from them depending on the manufacturing process. The influence of formulation and process on polymorphism of active ingredients, on colours and on 'Ostwald ripening' leading to crystal growth in suspensions are important aspects of manufacture requiring close control (A. L. Smith (Ed.), 'Particle Growth in Suspensions', *Soc. Chem. Ind.*, Monograph 28).

Dosage Form Specifications

Specifications for dosage forms must not only embrace those of the individual constituents, but also be such as to provide for a reasonable measure of control over the product as a whole. It is essential, therefore, that they should provide for the identification and control of the content of active ingredients, to ensure that they are present within reasonable working tolerances in amounts which accord with the declared label strength. Tolerances must, therefore, take into account not only the precision of the assay, covering inherent errors of methodology, instrumentation and operator, but also the precision of the actual manufacturing process.

Allowing for the normal variation of patient response, it is generally considered that unit dosage variations of up to ±10 per cent are unlikely to produce observable differences in the clinical efficacy of the great majority of drugs, particularly if they are administered in repetitive doses. For a few compounds, however, which are employed in microgram or milligram doses in clinically critical situations, uniformity of dosage is essential and must be strictly controlled. Typical examples include the heart stimulant, digoxin, for which the normal adult dose is one or two 250 mcg tablets, and the contraceptive pill in which the small oestrogen content requires careful control at 50 mcg per tablet, both for efficacy and safety. These quantities are very small in relation to the total weight of the tablets, and special care in manufacture is essential to ensure even distribution throughout the batch. Moreover, the normal tablet assay procedure based on examination of a group of 20 tablets would not reveal gross inter-tablet variation. Specifications for such products must, therefore, additionally incorporate control procedure based on the assay of individual tablets to ensure uniformity of content.

Single dose injection solutions containing small amounts of potent medicaments seldom present problems, since it is easy to ensure uniformity of content with a solution. Uniformity of dosage may, however, need to be monitored for single dose injectables where the active ingredient is present at microgram level together with a diluent such as lactose to visualise the contents of the ampoule, as for example in Vincristine Injection.

Metered-dose pressurised aerosols and powders for insufflation directly into the lungs also present problems in the control of unitary dosage. Aerosol specifications should control the total weight of active ingredient per can, measured as an average of the weight found in samples of at least ten cans drawn from each batch. The total number of puffs per can should also be specified, and the average weights of aerosol and active ingredient per puff. The latter, too, should remain reasonably constant throughout the life of the can, and specifications should, therefore, provide for checks on the active ingredient per dose both when cans are full and almost empty. Even with this close control of the dosage form, the effective dose in the lungs depends very much on particle size of the drug substance, and the characteristics of the can adapter used during administration. Specifications must, therefore, provide for strict control of these factors.

Micronised powders for insufflation into the lungs are packed in hard gelatin capsules, but the terminology, cartridge, is used to distinguish these products from hard capsules for internal administration. They are usually used where the weight of powder to be administered is rather greater than can be conveniently projected into the lungs by an aerosol. The capsules are used in special inhalers, which incorporate a mechanism to puncture the cartridge, and allow its contents to be drawn into the lungs by a miniature breath-actuated turbo-fan or other appropriate device. As with aerosol preparations, control of particle size, cartridge content, and inhaler and adapter characteristics are critical factors in standardisation of dosage, and must be tightly specified.

Bio-availability

Evidence that the formulation is capable of releasing the active constituent at a clinically effective rate is an essential aspect of all product licence applications. Presentation of correlated dissolution, absorption and excretion kinetics both in animals and in man is, therefore, imperative. The metabolic pathway and extent of metabolism should be determined, but if this work is based on the use of radioactive tracers, the position of the label within the molecule must be indicated so that it is clearly apparent that the label has been incorporated in a metabolically stable position.

Ready and consistent bio-availability may well be self-evident in the case of water-soluble injections and even solid dosage forms of readily water-soluble compounds, but formulations of insoluble compounds require careful evaluation. Tablets, capsules and suppositories must be capable of rapid disintegration and should show evidence of appropriate release rates into solution in suitably designed *in vitro* tests. Plasma levels, provided they correlate reasonably with clinical response, however, are more likely to give a meaningful indication that the active principle is effectively absorbed from the point of administration, whether this be mouth, skin, lung, or other body cavity. Relatively few preparations require compilation of standards for batchwise monitoring of correlated release rates. They fall into two categories, tablets of insoluble, highly potent medicaments administered in small doses for the treatment of vital medical conditions, and tablets specially designed to give controlled release over a prolonged period, from which too rapid or too slow release could be hazardous to the patient.

Stability Studies

Product stability of material produced in full-scale production batches and stored in the final containers is also of paramount importance. This is obviously related in some degree to the inherent stability or otherwise of the active ingredients, but application for product licences must additionally include evidence to show that the product retains an acceptable level of potency on storage, and equally importantly, that toxic decomposition products are not produced in significant amount. Accelerated storage tests, conducted at elevated temperatures and under other conditions likely to give a clear indication of all readily conceivable hazards, should be used to determine an acceptable shelf-life for each product. The product should be examined under all conditions likely to be met during its storage in practice. For example, a large volume pack of a syrup containing an oxidisable phenothiazine such as chlorpromazine, which may be kept in broken bulk as stock on the dispensary shelf, should be examined for sulphoxide formation in partially filled containers, even if the product is to be issued in well-filled containers. Fluctuating temperatures with partly filled containers should be used to simulate conditions arising in patient use. Similarly, the unusually large pack of a sterile cream for the treatment of burns, which the patient or nurse might be tempted to retain for use on a second or third occasion, even if these be within a matter of hours, should be tested under appropriate conditions to demonstrate the efficacy of any preservatives it may contain. Nonetheless, if stability risks are shown to exist, it is preferable that they even be avoided by the use of unit dosage forms. The choice of container and pack should, therefore, reflect the stability characteristics of the product. Light-sensitive materials should be packed to exclude light—e.g. tablets in metal foil strip packs; amber glass is no longer considered effective and injectables or other liquid preparations should be packed in clear glass vessels, if necessary wrapped individually in foil, but otherwise in packages capable of excluding light.

All product stability data should be referable to a particular production batch, so that information is on file giving the Batch Number, size of batch and date of manufacture. This information is essential to enable the origin of particular ingredients to be traced if a query should arise.

Containers

The type of container, and the nature of the container material must be so specified to ensure that it is adequate for the purpose.

Package insertions, whether these be space fillers to prevent breakage of friable tablets in the course of normal container handling or desiccant packs to exclude moisture, have been recognised as potential hazards to unsuspecting patients. Any such insertion must be clearly distinguishable, by a suitable combination of size, shape, weight, colour and texture from the product itself, to ensure immediate distinction by sight and touch particularly in the hands of handicapped patients such as blind persons. The pack should be labelled with instructions, that it is not to be eaten. The need for inclusion of a desiccant pack should be clearly demonstrated by presentation of evidence of moisture-induced physico-chemical changes. The compatibility of desiccant and product must be established, and the desiccant must be such as to ensure satisfactory desiccant action without leakage from the pack. The quantity of desiccant should also be related to the shelf-life of the product.

Reports of Experimental and Biological Studies

Whilst it is generally considered sufficient for animal toxicity and related biological studies to be carried out on the individual active ingredients and excipients, it is important to consider the possible influence of formulation and route of administration on the significance of results so obtained. Since both these parameters can markedly influence the extent of absorption, the pattern of distribution, and consequently the kinetics of metabolism and excretion, it is clearly important that chronic toxicity, carcinogenicity, teratogenicity and fertility studies should be carried out on either the dosage form actually proposed or something very close to it administered by the same route as that proposed for clinical use.

Reports of Clinical Trials

Reports of clinical trials must be of such nature and sufficiently well documented to provide adequate evidence of efficacy and safety when administered to patients by the proposed route in the dosage indicated for the treatment of the indications proposed by the applicant. The results must be reported so as to clearly indicate the number of patients at the commencement and on completion of each trial; the range and mean dosage employed, the results obtained, and any adverse reactions which have been observed. In all applications for product licences, it is essential that some evidence of clinical efficacy in relation to the proposed *indication* for that medicine be demonstrated. Failing such evidence, the only reasonable course of action on the part of an applicant is an appropriate modification in the claims made for the product.

On matters of safety, product licence applications should take due note of established hazards and recommended practice in products containing substances for which particular hazards are known to exist. Well known examples, which have received considerable publicity, include products containing Aspirin, Hexachlorophane, Monoamine Oxidase Inhibitors and the oestrogen content of oral contraceptives. The formulation of such products and recommendation for their use should conform to established safety standards.

3 The Theoretical Basis of Quantitative Analysis

ACID-BASE TITRATIONS

Electrolytic Dissociation

Certain substances known as electrolytes dissolve in water to yield solutions which will conduct electricity. In 1887, Arrhenius suggested that this ability to conduct electricity was due to the fact that, in solution, electrolytes undergo dissociation into positively and negatively charged fragments which are called ions. Positive ions move towards a negative electrode, and negative ions towards a positive electrode. This passage of ions, and the subsequent neutralisation of the ionic charge at the electrode, brings about conduction of electric current through the solution.

Ions are usually solvated but, for simplicity, the effects of ion solvolysis have been ignored in the subsequent discussion of the behaviour of electrolytes in solution.

The Law of Mass Action

This law, which was first stated by Guldberg and Waage in 1867, may be expressed in the following form:

'The rate of a chemical reaction is proportional to the active masses of the reacting substances.'

In dilute solution where conditions approach the ideal state, 'active mass' may be represented by the concentration of the reacting species, i.e. gram-molecules or gram-ions per litre. The constant of proportionality is known as the velocity constant, so that in the simple reaction, $A \rightarrow B$, the rate of reaction $= k[A]$, where $[A]$ is the concentration of A, and k is the velocity constant.

Consider now the homogeneous, reversible reaction:

$$A + B \rightleftharpoons C + D$$

According to the law of mass action

$$v_f = k_1[A] \cdot [B] \quad \text{and} \quad v_b = k_2[C] \cdot [D]$$

where v_f = velocity of the forward reaction; v_b = velocity of the backward reaction; $[A]$ denotes molar concentration of A; and k_1 and k_2 are constants.

At equilibrium, $v_f = v_b$

$$\therefore \quad k_2[C] \cdot [D] = k_1[A] \cdot [B]$$

$$\frac{k_1}{k_2} = \frac{[C] \cdot [D]}{[A] \cdot [B]}$$

Since k_1 and k_2 are both constants, the fraction k_1/k_2 must also be a constant.

Hence

$$K = \frac{[C] \cdot [D]}{[A] \cdot [B]}$$

where K = the *equilibrium constant* of the reaction (constant at a given temperature).

In extension, the equilibrium constant for the general reversible reaction:

$$aA + bB + cC + \ldots \rightleftharpoons pP + qQ + rR + \ldots$$

is $$K = \frac{[P]^p \cdot [Q]^q \cdot [R]^r}{[A]^a \cdot [B]^b \cdot [C]^c}$$

where a, b, c, and p, q, r are the number of molecules of the reacting species.

Application of the Law of Mass Action to Solutions of Weak Electrolytes

Electrolytes may be classified as either strong or weak electrolytes depending upon the extent to which they are dissociated into ions in solution. Strong electrolytes are almost completely dissociated even in moderately concentrated solutions, and hence do not constitute equilibrium systems. Weak electrolytes, on the other hand, are only incompletely dissociated even in the favourable ionisation conditions of dilute solution; therefore an equilibrium, which can be considered in terms of the law of mass action, is reached between undissociated molecules and ions.

The dissociation of water. Water is an extremely weak electrolyte and is only very slightly dissociated into its ions:

$$H_2O \rightleftharpoons H^+ + OH^-$$

From the law of mass action:

$$K = \frac{[H^+] \cdot [OH^-]}{[H_2O]}$$

In pure water and in dilute aqueous solutions the concentration of free water may be considered constant and hence

$$K_w = [H^+] \cdot [OH^-]$$

where K_w is known as the *ionic product of water*. The latter varies with temperature, but under ordinary experimental conditions (about 25°), its value may be taken as 1×10^{-14}, when the concentrations of hydrogen and hydroxyl ions are expressed in grams-ions per litre.

In pure water,

$$[H^+] = [OH^-]$$

and hence

$$[H^+] = \sqrt{K_w}$$

$$= 10^{-7} \text{ gram-ions per litre.}$$

Solutions in which the hydrogen ion concentration is greater than 10^{-7} are *acid*; when it is less than 10^{-7} the solution is *alkaline*.

The hydrogen ion exponent (pH). Because of the very great variations in hydrogen ion concentration met with in practice, it is often convenient to adopt the pH notation first introduced by Sörensen. pH is defined as the negative logarithm (to base 10) of the concentration of hydrogen ions in solution:

$$pH = -\log_{10}[H^+] = \log_{10}\frac{1}{[H^+]}$$

This method of stating hydrogen ion concentration has the advantage that all degrees of acidity and alkalinity between that of a solution molar (or normal) with respect to hydrogen and hydroxyl ions can be expressed by a series of positive numbers between 0 and 14. A neutral solution is one in which $pH = 7$, an acid solution one in which $pH < 7$ and an alkaline solution one in which $pH > 7$.

The dissociation of weak acids and bases. Consider an aqueous solution of a weak acid HA, in which the following equilibrium between ions and undissociated molecules obtains

$$HA \rightleftharpoons H^+ + A^-$$

From the law of mass action:

$$K_a = \frac{a_{H^+} \times a_{A^-}}{a_{HA}} \tag{1}$$

where K_a = the ionisation constant or dissociation constant (at constant temperature);* a = the activity of the various species present.

In dilute solution the activity terms can be equated to concentrations, so that (1) may be re-written in the form:

$$k_a = \frac{[H^+] \cdot [A^-]}{[HA]} \tag{2}$$

where k_a is the approximate dissociation constant.

Now if v is the volume of solution (in litres) which contains 1 gram equivalent of HA ($v = 1/c$, where c is the concentration in gram equivalents per litre), and α is the degree of dissociation, then there will be present at equilibrium $(1 - \alpha)$ gram equivalents of unionised acid, HA, and α gram equivalents of each of the ions H^+ and A^-. The corresponding concentrations of unionised acid HA will then be $(1 - \alpha)/v$, and of the ions H^+ and A^- will be α/v, so that equation (2) may be expressed in the form:

$$k_a = \frac{\alpha^2}{(1-\alpha)v} \tag{3}$$

For very weak electrolytes, α may be neglected in comparison with unity, and the expression (3) then reduces to

$$k_a = \frac{\alpha^2}{v} = \alpha^2 c$$

$$\therefore \quad \alpha^2 = k_a/c \quad \text{and} \quad \alpha = \sqrt{(k_a/c)}$$

_____

* K is used (as opposed to k) in this text to denote thermodynamic constants.

This relationship can now be used to derive an expression from which the pH of the solution can be determined as follows:

$$[H^+] = \alpha c$$
$$\therefore \quad [H^+] = c\sqrt{(k_a/c)}$$
$$= \sqrt{(k_a c)}$$

Taking logarithms

$$\log [H^+] = \tfrac{1}{2}\log k_a + \tfrac{1}{2}\log c$$
$$\therefore \quad -\log[H^+] = -\tfrac{1}{2}\log k_a - \tfrac{1}{2}\log c$$
$$\therefore \quad pH = \tfrac{1}{2}(pk_a - \log c)$$

Similar equations may be derived for aqueous solutions of a weak base, the equilibrium system in this case being:

$$B + H_2O \rightleftharpoons BH^+ + OH^-$$

B represents the weak base and BH^+ the conjugate acid, whence

$$k_b = \frac{[BH^+] \cdot [OH^-]}{[B]}$$

$$= \frac{\alpha^2}{(1-\alpha)v} \simeq \frac{\alpha^2}{v}$$

where k_b = dissociation constant of the base at constant temperature
  α = degree of ionisation
  v = volume in litres containing one gram equivalent of the weak base.

Since

$$k_b = \alpha^2/v = \alpha^2 c$$
$$\alpha^2 = k_b/c$$

and

$$\alpha = \sqrt{(k_b/c)}$$

Now

$$[OH^-] = c\alpha$$
$$\therefore \quad [OH^-] = c\sqrt{(k_b/c)}$$
$$= \sqrt{(k_b c)}$$

But

$$[OH^-] = K_w/[H^+]$$
$$\therefore \quad K_w/[H^+] = \sqrt{(k_b/c)}$$
$$\therefore \quad [H^+] = K_w/\sqrt{(k_b c)}$$

Taking logarithms

$$\log [H^+] = \log K_w - \tfrac{1}{2}\log k_b - \tfrac{1}{2}\log c$$
$$\therefore \quad pH = pK_w - \tfrac{1}{2}(pk_b - \log c)$$

The Strength of Acids and Bases

An acid may be defined as a substance which ionises to yield hydrogen ions or protons, and a base as a substance which combines with hydrogen ions. An acid is accordingly a proton donor and a base a proton acceptor. Both can be defined by the expression

$$A \rightleftharpoons H^+ + B$$

where A is the acid and B is the base.

The strength of an acid is related to the concentration of hydrogen ions which it yields upon ionisation and will depend upon the value of the degree of dissociation α, at any given concentration. The acid dissociation constant k_a gives a relationship between α and the concentration, and accordingly is a measure of the acid strength. The strength of a base is likewise related to its dissociation constant.

The relationship between the dissociation constants of an acid (HA) and its conjugate base (A^-) can be derived from the equilibria concerned, i.e.

$$HA \rightleftharpoons H^+ + A^-$$
$$A^- + H_2O \rightleftharpoons HA + OH^-$$

Since

$$k_a = \frac{[H^+] \cdot [A^-]}{[HA]} \quad \text{and} \quad k_b = \frac{[HA] \cdot [OH^-]}{[A^-]}$$

$$k_a \cdot k_b = \frac{[H^+] \cdot [A^-]}{[HA]} \cdot \frac{[HA] \cdot [OH^-]}{[A^-]}$$

$$\therefore \quad k_a \cdot k_b = [H^+] \cdot [OH^-]$$

$$\therefore \quad k_a \cdot k_b = K_w$$

It follows from the expression that the dissociation constants of a weak acid and its conjugate base are complementary, i.e. the stronger the acid the weaker its conjugate base, and vice versa. For convenience in practice, the strength of a base is often expressed in terms of the dissociation constant of its conjugate acid.

The dissociation constant exponent (pK). Dissociation constants of weak acids and bases are numerically small, and a logarithmic notation is therefore convenient. The dissociation constant exponent, pK, is derived from the dissociation constant in a manner analogous to the derivation of pH from hydrogen ion concentration. Hence

$$pK_a = -\log_{10} K_a$$

It follows from this expression that the higher the value of K_a the smaller the value of pK_a, so that the stronger the acid the smaller the pK_a value.

The Hydrolysis of Salts

When salts are dissolved in water, interaction may occur with the ions of water and the resultant solution may be neutral, acid or alkaline according to the nature of the salt. Such interaction is termed hydrolysis.

With an aqueous solution of a salt of the strong acid-strong base type, e.g. sodium chloride, neither the anions have any tendency to combine with the hydrogen ions nor the cations with the hydroxyl ions of water, since the related acids and bases are strong electrolytes and are themselves completely dissociated. The equilibrium between hydrogen and hydroxyl ions in water is therefore not disturbed and the solution remains neutral.

Weak base—strong acid salts. Consider the salt of a weak base and a strong acid, e.g. ammonium chloride. The chloride ions do not react significantly with hydrogen ions. The ammonium ions, on the other hand, are cations of a weak base and have a tendency to combine with hydroxyl ions to form undissociated ammonium hydroxide. Consequently the hydroxyl ion concentration of the water will be decreased and the hydrogen ion concentration increased since the product $[H^+] \cdot [OH^-]$ must remain constant.

$$NH_4Cl \rightleftharpoons NH_4^+ + Cl^-$$
$$NH_4^+ + H_2O \rightleftharpoons NH_3 + H_3O^+$$

Its pH can be calculated from the following expression which relates the pH of a solution of a weak acid with its dissociation constant, namely:

$$pH = \tfrac{1}{2}(pk_a - \log c)$$

by substituting in this expression for pk_a which equals $pK_w - pk_b$

$$pH = \tfrac{1}{2}(pK_w - pk_b - \log c)$$

Weak acid—strong base salts. Similarly, the salt of a strong base and a weak acid, e.g. potassium cyanide, will have an alkaline reaction because of hydrolysis represented by the equilibria:

$$KCN \rightleftharpoons K^+ + CN^-$$
$$CN^- + H_2O \rightleftharpoons HCN + OH^-$$

The pH of the solution can be calculated from the expression which relates the pH of a solution of a weak base to its dissociation constant, namely

$$pH = pK_w - \tfrac{1}{2}(pk_b - \log c)$$

Substituting in this expression for pk_b which equals $pK_w - pk_a$ we obtain the expression:

$$pH = pK_w - \tfrac{1}{2}pK_w + \tfrac{1}{2}pk_a + \tfrac{1}{2}\log c$$
$$\therefore \quad pH = \tfrac{1}{2}(pK_w + pk_a + \log c)$$

Weak acid—weak base salts. Both ions derived from the salt of a weak acid and a weak base such as ammonium acetate undergo hydrolysis in aqueous solution according to the equilibria:

$$NH_4^+ \rightleftharpoons H^+ + NH_3$$

$$CH_3COO^- + H_2O \rightleftharpoons CH_3COOH + OH^-$$

Summation of these equilibria gives the expression:

$$NH_4^+ + CH_3COO^- + H_2O \rightleftharpoons NH_3 + CH_3COOH + H^+ + OH^-$$

Provided the dissociation constants of the acid and base are not widely different, hydroxyl and hydrogen ions will be produced in approximately equal amounts. It is, therefore, permissible to substract the water equilibrium $H_2O \rightleftharpoons H^+ + OH^-$ from the overall expression, to give the simplified expression:

$$NH_4^+ + CH_3COO^- \rightleftharpoons NH_3 + CH_3COOH$$

$$(1-x)c \qquad (1-x)c \qquad xc \qquad xc$$

$$\therefore \quad k_h = \frac{[NH_3] \cdot [CH_3COOH]}{[NH_4^+] \cdot [CH_3COO^-]}$$

where k_h is the hydrolysis constant.

Introduction of the ionic product of water into this expression gives

$$k_h = \frac{[CH_3COOH]}{[H^+] \cdot [CH_3COO^-]} \times \frac{[NH_3]}{[NH_4^+] \cdot [OH^-]} \times [H^+] \cdot [OH^-]$$

$$\therefore \quad k_h = 1/k_a \cdot 1/k_b \cdot K_w$$

$$= K_w/k_a k_b$$

Consider a solution of ammonium acetate containing c g-mol. per litre. If the degree of hydrolysis is x, then

$$[CH_3COOH] = [NH_3] = xc$$

and

$$[CH_3COO^-] = [NH_4^+] = (1-x)c$$

Substituting in the expression

$$k_h = \frac{[NH_3] \cdot [CH_3COOH]}{[NH_4^+] \cdot [CH_3COO^-]}$$

$$k_h = \frac{xc \cdot xc}{(1-x)c \cdot (1-x)c}$$

$$\therefore \quad k_h = \frac{x^2}{(1-x)^2}$$

The pH of the solution may be calculated from a knowledge of the concentration of hydrogen ions in equilibrium with acetic acid as follows:

$$k_a = \frac{[H^+] \cdot [CH_3COO^-]}{[CH_3COOH]}$$

$$\therefore \quad H^+ = \frac{k_a \cdot [CH_3COOH]}{[CH_3COO^-]}$$

$$= \frac{k_a \cdot xc}{(1-x)c}$$

Substituting for $x/(1-x)$ which equals $\sqrt{k_h}$:

$$[H^+] = k_a \cdot \sqrt{k_h}$$

and substituting for k_h which equals $K_w/k_a k_b$:

$$[H^+] = k_a \sqrt{(K_w/k_a k_b)}$$
$$= \sqrt{(K_w k_a / k_b)}$$

whence:

$$\log[H^+] = \tfrac{1}{2}\log K_w + \tfrac{1}{2}\log k_a - \tfrac{1}{2}\log k_b$$
$$\therefore \quad pH = \tfrac{1}{2}(pK_w + pk_a - pk_b)$$

Buffer Solutions

The resistance of a solution to changes in hydrogen ion concentration upon addition of small amounts of acid or alkali is termed *buffer action*; a solution which possesses such properties is known as a buffer solution. Buffer solutions usually consist of solutions containing a mixture of a weak acid or base and its salt. Buffer action in a solution of a weak acid and its salt is explained by the fact that hydrogen ions are removed by the anions of a weak acid to form unionised molecules, thus:

$$H^+ + A^- \rightarrow HA$$

Hydroxyl ions are also removed by neutralisation, according to the equation:

$$OH^- + HA \rightleftharpoons H_2O + A^-$$

The concentrations of hydrogen ion (relative to those of the weak acid and its salt) in such a buffer solution will be determined by the dissociation constant of the acid according to the expression:

$$k_a = \frac{[H^+] \cdot [A^-]}{[HA]}$$

$$\therefore \quad \log k_a = \log[H^+] + \log\frac{[A^-]}{[HA]}$$

$$\therefore \quad -\log k_a = -\log[H^+] - \log\frac{[A^-]}{[HA]}$$

$$\therefore \quad pk_a = pH - \log\frac{[A^-]}{[HA]}$$

$$\therefore \quad pH = pk_a + \log\frac{[A^-]}{[HA]}$$

Now, the acid HA is weak, and only slightly ionised. Moreover, its ionisation is repressed by the relatively large concentration of anions A^- from the fully dissociated salt.

Hence, [HA] is numerically equal to the initial concentration of acid, and $[A^-]$ is numerically equal to the initial concentration of salt.

$$pH = pk_a + \log\frac{[salt]}{[acid]}$$

This is known as the Henderson equation.

If [salt] = [acid] the expression becomes

$$pH = pk_a + \log 1$$

$$\text{i.e.} \quad pH = pk_a$$

A tenfold increase or decrease of the ratio [salt]/[acid] would raise or lower the pH of the solution by one pH unit. The resistance of a buffer solution to such a pH change is a measure of the 'buffer capacity'. This is defined as the number of gram equivalents of strong acid or strong alkali necessary to produce a change of 1 pH unit in 1 litre of the solution.

Consider a solution containing 0.5 g equivalents/litre of acid and 0.5 g equivalents/litre of salt. An increase of 1 pH unit will be brought about when the salt concentration has been raised to approximately 0.91 g equivalents/litre and the acid concentration reduced to approximately 0.09 g equivalents/litre. This change would require the addition of 0.41 g equivalents of strong base. Hence the buffer capacity is 0.41 g equivalents.

In a buffer solution which consists of a mixture of a weak base and its salt, hydroxyl ions are removed by the salt cations (BH^+) to form unionised molecules:

$$OH^- + BH^+ \rightarrow B + H_2O$$

Hydrogen ions are also removed by neutralisation,

$$H^+ + B \rightarrow BH^+$$

The concentrations of weak base and its salt relative to that of hydroxyl ion in the solution will be determined by the dissociation constant of the base, i.e.

$$k_b = \frac{[BH^+] \cdot [OH^-]}{[B]}$$

$$\therefore \quad \log k_b = \log [OH^-] + \log \frac{[BH^+]}{[B]}$$

$$\therefore \quad -\log k_b = -\log [OH^-] - \log \frac{[BH^+]}{[B]}$$

$$\therefore \quad pk_b = pOH - \log \frac{[BH^+]}{[B]}$$

but

$$pOH = pK_w - pH$$

$$\therefore \quad pk_b = pK_w - pH - \log \frac{[BH^+]}{[B]}$$

Hence,

$$pH = pK_w - pk_b - \log \frac{[BH^+]}{[B]}$$

Now the base B is weak, and only slightly ionised. Also its ionisation is repressed by the relatively large concentration of cations BH^+ from the fully dissociated salt. Hence [B] is numerically equal to the initial concentration of base and $[BH^+]$ is numerically equal to the initial concentration of salt,

$$\therefore \quad pH = pK_w - pk_b - \log \frac{[salt]}{[base]} \qquad (4)$$

That this is merely a restatement of the Henderson equation is shown by substitution for $pK_w - pk_b = pk_a$ in eq. (4) when

$$pH = pk_a - \log \frac{[acid]}{[salt]}$$

and

$$pH = pk_a + \log \frac{[salt]}{[acid]}$$

This follows from the fact that BH^+ is the conjugate acid of base B, and that the latter, as before, can be equated to the concentration of salt.

Neutralisation Indicators

It follows from the foregoing discussion of the behaviour of weak acids and bases, that the equivalence point in the titration of standard acids and alkalis will not always be the point of exact neutrality (pH 7.0). Coincidence of equivalence point and exact neutrality is attained only in strong acid—strong base titrations, since if either base or acid is weak the resulting salt will be hydrolysed, and the solution will become either acid or alkaline respectively. The actual pH of the solution at the end point can be determined potentiometrically, or by means of a neutralisation indicator which changes colour according to the hydrogen ion concentration of the solution.

Indicators are weak acids or weak bases which have different colours in their conjugate base and acid forms (two-colour indicators); others are one-colour indicators, and have one form coloured with a colourless conjugate form. Most indicators in common use are intensely coloured, and can be used in dilute solution in such small quantities that the acid-base equilibrium which is under examination is not disturbed by the addition of the indicator. As weak acids or weak bases, they are able to reach instantaneous equilibrium with the system, and the colour of the solution will range between the extreme colours of the two forms as the proportion of acidic and basic forms automatically adjusts itself to the pH of the solution.

The following equilibrium will apply for an indicator functioning as a weak acid:

$$HIn \rightleftharpoons H^+ + In^-$$

$$\begin{array}{ccc} \text{unionised} & & \text{ionised} \\ \text{colour} & & \text{colour} \end{array}$$

In acid solution, the excess of H^+ ions will depress the ionisation of the indicator. The concentration of In^- will be small, and of HIn large, and the colour will be that of the unionised form. Alkali will promote removal of hydrogen ions from the system with an increase in the concentration of the ionised form (In^-), so that the solution acquires the ionised colour.

Then

$$k_{In_a} = \frac{[H^+] \cdot [In^-]}{[HIn]}$$

$$[H^+] = k_{In_a} \frac{[HIn]}{[In^-]}$$

$$\therefore \quad -\log[H^+] = -\log k_{In_a} - \log \frac{[HIn]}{[In^-]}$$

$$\therefore \quad pH = pk_{In_a} + \log \frac{[In^-]}{[HIn]}$$

Similarly for an indicator functioning as a weak base, the following equilibrium will apply:

$$\text{In} \quad + \quad H_2O \quad \rightleftharpoons \quad InH^+ + OH^-$$

unionised   ionised
colour   colour

and

$$k_{In_b} = \frac{[InH^+] \cdot [OH^-]}{[In]}$$

$$\therefore \quad pH = pK_w - pk_{In_b} - \log \frac{[InH^+]}{[In]}$$

Tautomeric neutralisation indicators. Although the behaviour of indicators can be explained in terms of ionisation of weak acids and bases, as above, the equilibrium is actually more complex, the colour changes being brought about by tautomeric changes in the structure of the molecule. This is illustrated by the behaviour of phenolphthalein in solution:

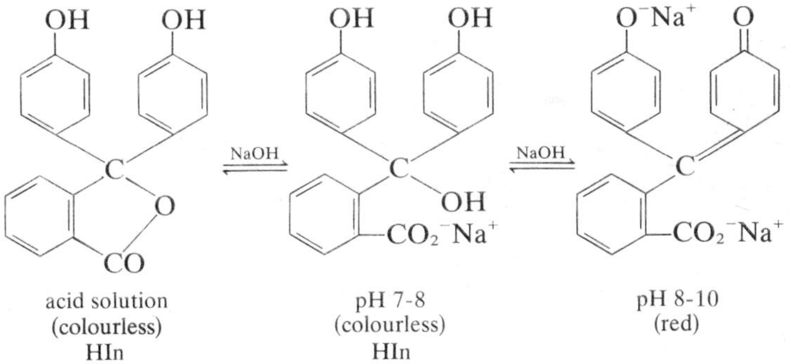

acid solution	pH 7-8	pH 8-10
(colourless)	(colourless)	(red)
HIn	HIn	

The red colour in alkaline solution is due to the quinonoid structure, with the resulting increased possibilities for resonance between the various ionic forms as:

coloured (In⁻)   colourless
                                                                  pH 12

Table 10. Indicator ranges and colour changes

INDICATOR	1	2	3	4	5	6	7	8	9	10	11	12	13
ALIZARIN YELLOW G							Yellow			Orange		Red-Orange	
BROMOCRESOL GREEN	Yellow			Green		Blue							
BROMOCRESOL PURPLE			Yellow		Grey		Purple						
BROMOPHENOL BLUE	Yellow		Grey			Blue							
BROMOTHYMOL BLUE				Yellow		Green		Blue					
CONGO RED	Blue		Violet		Red								
CRESOL RED	Red Orange	Yellow						Pink	Violet-Red				
DIMETHYL YELLOW		Red	Orange	Yellow									
LITMUS				Red		Violet			Blue				
METHYL ORANGE		Red		Orange	Yellow								
METHYL RED			Red	Orange			Yellow						
α-NAPHTHOLPTHALEIN					Pale Red			Violet	Blue				
NEUTRAL RED						Red	Orange		Orange				
PHENOLPHTHALEIN					Colourless			Pink		Red			
PHENOL RED					Yellow		Pink		Red				
THYMOL BLUE	Red	Orange		Yellow					Grey	Blue			
THYMOLPHTHALEIN							Colourless		Blue		Blue		
TITAN YELLOW										Yellow	Orange	Red	
TROPAEOLIN OO	Red	Orange		Yellow									
TROPAEOLIN O								Yellow		Orange	Red-Orange		

However, despite the complexity of equilibria such as these, for most practical purposes the expression

$$pH = pk_{In_a} + \log \frac{[In^-]}{[HIn]}$$

can be adopted.

The range of indicators. The observed colour of a two-colour indicator is determined by the ratio of the concentrations of ionised and unionised forms. Observable colour changes are, however, limited by the ability of the human eye to detect changes of colour in mixtures. This is particularly difficult where one colour predominates, and in practice is almost impossible when the ratio of the two forms exceeds 10 to 1. Thus the limit of visible colour change will be represented by the introduction of the term $\pm\log 10$ for $+\log [In^-]/[HIn]$ in the above expression so that

$$pH = pk_{In_a} \pm 1$$

The average colour-change interval (range) of an indicator is, therefore, about two pH units. The observed colour changes within the indicator range are seen as a gradual change of tint or shade which ranges from one extreme colour to the other. The shade of colour is independent of the amount of indicator present, but the use of too much indicator should be avoided as slight changes are then more difficult to detect.

With a single colour indicator, such as phenolphthalein, the intensity of colour is important, and not shade difference. The actual concentration of indicator is therefore significant, and should be carefully controlled.

Since the useful range of an indicator only extends over approximately two pH units, it is essential to have a series of indicators available to cover the complete pH scale. A list of such indicators in common use, together with their colour changes is given in Table 10.

PRECIPITATION AND COMPLEX FORMATION

Solubility Product

Consider a solution of a slightly soluble salt, BA, which is in equilibrium with the solid phase BA:

$$BA \rightleftharpoons B^+ + A^-$$
$$\text{(solid)} \qquad \text{(solution)}$$

Applying the law of mass action to this system,

$$K = \frac{[B^+] \cdot [A^-]}{[BA]}$$

The concentration of BA in solution will be constant in the presence of undissolved BA.

$$\therefore \quad K \times \text{constant} = [B^+] \cdot [A^-]$$
$$= S_{BA}$$

where S_{BA} is a constant (at constant temperature) and is called the solubility product of the salt BA.

If the sparingly soluble salt has the general formula $B_m A_n$ each molecule will furnish m cations and n anions:

$$B_m A_n \rightleftharpoons m B^+ + n A^-$$

and

$$S_{B_m A_n} = [B^+]^m \cdot [A^-]^n$$

In a saturated solution of the slightly soluble salt BA in water:

$$[B^+] = [A^-] = S$$

where S is the molar solubility of the salt, hence

$$S = \sqrt{S_{BA}}$$

Common ion effect. In any system in which solid is in equilibrium with its solution, the product of the ion concentrations is determined by the solubility product. Thus if an excess of silver ions is added to a saturated solution of silver chloride in water, the solubility product [Ag][Cl] is exceeded and consequently some silver chloride will be precipitated, equilibrium being reached when the product of the silver and chloride ion concentrations becomes equal to the solubility product. A similar effect will occur if an excess of chloride ions is added to a saturated silver chloride solution. A compound having an ion in common with a slightly soluble salt decreases the solubility of the latter (common-ion effect). The extent of the depression of the solubility can be calculated if the excess of the common ion is known.

Example. Calculate the solubilities of silver chloride in 0.001M, 0.01M and 0.1M potassium chloride respectively.

In 0.001M potassium chloride $[Cl^-] = 10^{-3}$.

$$\therefore \quad [Ag^+] = \frac{S_{AgCl}}{[Cl^-]} = \frac{10^{-10}}{10^{-3}} = 10^{-7}$$

In 0.01M potassium chloride

$$[Ag^+] = \frac{10^{-10}}{10^{-2}} = 10^{-8}$$

In 0.1M potassium chloride

$$[Ag^+] = \frac{10^{-10}}{10^{-1}} = 10^{-9}$$

In the above equations the concentration of Cl^- ions furnished by the silver chloride itself is neglected, since it is very small compared with the concentration of the excess Cl^- ions added.

Depression of solubility by the common-ion effect is of fundamental importance in gravimetric analysis. Addition of a suitable excess of a precipitating agent usually decreases the solubility of a precipitate to such an extent that the loss by washing is negligible. On the other hand a large excess of precipitant must be avoided, one reason being that the salt effect counteracts the common ion effect.

Influence of temperature upon solubility. The solubility of the precipitates encountered in quantitative analysis increases in varying degrees with rises of temperature. In many instances the common-ion effect reduces the solubility to so small a value that the temperature effect, which might otherwise be appreciable, becomes very small. It is, however, significant with magnesium ammonium phosphate hexahydrate and silver chloride, which are usually filtered at room temperature to avoid appreciable solubility loss.

Complex Ions

The increase in solubility of a precipitate upon the addition of excess of the precipitating agent is frequently due to the formation of a complex ion. A complex ion is formed by the union of a simple ion with either ions of opposite charge or with neutral molecules. Thus when potassium cyanide is added to a solution of silver nitrate, a white precipitate of silver cyanide is first formed because the solubility product of silver cyanide is exceeded. The precipitate dissolves on the addition of excess of potassium cyanide due to the formation of the complex ion $[Ag(CN)_2]^-$.

$$AgCN \text{ (solid)} + CN^- \text{ (excess)} \rightleftharpoons [Ag(CN)_2]^-$$

That the complex ion itself dissociates, is shown by the fact that silver sulphide can be precipitated from the solution with hydrogen sulphide, dissociation occurring as follows:

$$[Ag(CN)_2]^- \rightleftharpoons Ag^+ + 2CN^-$$

Application of the law of mass action to the equilibrium gives:

$$K = \frac{[Ag^+] \cdot [CN^-]^2}{[Ag(CN)_2]^-}$$

in which K is the dissociation or instability constant of the complex ion. The experimentally determined value of K is very small ($K = 1.0 \times 10^{-21}$); also the dissociation of the complex ion is repressed by the excess of cyanide ions present in the solution. Hence the silver ion concentration is reduced to such a low level that the solubility product of silver cyanide is not exceeded.

Indicators in Argentimetric Titrations

Potassium chromate (Mohr method). Potassium chromate is used as an indicator in the titration of chloride ions with standard silver nitrate in neutral solution, giving a precipitate of red silver chromate at the end point. The process is one of fractional precipitation of a pair of sparingly soluble salts—silver chloride ($S_{AgCl} = 1.56 \times 10^{-10}$ at 25°) and silver chromate ($S_{Ag_2CrO_4} = 9 \times 10^{-12}$ at 25°).

As the titration proceeds, silver chloride, the least soluble salt, will be precipitated so long as the chloride ion concentration is significant. The end point is reached when chloride ions are still present ($[Cl^-] = \sqrt{S_{AgCl}} = 1.249 \times 10^{-5}$) since silver chromate will commence to precipitate as soon as its solubility product is exceeded. This occurs when:

$$[Ag^+] \cdot [Cl^-] = S_{AgCl} = 1.56 \times 10^{-10}$$

$$[Ag^+]^2 \cdot [CrO_4^{2-}] = S_{Ag_2CrO_4} = 9 \times 10^{-12}$$

$$\therefore \quad [Ag^+] = S_{AgCl}/[Cl^-] = \sqrt{\{S_{Ag_2CrO_4}/[CrO_4^{2-}]\}}$$

$$\therefore \quad \frac{[Cl^-]}{\sqrt{[CrO_4^{2-}]}} = \frac{S_{AgCl}}{\sqrt{S_{Ag_2CrO_4}}}$$

$$= \frac{1.56 \times 10^{-10}}{\sqrt{(9 \times 10^{-12})}}$$

$$= 5.2 \times 10^{-5}$$

$$\therefore \quad [CrO_4^{2-}] = \left(\frac{[Cl^-]}{5.2 \times 10^{-5}}\right)^2 = \left(\frac{1.249 \times 10^{-5}}{5.2 \times 10^{-5}}\right)^2$$

$$= 5.77 \times 10^{-2}$$

Hence potassium chromate must be present at a concentration of at least 0.058M if precipitation is to occur at the end point. This calculation is based on solubility product figures determined at 25°; if the laboratory temperature is lower these will be significantly smaller, the corresponding potassium chromate concentration to ensure precipitation at 15° being 0.015M.

In practice, two other factors also operate. High concentrations of potassium chromate tend to obscure the end point and it is also necessary to add a small excess of silver nitrate (ca. 0.05 ml 0.1N AgNO₃) before the eye can detect a change. The errors involved can be corrected by a blank determination carried out with the same volume of indicator (usually 1 ml of 5 per cent solution) in sufficient water to reproduce the indicator concentration which obtains at the end point of the actual determination. The blank titration volume is subtracted from the determination titre.

Ferric thiocyanate (Volhard method). Ferric alum (ferric ammonium sulphate) is used as an indicator for the titration of silver ions with ammonium thiocyanate solution in the presence of nitric acid. This method forms the basis of the Volhard determination of chlorides, bromides and iodides in acid solution. When excess of standard silver nitrate solution is added to a solution of the halide, precipitation of silver halide occurs:

$$MX + AgNO_3 \rightarrow MNO_3 + AgX \downarrow$$

The excess of silver nitrate is then back titrated with ammonium thiocyanate:

$$AgNO_3 + NH_4SCN \rightarrow AgSCN \downarrow + NH_4NO_3$$

At the end point, excess thiocyanate reacts with ferric iron to give the red colour due to formation of ferric ferrithiocyanate:

$$2Fe^{3+} + 6SCN^- \rightleftharpoons Fe^{3+}[Fe(SCN)_6]^{3-}$$

In the determination of a chloride by this method, it is necessary to consider only the equilibria:

$$Ag^+ + Cl^- \rightleftharpoons AgCl \downarrow$$

$$Ag^+ + SCN^- \rightleftharpoons AgSCN \downarrow$$

Both 'insoluble' salts are in equilibrium with the solution, so that

$$[Ag^+] \cdot [Cl^-] = S_{AgCl} = 1.56 \times 10^{-10}$$
$$[Ag^+] \cdot [SCN^-] = S_{AgSCN} = 1.16 \times 10^{-12}$$
$$\therefore \quad [Cl^-]/[SCN^-] = S_{AgCl}/S_{AgSCN}$$
$$= 1.56 \times 10^{-10}/1.16 \times 10^{-12}$$
$$= 134$$

Because silver thiocyanate is less soluble than silver chloride, it follows that when the equivalence point is reached, further addition of thiocyanate would result in reaction with the precipitated silver chloride, and the reaction would proceed until the $[Cl^-]/[SCN^-]$ ratio reached 134, before any reaction occurred with the indicator. This is prevented, in practice, either by filtering out the precipitated chloride ions before back-titration, or alternatively by addition of a small quantity of nitrobenzene (*ca.* 1 ml) which assists coagulation of the precipitate, and coats the coagulated particles with a film of oil, thus preventing reaction with the thiocyanate ions.

The corresponding ratio of $[Br^-]/[SCN^-]$ which controls the end point in the Volhard titration of bromides is very much smaller being only 0.66 at 25°. There is little or no titration error in practice and there is, in consequence, no need for removal or treatment of the silver bromide precipitate. Similarly, the titration error with iodides is negligible.

Adsorption indicators. Adsorption indicators are acidic or basic dyes which change colour on adsorption on to the precipitate at the end point. Silver chloride precipitated during the titration of sodium chloride by silver nitrate adsorbs chloride ions on to the surface of the precipitate to form a layer of adsorbed ions. This layer of negatively charged chloride ions in turn promotes secondary adsorption of oppositely charged ions (cations) present in solution as shown in Fig. 4(*a*). At the end point some of the silver ions, now present in excess, will also be adsorbed on other particles of precipitate and will also constitute a primary adsorption layer, as in Fig. 4(*b*). The formation of the secondary adsorption layer involves competition between the anions present in the solution and the dye (usually fluorescein or dichlorofluorescein) which is adsorbed preferentially. Combination of the dye and Ag^+ ions on the surface of the precipitate gives the characteristic orange end point colour.

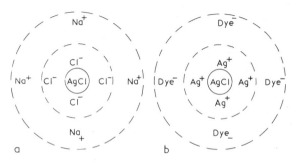

Fig. 4

Recommended adsorption indicators. Fluorescein (adsorbed colour pink) is suitable for the titration of chlorides, but since it is a very weak acid ($pK_a = 8.0$) it cannot be used in acid solution. Dichlorofluorescein (adsorbed colour orange) which is a stronger acid, is suitable for use in weakly acid solution (pH > 4.5).

Phenosafranine is suitable for the titration of both chlorides and bromides. It acts by a mechanism different from that of other adsorption indicators, in that the indicator is adsorbed throughout the titration, the adsorbed colour changing at the end point from pink to pale lilac (chlorides) and to deep lilac (bromides).

OXIDATION-REDUCTION TITRATIONS

Oxidation and reduction can be defined in terms of loss or gain of electrons. A definite equilibrium exists in any oxidation-reduction system if both oxidised and reduced forms are present. This is shown in the tendency of the system to gain or lose electrons from an inert metal (Pt) electrode in contact with the system. Such an electrode would assume a definite potential, losing electrons to become positively charged in an oxidising system, and gaining electrons to become negatively charged in a reducing system.

Standard Oxidation-Reduction Potentials

Oxidation-reduction potentials may be calculated by measuring the potential difference of a cell in which the oxidation-reduction half cell is coupled with a standard reference electrode. This is usually the normal hydrogen electrode, the potential of which is taken as zero. Thus for the ferrous-ferric system, the cell would be:

$$\text{Pt} \left| \begin{matrix} \text{Fe}^{3+} \\ \text{Fe}^{2+} \end{matrix} \right| \left| \text{H}^+ \right| \text{Pt, H}_2$$

In this system the cell reaction is $\text{Fe}^{3+} + \frac{1}{2}\text{H}_2 \rightleftharpoons \text{Fe}^{2+} + \text{H}^+$ and application of the van't Hoff isotherm gives the expression:

$$-\Delta G = nFE = RT \log_e K - RT \log_e \frac{a_{\text{Fe}^{2+}} \cdot a_{\text{H}^+}}{a_{\text{Fe}^{3+}} \cdot (a_{\text{H}_2})^{\frac{1}{2}}}$$

$$\therefore \quad E = \frac{RT}{nF} \log_e K - \frac{RT}{nF} \log_e \frac{a_{\text{Fe}^{2+}} \cdot a_{\text{H}^+}}{a_{\text{Fe}^{3+}} \cdot (a_{\text{H}_2})^{\frac{1}{2}}}$$

If the potential of the hydrogen electrode is taken as zero, then the e.m.f. of the cell is equal to the oxidation-reduction potential of the particular ferrous-ferric system under examination.

The expression

$$\frac{RT}{F} \cdot \log_e K - \frac{RT}{F} \log_e \frac{a_{\text{H}^+}}{(a_{\text{H}_2})^{\frac{1}{2}}}$$

is constant (n being equal to 1, since one electron only is involved in the reaction). This expression is equal to $E°$, the *standard oxidation-reduction potential*.

Table 11. Standard oxidation-reduction potentials ($E°$) at 25°

Electrode system	Electrode reaction	$E°$ (volts)
MnO_4^-, MnO_2/Pt	$MnO_4^- + 4H^+ + 3e \rightleftharpoons MnO_2 + 2H_2O$	+1.59
MnO_4^-, Mn^{2+}/Pt	$MnO_4^- + 8H^+ + 5e \rightleftharpoons Mn^{2+} + 4H_2O$	+1.52
Ce^{4+}, Ce^{3+}/Pt	$Ce^{4+} + e \rightleftharpoons Ce^{3+}$	+1.45
$Cr_2O_7^{2-}$, Cr^{3+}/Pt	$Cr_2O_7^{2-} + 14H^+ + 6e \rightleftharpoons 2Cr^{3+} + 7H_2O$	+1.36
IO_3^-, I_2/Pt	$IO_3^- + 6H^+ + 5e \rightleftharpoons I_2 + 3H_2O$	+1.20
Fe^{3+}, Fe^{2+}/Pt	$Fe^{3+} + e \rightleftharpoons Fe^{2+}$	+0.77
H_3AsO_4, H_3AsO_3/Pt	$H_3AsO_4 + 2H^+ + 2e \rightleftharpoons H_3AsO_3 + H_2O$	+0.56
I_2, I^-/Pt	$I_2 + 2e \rightleftharpoons 2I^-$	+0.53
$[Fe(CN)_6]^{3-}$, $[Fe(CN)_6]^{4-}$/Pt	$[Fe(CN)_6]^{3-} + e \rightleftharpoons [Fe(CN)_6]^{4-}$	+0.36
H^+, H_2/Pt	$H^+ + e \rightleftharpoons \frac{1}{2}H_2$	0.00
Ti^{4+}, Ti^{3+}/Pt	$Ti^{4+} + e \rightleftharpoons Ti^{3+}$	−0.06
S, S^{2-}/Pt	$S + 2e \rightleftharpoons S^{2-}$	−0.51

Then at temperature T,

$$E_T = E° + \frac{RT}{F} \cdot \log_e \frac{a_{Fe^{3+}}}{a_{Fe^{2+}}}$$

and when both ferrous and ferric ions are present in equal concentrations the second term disappears, and $E_T = E°$. That is, the e.m.f. of the cell is a direct measure of the standard oxidation-reduction potential when the concentration ($=$ activity) of oxidised and reduced forms of the system are equal. In practice, accurate measurement of the standard O/R potential of the ferrous-ferric system must be carried out in the presence of hydrochloric acid to depress hydrolysis. This necessitates extrapolation of the results to zero acid concentration. The standard oxidation-reduction potentials of a number of systems in common use for volumetric analysis are given in Table 11.

The greater the value of the oxidation-reduction potential, the more powerful the oxidising agent, and conversely, the lower the value the more powerful the reducing agent. Thus the position in the table of an oxidation-reduction system is an indication of its ability to oxidise or reduce other systems. Any system will oxidise any other system which occurs below it in the table and similarly will reduce any system situated above it in the table.

It is important to distinguish between the strength of an oxidising or reducing agent as expressed by its oxidation-reduction potential, and its actual capacity to oxidise or reduce. Thus strong oxidising or reducing agents, as implied by the value of E° for the system, may exhibit only limited ability, actually to oxidise or reduce, due to buffering of the sytem. Such systems are said to be well-poised.

The General Equation for the Calculation of Oxidation-Reduction Potentials

In a single oxidation-reduction system represented by the equilibrium:

$$\text{Oxidised form} + ne \rightleftharpoons \text{Reduced form}$$
$$\text{(ox)} \qquad\qquad\qquad \text{(red)}$$

where n is the number of electrons involved in the process, the electrode potential is given by the expression:

$$E_T = E° - \frac{RT}{nF} \log_e \frac{a_{red}}{a_{ox}}$$

$$= E° + \frac{RT}{nF} \log_e \frac{a_{ox}}{a_{red}}$$

where E_T is the observed potential at $T°$ absolute, and $E°$ the standard-oxidation potential of the system.

Substituting in this expression for R (8.313 joules), F (96 500 coulombs), taking T at 298° A (25°), converting to common logarithms and expressing activities in terms of concentration then

$$E_{25} = E° + \frac{0.0592}{n} \cdot \log_{10} \frac{[ox]}{[red]}$$

The actual oxidation-reduction potential of a system can be calculated from a knowledge of the standard potential and the percentage of oxidised and

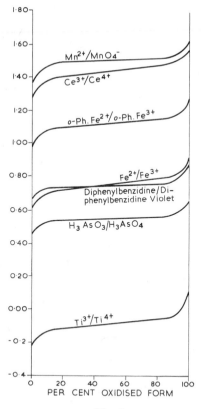

Fig. 5

reduced forms which are present, by substituting in the general equation:

$$E_T = E° + \frac{RT}{nF} \cdot \log_e \frac{[ox]}{[red]}$$

The results of such calculations can be expressed graphically as shown in Fig. 5 which shows the relationship between the calculated oxidation-reduction potential of a number of systems and the percentage of oxidised forms present. The curves for systems such as MnO_4^-/Mn^{2+} which are dependent upon the hydrogen ion concentration are calculated on the basis that it is molar. An increase or decrease of hydrogen ion concentration could either raise or lower the position of the curve in the table.

Oxidation of ferrous iron by permanganate. Ferrous iron is oxidised to the ferric state by potassium permanganate in acid solution according to the following equation:

$$MnO_4^- + 5Fe^{2+} + 8H^+ \rightleftharpoons Mn^{2+} + 5Fe^{3+} + 4H_2O$$

Since the reaction is carried out in dilute aqueous solution the water term of the equilibrium is constant and the equilibrium constant, K is given by the expression:

$$K = \frac{[Mn^{2+}] \cdot [Fe^{3+}]^5}{[MnO_4^-][Fe^{2+}]^5[H^+]^8}$$

The reaction can now be considered as the summation of two separate oxidation-reduction systems as follows:

$$MnO_4^- + 8H^+ + 5e \rightleftharpoons Mn^{2+} + 4H_2O$$
$$5[Fe^{2+} \rightleftharpoons Fe^{3+} + e]$$

An equation for the oxidation-reduction potential of each system can be written:

$$E = 1.52 + \frac{0.0592}{5} \cdot \log \frac{[MnO_4^-][H^+]^8}{[Mn^{2+}]}$$

and

$$E = 0.77 + \frac{0.0592}{5} \cdot \log \frac{[Fe^{3+}]^5}{[Fe^{2+}]^5}$$

$E°$ being 1.52 and 0.77 for the two systems respectively.

Considering these two electrode systems as being combined together in a cell, the e.m.f. at equilibrium will be zero and the two equations can be equated,

$$\therefore \quad 1.52 + \frac{0.0592}{5} \cdot \log \frac{[MnO_4^-][H^+]^8}{[Mn^{2+}]} = 0.77 + \frac{0.0592}{5} \cdot \frac{[Fe^{3+}]^5}{[Fe^{2+}]^5}$$

$$\therefore \quad \log \frac{[Mn^{2+}][Fe^{3+}]^5}{[MnO_4^-][H^+]^8[Fe^{2+}]^5} = \frac{5(1.52-0.77)}{0.0592}$$

$$\therefore \quad \log K = 63.34 \quad \text{and} \quad K = 2.19 \times 10^{63}$$

It is clear from the large value of K, the equilibrium constant, that the reaction will proceed to completion, the residual ferrous iron concentration being of negligible proportion.

Oxidation of ferrous iron by ceric sulphate. Ferrous iron can also be oxidised to the ferric state by ceric sulphate according to the equation:

$$Ce^{4+} + Fe^{2+} \rightleftharpoons Ce^{3+} + Fe^{3+}$$

and

$$K = \frac{[Ce^{3+}] \cdot [Fe^{3+}]}{[Ce^{4+}] \cdot [Fe^{2+}]}$$

The two separate oxidation-reduction systems are the ferrous-ferric and the ceric-cerous electrode reactions, which may be expressed as follows:

$$E_{Fe^{3+}/Fe^{2+}} = 0.77 + 0.0592 \log \frac{[Fe^{3+}]}{[Fe^{2+}]}$$

and

$$E_{Ce^{4+}/Ce^{3+}} = 1.45 + 0.0592 \log \frac{[Ce^{4+}]}{[Ce^{3+}]}$$

Hence at equilibrium:

$$\log \frac{[Ce^{3+}] \cdot [Fe^{3+}]}{[Ce^{4+}] \cdot [Fe^{2+}]} = \frac{1.45 - 0.77}{0.0592}$$

$$\therefore \quad \log K = 11.62 \quad \text{and} \quad K = 4.17 \times 10^{11}$$

Again the large value of K indicates that the reaction will proceed to completion.

Reduction of ferric iron by titanous chloride. The reduction of ferric iron to the ferrous state by titanous chloride (or sulphate) may be expressed by the following equation:

$$Fe^{3+} + Ti^{3+} \rightleftharpoons Fe^{2+} + Ti^{4+}$$

and

$$K = \frac{[Fe^{2+}] \cdot [Ti^{4+}]}{[Fe^{3+}] \cdot [Ti^{3+}]}$$

The separate oxidation-reduction systems may be expressed:

$$E_{Fe^{3+}/Fe^{2+}} = 0.77 + 0.0592 \log \frac{[Fe^{3+}]}{[Fe^{2+}]}$$

and

$$E_{Ti^{4+}/Ti^{3+}} = -0.06 + 0.0592 \log \frac{[Ti^{4+}]}{[Ti^{3+}]}$$

Hence at equilibrium:

$$\log \frac{[Fe^{2+}] \cdot [Ti^{4+}]}{[Fe^{3+}] \cdot [Ti^{3+}]} = \frac{0.77 + 0.06}{0.0592}$$

$$\therefore \quad \log K = 14.03 \quad \text{and} \quad K = 1.07 \times 10^{14}$$

As in the above examples the large value of K indicates that the reaction will proceed to completion.

Oxidation of arsenic trioxide by iodine. Consider the oxidation of arsenious acid ($As_2O_3 \equiv H_3AsO_3$) to arsenic acid ($As_2O_5 \equiv H_3AsO_4$) by iodine:

$$H_3AsO_3 + I_2 + H_2O \rightleftharpoons H_3AsO_4 + 2H^+ + 2I^-$$

Assuming the concentration of water to be constant, then

$$K = \frac{[H_3AsO_4][I^-]^2[H^+]^2}{[H_3AsO_3][I_2]}$$

The reaction can now be considered as the summation of the two separate oxidation-reduction systems:

$$H_3AsO_4 + 2H^+ + 2e \rightleftharpoons H_3AsO_3 + H_2O$$
$$I_2 + 2e \rightleftharpoons 2I^-$$

Hence

$$E = 0.56 + \frac{0.0592}{2} \cdot \log \frac{[H_3AsO_4][H^+]^2}{[H_3AsO_3]}$$

and

$$E = 0.53 + \frac{0.0592}{2} \cdot \log \frac{[I_2]}{[I^-]^2}$$

$$\therefore \quad 0.56 + \frac{0.0592}{2} \cdot \log \frac{[H_3AsO_4][H^+]^2}{[H_3AsO_3]} = 0.53 + \frac{0.0592}{2} \cdot \log \frac{[I_2]}{[I^-]^2}$$

$$\therefore \quad \log \frac{[H_3AsO_4][H^+]^2[I^-]^2}{[H_3AsO_3][I_2]} = \frac{2(0.53 - 0.56)}{0.0592}$$

$$\therefore \quad \log K = -\frac{0.06}{0.0592}$$

$$= -1.014$$

and

$$K = 9.68 \times 10^{-2}$$

The very small value of K indicates that the reaction does not proceed to completion, and in fact the equilibrium is displaced from right to left. The reaction can be displaced completely from left to right if it is carried out in the presence of sodium bicarbonate to remove hydrogen ions.

Irreversible Oxidations and Reductions

Many oxidations and reductions are not based on reversible electrode reactions. This applies generally to the oxidation and reduction of covalent compounds, when the necessary loss or gain of electrons takes place very much less readily than with similar ionic processes. The loss of a volatile oxidation (or reduction) product, such as carbon dioxide in the oxidation of oxalic acid by potassium permanganate similarly provides for an irreversible form of reaction:

$$5[(COOH)_2 \rightarrow 2CO_2 + 2H^+ + 2e]$$
$$2[MnO_4^- + 8H^+ + 5e \rightleftharpoons Mn^{2+} + 4H_2O]$$

Summation of these two equations gives:

$$2MnO_4^- + 5(COOH)_2 + 6H^+ \rightarrow 2Mn^{2+} + 10CO_2 + 8H_2O$$

This last equation shows that the oxalic acid alone is unable to supply sufficient hydrogen ions for the reaction, and this explains the necessity for conducting the titration in the presence of dilute sulphuric acid.

Since the overall reaction is irreversible, the law of mass action does not apply and no equilibrium constant can be calculated.

Other examples of non-reversible oxidation and reduction used in quantitative analysis include the oxidation of iodides with potassium iodate, the reduction of iodine with sodium thiosulphate, and the oxidation of sulphydryl (-SH) compounds with ferrous iron or iodine.

The Speed of Oxidation-Reduction Reactions

It is possible to calculate the equilibrium constant of reversible oxidation-reduction reactions but this gives no indication of the speed at which equilibrium will be reached. In many cases the reaction is quite slow and can only be used as the basis for quantitative titration in the presence of a catalyst which increases the reaction velocity. Thus the oxidation of oxalic acid by potassium permanganate is extremely slow at room temperature and in the absence of manganous ions. The reaction velocity is increased to a reasonable rate by heating to 60°. Even so, the reaction is slow to start, and potassium permanganate should be added cautiously at first until the necessary catalytic quantities of manganous ion can be built up. Similarly, in the standardisation of ceric sulphate by arsenic trioxide, the reaction is slow at room temperature but can be catalysed by the addition of a trace of osmium tetroxide.

Oxidation-Reduction Indicators

The oxidised or reduced form is self-indicating. Potassium permanganate is a good example; its solutions are so intensely coloured that a single drop will impart a definite pink colour at the end point of a titration to a comparatively large volume of solution. Ceric sulphate, which is yellow, and iodine (brown), both reduce to colourless ions, and are themselves sufficiently deeply coloured to provide good visual end points. The only objection to the use of self-indicating titrants is that the visual end point represents a slight over-titration.

External indicators. These are little used now. Some depend on the fact that excess titrant present immediately after the end point gives a visible reaction with some specific reagent. Others are based on some visible reaction of the titrated substance with a suitable reagent, so that the end point is marked by failure to elicit the reaction. An example of the latter type is the use of potassium ferricyanide as external indicator in the titration of ferrous iron by potassium dichromate. Drops of the solution removed to a spotting tile during the titration will give a deep prussian blue colour with potassium ferricyanide because ferrous ions are still present. At the end point, ferric iron only is present and this does not give a colour with potassium ferricyanide.

Internal indicators. Consider the oxidation of ferrous iron by potassium permanganate in molar acid solution. As the titration proceeds, the changes in the oxidation potential of the solution will follow a curve (Fig. 5) which is given by the sum of the separate curves of the two systems. At the end point, when all the iron has been oxidised to the ferric state, there is a steep rise in potential and this can be observed either potentiometrically or by the use of oxidation-reduction indicators.

Oxidation-reduction indicators are substances which have different colours in their oxidised and reduced form. Ideally, the reactions should be reversible and should give a precise and easily observable colour change at the end point. The two forms of the indicator comprise an oxidation-reduction system, the standard oxidation-reduction potential of which is intermediate between that of the titrated system and the titrant:

$$In_{ox} + ne \rightleftharpoons In_{red}$$

Most oxidation-reduction indicators are dyes, the reduced or leuco forms of which are colourless. Since dyes are intensely coloured, the indicator can be used at such a low concentration that there is no interference with the system under examination.

Consider an indicator at a potential of 0.06 volt above that of its standard oxidation potential $(E°)$

$$\therefore \quad E_T - E° = 0.06 = 0.0592 \log_{10} \frac{[ox]}{[red]}$$

$$\therefore \quad \log_{10} \frac{[ox]}{[red]} = \frac{0.06}{0.0592}$$

$$= 1.014$$

$$\therefore \quad \frac{[ox]}{[red]} = \frac{10.33}{1}$$

The indicator is therefore over 90 per cent oxidised and its colour will be indistinguishable from that of the oxidised form. Similarly, when the indicator is at a potential 0.06 volt lower than that of its $E°$, over 90 per cent of the indicator will be reduced and its colour, for all practical purposes, will be that of the reduced form. Such an indicator will change colour over a potential range which is given by $E_{In}°$ ±0.06 volt. If the indicator is to be of practical use in a particular titration, this potential range must not overlap that of either oxidation-reduction equilibrium concerned in the reaction. This is ensured if the potential differences $(E_1° - E_{In}°)$ and $(E_{In}° - E_2°)$ are not less than 0.15 volt. The above calculation is based on a working temperature of 25°. Sharper changes over smaller potential ranges will be obtained at lower temperatures since the value of the denominator in the above expression will be smaller.

Diphenylamine; sodium diphenylaminesulphonate; diphenylbenzidine. These three indicators, which are all used as solutions in concentrated sulphuric acid, are considered together since their mechanism of action is similar. The first step in the oxidation of diphenylamine (I) is its irreversible conversion into diphenylbenzidine (II)

Diphenylbenzidine (II) is colourless, but is reversibly oxidised into diphenylbenzidine violet (III):

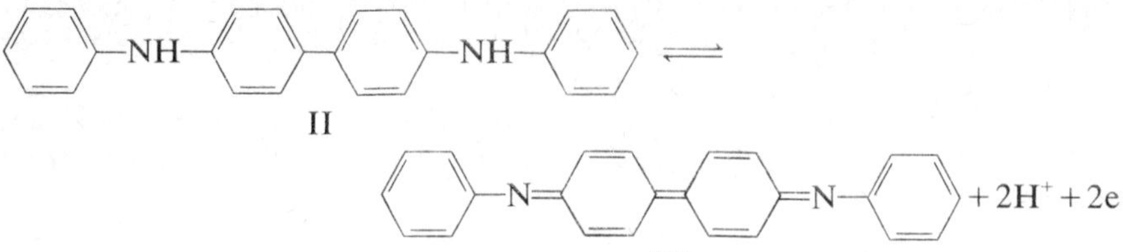

II

$+ 2H^+ + 2e$

III

The oxidation potential of this system is 0.76 volt and it is used as an indicator in the titration of ferrous iron by potassium dichromate. The standard oxidation-reduction potentials $E^{\circ}_{Cr_2O_7^{2-}/Cr^{3+}}$ and $E^{\circ}_{Fe^{3+}/Fe^{2+}}$ are 1.36 and 0.77 volt respectively, so that diphenylamine is only able to function as an indicator in this reaction when phosphoric acid is present in the solution. This decreases the concentration of ferric ions in the solution by complex formation, and hence reduces the actual potential of the ferrous-ferric system to a level which is sufficiently low to permit the indicator to function.

Sodium diphenylaminesulphonate acts by a similar mechanism, its oxidation-reduction potential (E°) being 0.83 volt, but unlike diphenylamine and diphenylbenzidine, it is readily soluble in water.

ortho-*Phenanthroline-ferrous iron (Ferroin)*. ortho-Phenanthroline-ferrous iron is a bright red complex formed by combination of the base *ortho*-phenanthroline with ferrous ions:

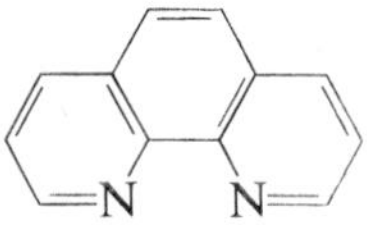

$+ Fe^{2+} \longrightarrow [(C_{12}H_8N_2)_3Fe]^{2+}$

Table 12. Oxidation-reduction indicators and their colour changes

INDICATOR	E° (at pH 0)												
	0·2	0·3	0·4	0·5	0·6	0·7	0·8	0·9	1·0	1·1	1·2	1·3	
2,6-DIBROMOPHENOLINDOPHENOL	Colourless	0·26	Blue										
DIPHENYLAMINE					Colourless		0·76	Blue					
DIPHENYLAMINESULPHONATE					Colourless			0·85	Blue				
DIPHENYLBENZIDINE					Colourless		0·76	Red Violet					
ERIOGLAUCINE							Orange		1·00	Green Yellow			
INDIGOMONOSULPHONATE	Colourless	0·26	Blue										
METHYLENE BLUE		Colourless		0·52	Blue								
NITRO-o-PHENANTHROLINE FERROUS SULPHATE										Red	1·14	Magenta	
o-PHENANTHROLINE FERROUS SULPHATE									Red	1·14	Blue		
PHENOSAFRANINE	Colourless	0·28	Red										
N-PHENYLANTHRANILIC ACID							Colourless		1·08	Red Violet			

This complex is readily oxidised reversibly to the corresponding *ortho*-phenanthroline-ferric iron complex, which is pale blue in colour:

$$[(C_{12}H_8N_2)_3Fe]^{2+} \rightleftharpoons [(C_{12}H_8N_2)_3Fe]^{3+} + e$$

The oxidation-reduction potential of this system is 1.14 volts. The complex is used as an indicator in the titration of ferrous iron ($E^{\circ}_{Fe^{3+}/Fe^{2+}} = 0.77$ V) by ceric sulphate ($E^{\circ}_{Ce^{4+}/Ce^{3+}} = 1.45$ V), the complete colour range of the indicator lying well within the potential gap between the two systems.

Table 12 gives a list of oxidation-reduction indicators, showing the E° values at pH = 0 and the approximate potential range of visible colour change.

4 Technique of Quantitative Analysis

GENERAL INFORMATION

Attention to detail is absolutely essential if success in quantitative analysis is to be achieved. A number of the more important general points which should be observed are enumerated below.

Cleanliness

The bench and apparatus must be kept scrupulously clean. A bench-cloth and a glass-cloth are required. All glassware should be rinsed with water (distilled or de-ionised) before use if it has been standing in the cupboard. The outside of vessels should be wiped dry with the glass-cloth, but the latter should not be used on the inside of vessels. Glass vessels should be free from grease. In general, soap or detergent solution is satisfactory for cleaning glassware. The apparatus should then be *thoroughly* rinsed with tap water followed by Water. On some occasions, however, a stronger cleaning agent may be required; this can be prepared by adding 15 g of powdered sodium dichromate to 500 ml of concentrated sulphuric acid. (NOTE: This cleaning agent is extremely corrosive.) A little of the 'cleaning mixture' is poured into the vessel to be freed from grease, and allowed to come into contact with the whole of the interior surface. Several hours contact is desirable. Any surplus should then be drained into the stock 'cleaning mixture' bottle, and the vessel rinsed successively with tap water and then Water. In the case of graduated apparatus, the first washing should be done with a large volume of cold water added quickly to prevent the apparatus becoming too hot as the solution and the water are mixed.

Tidiness

All reagent bottles must be returned to their *correct* positions on the reagent shelf *immediately* after use. Stoppers of reagent bottles must not be placed on the bench; they should be held in the left hand and replaced in the bottle after the reagent has been used. The apparatus should be arranged on the bench and in the cupboard in an orderly manner.

Labelling

All solutions, filtrates, precipitates etc. should be labelled systematically throughout the analytical procedure. If any liquid other than water is introduced into a wash-bottle, the bottle should be labelled appropriately and *immediately*.

Planning

Before commencing a determination, the directions for the assay to be performed must be read carefully. Care should be taken to ensure that all the details of the technique and the underlying basic principles are understood. The work should be planned so that there is no hold-up at any stage, e.g. the Gooch crucibles should be prepared and dried before the particular solution and precipitate is ready for filtration. Two or more operations should be kept going at the same time e.g. while a precipitate is cooling in a desiccator, a weighing or a titration should be carried out.

Graduated Apparatus

Graduated apparatus must not be heated and must not be used as containers for hot liquids, because the glass will expand and may not contract to its initial volume on cooling to room temperature.

Determinations in Duplicate

In general, determinations should be performed in duplicate. Good agreement between the two results engenders confidence; bad agreement provides evidence of incorrect work. Two results should not be averaged unless they are within 0.4 per cent. Burette readings should agree within 0.05 ml for titration figures of between 20 to 30 ml when equal volumes of solution are being titrated. A further titration (or titrations) is necessary if this agreement is not obtained.

Records

All data should be recorded *directly* into a laboratory note-book and *not* on pieces of paper.

It is recommended in a student's course in quantitative analysis, that the experimental observations e.g. weighings and burette readings should be recorded on the left-hand page, and the equations, description of the determination, calculations, and conclusion should be recorded on the right-hand page of the laboratory note-book. Examples of typical laboratory records are given on p. 129.

BALANCES

The Analytical Balance

The analytical balance possesses a high degree of sensitivity in order to give the true weight of samples (weight in this context being synonymous with mass). Factors affecting these requirements are

    (i) the length of the balance arms
   (ii) co-planarity of the knife edges
  (iii) the weight of the beam
  (iv) the position of centre of gravity of the beam in relation to the central knife edge or pivot.

The influence of these factors is shown by considering the hypothetical balance beam ABC (Fig. 6) of weight W with knife edges at A, B and C and centre of gravity at G. Figure 6 shows the position of the beam when loads W_1 and W_2 have caused a small deflection θ.

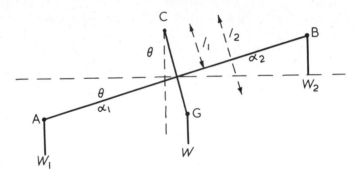

Fig. 6. Hypothetical balance beam

At equilibrium it may be shown that

$$\tan \theta = \frac{W_1 a_1 - W_2 a_2}{(W_1 + W_2)l_1 + Wl_2} \tag{1}$$

When equal weights are on each arm, the balance beam should be horizontal, i.e. $\theta = 0$; this applies when the lengths (a_1 and a_2) of the arms are equal. If they are not, a true weight may be obtained by weighing the sample first on one pan to give an apparent weight W_3 and then on the other pan to give an apparent weight W_4. The true weight is calculated from the formula True Weight $= \sqrt{(W_3 \cdot W_4)}$.

The sensitivity of a balance can be defined as the deflection of a pointer caused by a standard difference of weight between the two sides, e.g. the deflection, in scale divisions, per mg weight difference. For small deflections, $\tan \theta$ may be taken as θ expressed in radians and, incorporating the condition of equal balance arms in equation (1), the sensitivity is proportional to

$$\frac{\theta}{W_1 - W_2} = \frac{a}{(W_1 + W_2)l_1 + Wl_2} \tag{2}$$

where a represents the length of a balance arm. Increasing the length of each arm will increase the sensitivity of the balance only if it can be done without too great an increase in the weight W of the beam.

Sensitivity can be increased by reducing the magnitude of the denominator in (2) and manufacturers eliminate l_1 by making all three knife edges co-planar. Under this condition, the sensitivity is now proportional to a/Wl_2 and should remain constant with increase in load. If this co-planarity is lost due to excessive load and distortion of the beam, the sensitivity will decrease with increase in load. If, on the other hand, the central knife edge is below the other two, the sensitivity will increase with increase in load. The position of the centre of gravity can also be varied by means of a sensitivity bob. In this way G can be made to approach C and hence l_2 will be reduced leading to increased sensitivity. In the limit, when G coincides with C, a horizontal position of equilibrium is possible, but the slightest additional weight on one side causes the beam to swing to an extreme position; the balance is now unstable.

The ordinary technique in weighing is to determine the weight of sample by difference, and hence inequality of balance arms and loss of co-planarity of knife edges under load, with concomitant variations in sensitivity, may give rise to small errors. To avoid these errors, weighing by substitution has received considerable attention and is likely to achieve much wider use with the advent of synthetic sapphire for planes and knife edges. This hard material meets the objection of excessive wear on bearing surfaces at high loads when 'softer' materials are used. The technique involves both balance pans being fully loaded, and when a sample is placed upon the left-hand pan, weights must be *taken off* until equilibrium is attained. This is time-consuming on the conventional balance, but is convenient and rapid on certain types of single pan balance (see below).

Use and Care of the Balance

It is presumed that the student is already well acquainted with the principles of the balance and its use in weighing. The following points are intended to draw attention to certain important practical details, using the simple analytical balance for illustration.

(a) Material spilled on the balance pans or base must be cleaned up immediately.

(b) The balance should be tested to see that it swings freely and is in adjustment before each weighing is made.

(c) Sample must not be weighed directly upon the balance pans. A stoppered container must be used for weighing liquids and volatile, deliquescent or hygroscopic solids.

(d) Hot objects must be cooled to room temperature before introduction into the balance case.

(e) The balance door must be closed when the final weighing is made.

(f) The arrestment of the beam must be lowered *slowly* when setting the balance swinging.

(g) The beam must be raised and the pans arrested before an object or weight is added to or removed from the pans.

(h) The object to be weighed is usually placed on the left-hand pan and weights on the right-hand one.*

(i) All weights must be handled with forceps and *not* with the fingers. The forceps are manipulated with the right hand and the balance arrestment with the left hand. The heavy weight should be placed towards the centre of the balance pan. There are only two permissible places for weights to rest—on the balance pan or in the correct space in the box of weights.

(j) Before recording a weight, check the empty spaces in the box of weights as well as the weights on the balance pan.

(k) The balance must not be overloaded (maximum load usually 200 g).

(l) When a weighing has been completed, a check should be made to see that the beam is arrested, the weights are in their correct places in the box of weights, the 'rider' is on the sliding hook and the balance-case and box of weights are closed.

* A left-handed person may use the reverse only if the balance beam is graduated as shown in A.

Use of a 'rider'

Most laboratories now use aperiodic or single pan balances. Balances of this sort use a dial-operated system of weight addition and subtraction, with the smallest weights automatically visualised by means of a light pointer on an illuminated screen. The principle on which these modern balances operate, however, is fundamentally the same as that of the simple analytical balance, which requires the use of a beam rider for measurement of the smallest weights.

Weights smaller than 0.01 g are inconvenient to handle on the conventional analytical balance, therefore a 'rider' which is placed on a graduated scale along the top of the balance beam is used instead. The scale is divided into 20 or 10 equal parts (Fig. 7 A, B and C), each of the parts being subdivided.

If the beam is graduated as in A (Fig. 7), a 10 mg rider, suitably shaped to hang on the beam, is used. If the rider is placed at 'a' (Fig. 7A) the effect is equal to that of a 0.01 g weight placed on the right-hand pan of the balance; if it is hung at 'b', the effective weight is equal to 0.006 g. The distance between two small divisions as shown in Fig. 7A, represents a weight of 0.0002 g. The left-hand section of the scale (below zero) is seldom used; if the rider is placed on this scale it represents the effective weight of the rider applied to the left-hand pan. The rider is manipulated from outside the balance case by means of a sliding hook which moves above the balance beam. When the balance is not in use the rider is kept suspended from the rider hook away from contact with the beam.

In Fig. 7B another type of division of the beam is shown. A 5 mg rider is used for a balance with a beam of this type. The zero position of the rider is on the beam at division '0' and not on the hook. The effective weight of the rider

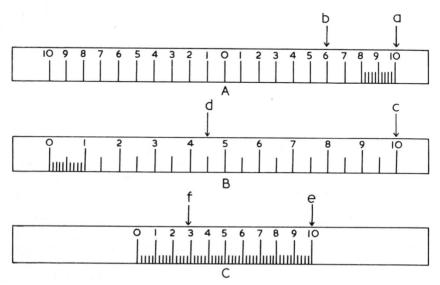

Fig. 7. Types of balance beam

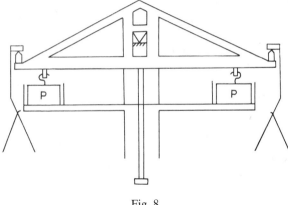

Fig. 8

is always on the right-hand pan. The rider at 'c' represents an effective weight of 0.01 g on the pan while at 'd' it represents an effective weight of 0.0045 g. The distance between two small divisions represents an effective weight of 0.0001 g.

A third type of beam division is shown in Fig. 7C. The graduations cover only half the effective length of the beam, and a 10 mg rider is used. The zero position of the rider is on the beam at division '0' and, as in B, the effective weight of the rider is always on the right-hand pan. The rider at 'e' represents an effective weight of 0.01 g on the pan, and at 'f' represents an effective weight 0.0028 g. The scale is subdivided in fifths and each small division represents an effective weight of 0.0002 g.

Aperiodic Balances

Aperiodic balances are those in which the swing of the beam is damped so that the beam comes to rest rapidly, usually in about 12 seconds or less. The damping is achieved by a piston arrangement on each arm of the balance as shown diagrammatically in Fig. 8.

The clearance between the pistons (P) and the cylinders, which are fixed, is small. Therefore, in setting up the balance, care should be taken to exclude dust, and to level the instrument accurately. Since the use of a rider would tend to increase weighing time, an optical system of some kind is usually incorporated to enable readings direct to 0.1 mg to be taken. The scale (S) is attached to the pointer of the balance and projected on to a suitable screen. It is of the utmost importance to ensure that the range of the scale, e.g. 0–10 (or 0–100) represents 10.0 (or 100.0) mg and this should be checked frequently at different loads.

Single Pan Balances

If the right-hand pan is enclosed completely, the left-hand pan of the aperiodic balance now becomes the single pan of the conventional single pan balance, i.e. one in which three knife edges are still retained. This is the variable load balance which is operated in the same way as an ordinary balance,

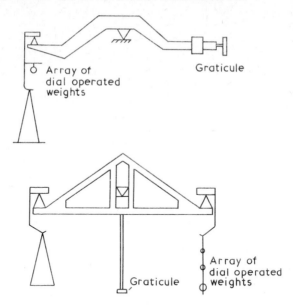

Fig. 9. Two types of single-pan balance (diagrammatic only)

except that mechanical addition of weights to the 'right-hand' side is achieved by an array of dial controls, as illustrated in Fig. 9.

When weighing by substitution is adopted, i.e. weighing under constant load, one of the outer knife edges may be replaced by a counterweight rigidly fixed to the balance beam. For maximum benefit from this design, the two arms of the balance are of unequal length as illustrated in Table 13, column 3. The inequality in the arms leads to a reduction in the weight necessary for a counterweight and hence to a reduction in the weight of the beam. Damping is achieved by movement of a damping disc in a cylinder much in the same way as described for aperiodic balances.

A summary of data on the various types of balance is reproduced in Table 13.

WEIGHING THE SAMPLE

An accurate weight of a sample may be obtained either by 'weighing by difference' or by 'weighing by addition'.

Weighing by Difference

Method A. Place sufficient sample for several analyses in a stoppered weighing bottle, and weigh accurately. Take the bottle from the balance with the *tips* of a *clean* dry thumb and fingers, and remove the stopper (left hand), holding the bottle over the mouth of the vessel in which the sample is to be placed. Pour out the estimated correct amount of substance by *carefully* tilting the bottle from the horizontal, and rotating it. Whilst still over the receiving vessel, tilt the bottle towards the upright position and

Table 13. Comparative information on different types of modern, direct-reading, knife-edge balance

SUBJECT \ TYPE	Normal three-knife two-pan balance	Three-knife single-pan balance	Two-knife single-pan balance
Accuracy	Usually best at loads less than maximum capacity		Limited to that at full knife loading
Variation of zero with air density	None	Slight	May be irksome
Sensitivity graticule	Only match for all loads on pan when adjustment of knives is ideal (involves professional skill)		Sensitivity independent of load on pan (because the load on the beam is always the same). Graticule matched by user
Variation of sensitivity with balance temperature	Slight		May be greater with this type
Stirrup error	Due to two stirrups		Due to one stirrup only
Weighings with counterpoise or tare	Both possible	Small tare or counterpoise may be possible	Neither possible
Adjustment of arm length	Equality required to very fine tolerances		Unnecessary
Other adjustments	All three knives to be adjusted parallel and co-planar		Only two knives to be made parallel
Wear of knives	Normal		Accelerated Knife loading always at maximum

rotate the bottle about its vertical axis. Tap the bottle lightly with the finger so that sample remaining on the inside of the bottle lip slides back in, and that on the outside falls into the receiving flask. Insert the stopper slowly and carefully to prevent powder from being blown out of the bottle. Reweigh the bottle and contents. The difference between the two weighings gives the weight of sample.

The disadvantage of this method lies in the difficulty of estimating the correct amount to be tipped out of the bottle. Since it is inadmissible to return any sample to the weighing bottle after it has been introduced into the receiving vessel, a number of additions of small portions are frequently required before the correct sample weight is obtained. The B.P. states that in assays 'the quantity actually used must not deviate by more than 10 per cent from that stated'. The following (Method B), is therefore recommended despite the fact that it involves *four accurate* weighings to obtain *two* analytical sample portions instead of the *three accurate* weighings which would be required by Method A.

Method B. Weigh the weighing bottle approximately to the nearest 0.01 g. (The approximate weight of the students's own bottle should be known.) Introduce into the bottle slightly more than the specified amount of sample. Alternatively, if the material to be weighed is stable, weigh slightly more than the specified amount on a rough balance and then introduce into the weighing bottle. Weigh the bottle and contents accurately. Tip out the contents of the bottle into the receiving vessel, using the technique described under Method A. When most of the solid has been transferred (no attempt should be made to transfer the last few particles which adhere to the inside of the bottle), replace the stopper, taking the precautions mentioned above, and reweigh the bottle and remaining contents *accurately*.

General Precautions for Weighing the Sample

(*a*) If the material is deliquescent or hygroscopic, exposure to the atmosphere should be reduced to a minimum and method B adopted.

(*b*) If the sample is being transferred to a conical flask or other vessel with a narrow aperture which is not narrow enough to require the use of a funnel, it is necessary to ensure that the mouth of the vessel is dry because of the possibility of contact with the weighing bottle. A funnel need not be used when transferring a solid from a weighing bottle to a conical flask in the 'weighing by difference' procedure.

(*c*) A record of the weighings must be made in the laboratory note-book *immediately*, in ink.

(*d*) Analytical samples of liquids should not be weighed by difference but by the method of 'weighing by addition' as outlined below.

Weighing by Addition

The empty dry vessel is first weighed *accurately*. The vessel can be a clock-glass, a weighing bottle, a *small* beaker, a *small* conical flask, a crucible etc. depending upon the weight and nature of the material to be weighed. The sample is then introduced into the vessel and the container and sample weighed again *accurately*. It is essential that the vessel should not be too large in order that changes due to adsorption of moisture and electrification during weighing may be kept to a minimum. Solids require slightly different weighing procedures from those used for liquids and the general methods are outlined below.

Method A. *Solids* (as in the preparation of standard solutions). Weigh the vessel (clock-glass, watch-glass or weighing bottle) *accurately*. Introduce the solid in portions until the correct weight has been added. Weigh the vessel and contents *accurately*. Hold the vessel over the receiver (beaker, or funnel in the mouth of the flask) and incline slightly so that the solid slides down slowly into the receiver. Wash the weighing vessel thoroughly with a jet of Water, collecting *all* the washings in the receiver.

On the semi-micro scale, the solid can be weighed directly in the small conical flask or beaker in which the titration or other subsequent treatment will be performed. *Great care* must be taken in weighing small quantities to ensure that the vessel into which the sample is to be weighed directly has been allowed to remain on the balance pan for some time so that a constant weight is obtained. The material to be weighed should be introduced directly into the weighed container while it remains on the balance pan, and then the container and its contents reweighed accurately.

Method B. *Water-miscible liquids*. Weigh accurately an empty, clean, stoppered weighing bottle. Remove the stopper, placing it on the balance pan. Add the liquid carefully from a teat pipette, until the correct weight has been added. Insert the stopper into the weighing bottle, and weigh the latter and its contents accurately. Liquid should not come into contact with the ground glass portion of the bottle in order that the stopper is not wetted. Take the bottle from the balance and, holding it at the base between the tips of the fingers and the thumb of the right hand, remove the stopper with the left hand, and pour the liquid gently down a guiding rod (also in the left hand) into the receiving vessel. A funnel need only be used if the receiver has a narrow neck. While the bottle is upturned and still held over the receiving vessel, transfer it to the left hand and wash down the bottle and guide rod with a jet of Water from a wash-bottle held in the right hand.

Method C. *Water-immiscible liquids*. If the titration (or other treatment) of the sample necessitates the use of a large flask or other container, the sample is usually weighed in a small vessel, e.g. a sample tube or other special container with a flattened end (see the Determination of Iodine values, p. 204).

Weigh the small vessel empty, introduce the sample, and then accurately weigh the vessel and its contents. Introduce the weighing vessel and its contents into the larger vessel in which the determination is to be carried out.

If the determination can be carried out in a small vessel, and the liquid is non-volatile, the liquid may be weighed directly into the vessel.

Accuracy of Weighings

It is not necessary to weigh a sample with a degree of accuracy which is appreciably greater than that which can be attained in the subsequent steps of the analysis.

In volumetric work, if the accuracy of the apparatus is ± 0.2 per cent and 1 g of sample is required for a particular assay, it would be unnecessary to weigh the sample to ± 0.0001 g since this would constitute an error of only ± 0.01 per cent in the weighing. For samples of such size in volumetric work, it is not necessary to weigh closer than ± 0.0005 g.

In gravimetric work, however, all weighings should be within ± 0.0002 g.

Constant Weight

In practice it is unusual for two consecutive weighings of the same object to be completely identical. This is particularly true where the object has been submitted to some physical treatment, such as heating and cooling between weighings. In gravimetric analysis, an object or sample is said to be at constant weight when two consecutive weighings after heating and then cooling in the desiccator differ by not more than 0.0003 g. The British Pharmacopoeia defines the term 'constant weight' used in relation to the determination of loss on drying or loss on ignition as meaning that two consecutive weighings do not differ by more than 1.0 mg per g of substance or residue for the determination, the second weighing being made after an additional hour of drying or after further ignition.

TECHNIQUE OF VOLUMETRIC ANALYSIS

Volumetric analysis requires the accurate measurement of *volumes* of interacting solutions. The graduated apparatus most commonly used in volumetric work are measuring (graduated) flasks, burettes, pipettes and measuring cylinders.

A Maker
BS 1792
in 100ml
20°C

Fig. 10.
Graduated
flask

Measuring (graduated) Flasks

Measuring flasks are usually round or pear-shaped, flat-bottomed, and with a long neck, which bears a single graduation mark extending right round the neck (Fig. 10). Such flasks, with one graduation, are made to *contain* a specified volume of liquid at 20°, when the level of the bottom of the meniscus (top, if convex liquid surfaces, as with mercury) coincides with the mark. The long narrow neck makes for accurate adjustment, the height of the liquid being sensitive to small changes of volume.

The litre is the standard unit of volume for all volumetric glassware and this is defined as 'the volume occupied by one kilogram of water at its temperature of maximum density (4°) and subjected to normal atmospheric pressure'. The cubic centimetre is the volume occupied by a cube of which each edge is 1 cm in length, and 1 litre equals 1000.028 cc. Thus the millilitre and cubic centimetre are not the same, although the difference is very small. Volumetric apparatus is therefore standardised in millilitres.

Volumetric apparatus is manufactured to two sets of tolerances (Grade A and Grade B respectively) which have been laid down by the National Physical Laboratory and the British Standards Institution (B.S. 1792). Volumetric flasks are required to be marked with the nominal capacity expressed in ml, the temperature at which standardised (20°C), the letters 'In' to indicate that the flask is graduated to contain, the letter 'A' or 'B' to indicate the class of accuracy to which the flask has been

Table 14. Tolerances in the capacity of measuring flasks

Capacity (ml)	Tolerance Class A (±ml)	Tolerance Class B (±ml)
5	0.02	0.04
10	0.02	0.04
25	0.03	0.06
50	0.05	0.10
100	0.08	0.15
200	0.15	0.30
250	0.15	0.30
500	0.25	0.50
1000	0.40	0.80
2000	0.60	1.20

graduated, the maker's name and the B.S. standard number (Fig. 10). Permitted tolerances for graduated flasks in common use are shown in Table 14.

The preparation of solutions of definite concentration
Transfer the appropriate weight of solid (weighed by addition, see p.112) quantitatively to a beaker and dissolve in Water (or other specified solvent).* Transfer the solution quantitatively by means of a glass rod, a funnel and a jet of Water from a wash-bottle as follows. Hold the beaker in the right hand and pour the liquid gently down the guiding rod, held in the left hand, into the funnel placed in the mouth of the graduated flask. Transfer the beaker to the left hand while still upturned, and still held over the funnel. Wash the beaker and guide rod with a jet of Water from the wash-bottle held in the right hand. Wash the funnel until the flask is two-thirds full, remove the funnel, swirl the flask to mix the contents and adjust to the mark. The final adjustment should be made by adding Water dropwise from a teat pipette; time must be allowed for water to drain down the inside of the neck of the flask. Finally shake the flask thoroughly to mix the contents.

If the solid is easily soluble in water, it may be added directly to a dry funnel in the mouth of the flask. The solid should slide easily down the funnel and the final trace of solid on the weighing vessel and funnel can be washed into the flask until the latter is about half full. The flask is then swirled gently until solution is effected and the volume made up to the mark as described above. The dry stopper is then inserted and the solution mixed thoroughly by inverting and rotating the flask.

For precise work, the temperature of the solution should be adjusted to 20° before making up to the mark (see later for effect of temperature).

Standard solutions, if they are to be stored, are usually transferred to stock bottles. If such a solution is to be transferred, the receiving vessel should be rinsed with two or three successive small quantities of the solution before the main bulk is added. When a standard solution is used some time after preparation, the container and its contents should be shaken well before any

---

* Water (distilled or de-ionised) is the solvent described in this account. Other solvents are also used and consequently when water is mentioned in this account it is intended that the appropriate solvent should be used as required.

of the solution is withdrawn. This shaking mixes the condensed water droplets, which have collected on the inside of the container above the solution
with the bulk of the solution.

Accurate Dilutions of Standard Solutions

Accurate dilutions can be prepared by pipetting the standard solution into a
graduated flask and diluting to the mark with Water. For example 100 ml of 0.2N
solution can be prepared from N solution by accurately pipetting 20 ml of the
latter into a 100 ml graduated flask and diluting to the mark with Water. The
factor for the dilution is the same as for the original standard solution.

Pipettes

Pipettes are of two kinds:
(a) *transfer* pipettes which have one mark and are used to *deliver* a specified
    volume of liquid under certain specified conditions, and
(b) *graduated* (or *measuring*) pipettes which have graduated stems and are
    employed to deliver various small volumes as required; they are not
    employed for measuring very exact volumes of liquids.
Permitted tolerances and delivery times as laid down by the British Standards
Institution (B.S. 1583:1961) for bulb transfer pipettes in common use are
shown in Table 15.

Table 15. Tolerances and delivery times for one-mark pipettes

Capacity (ml)	Tolerance (±ml)		Delivery times (secs)		
	Class A	Class B	Min., Class A	Min., Class B	Max., Classes A and B
1	0.007	0.015	7	5	15
2	0.01	0.02	7	5	15
3	0.015	0.03	10	7	20
4	0.015	0.03	10	7	20
5	0.015	0.03	15	10	25
10	0.02	0.04	15	10	25
15	0.025	0.05	20	15	30
20	0.03	0.06	25	20	40
25	0.03	0.06	25	20	40
50	0.04	0.08	30	20	50
100	0.06	0.12	40	30	60
200	0.08	0.16	50	40	70

All transfer pipettes have a single graduation mark. The capacity, and the
temperature at which it was graduated (Ex) and a reference to delivery time in
seconds is stated on the bulb. In addition, class A pipettes also state the time
of outflow (25 in Fig. 11). The method of calibration obviates the need to
specify drainage time, though a waiting time of 3 seconds after apparent
cessation of outflow is still essential. The stated times apply only for water and
aqueous solutions.

Use of the transfer pipette. Rinse the pipette with Water before use and allow to drain. Remove the drop of water remaining in the tip by touching against filter paper, and wipe the outside of the pipette to prevent dilution of the solution to be pipetted. Rinse the pipette with two or three small portions of the solution, each portion being used to wet the whole inside surface before being allowed to run out. Suck the liquid up into the pipette until just above the single graduation mark; close the upper end of the pipette with the tip of the index finger. Remove the pipette from the bulk solution and gently wipe the outside of the stem free from adhering liquid. Then, holding the pipette vertical and with the graduation mark at eye level, gently relax the pressure on the finger, so that the level of the liquid slowly falls. When the bottom of the meniscus coincides with the mark* increase pressure with the finger to prevent further escape of liquid, and remove the drop adhering to the tip by gently touching against a porcelain tile. Introduce the pipette into the receiving vessel, and allow the liquid to run out with the tip of the pipette *touching* the inside of the vessel at an angle of 60°, but *not* dipping into the delivered liquid. When all the liquid has run in, allow the pipette to drain in this position for 3 seconds (*waiting time*), and then remove the pipette. (*Note*, the small drop of liquid which remains in the tip of the emptied pipette is taken into account in the calibration and must *not* be added to the delivered liquid by blowing down the pipette.) A draining time is necessary in the case of liquids which are more viscous or have a much larger surface tension than water, e.g. strong solution of iodine.

After use, rinse the pipette with Water and allow to drain.

Fig. 11.
Transfer pipette

Automatic pipettes. These can be used to advantage with certain corrosive or toxic liquids. One type, shown in Fig. 12, is frequently used for iodine monochloride solution required in the determination of iodine values of oils. The pipette delivers its stated volume of liquid when filled with the liquid in question from tip (B) to tip (C) and allowed to run out and drain in the normal manner. D is connected to an aspirator which is fixed above the pipette so that the solution feeds in under gravity. E is a waste overflow.

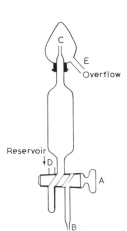

Fig. 12. Automatic pipette

Manipulation of the Automatic Pipette. Turn the two-way tap clockwise to open so that solution commences to flow into the pipette. When about 5 ml has run into the pipette, turn A clockwise through 180°, so that solution now flows from the pipette to fill the delivery tube B. As soon as B is full to the tip, again turn A clockwise through 180°, so that the body of the pipette is filled completely to the tip, closing the tap A by turning clockwise through 90° when solution commences to overflow at C. The pipette is now full from tip to tip and ready for use. Remove the drop of solution from tip B, run out and drain for 15 seconds in the usual way.

* The National Physical Laboratory describes a method of reading the meniscus in graduated glassware, viz. a dark horizontal line on a white background is placed 1 mm below the meniscus. A slight adjustment of the position of the dark line causes the meniscus to stand out sharply against the white background.

Burettes

A burette is a graduated vessel (Fig. 13) which is used for the accurate delivery of variable volumes of liquids. Burettes are made of varying overall capacity from 1 ml up to 100 ml; 50 ml burettes are of a size which is convenient for most volumetric operations. As with other forms of graduated glassware for volumetric work, they are produced to either Class A or Class B specifications (B.S. 846 : 1962). All Class A and some of Class B burettes have graduations which extend right round the barrel of the burette to reduce parallax errors in reading the burette. Class B burettes are usually graduated on one side only. Permitted tolerances on capacity for burettes in common use are shown in Table 16. The tolerance represents the maximum error allowed at any point and also the maximum difference allowed between the errors at any two points. Thus a tolerance of ±0.05 ml implies that the burette may be in error at any point by ±0.05 ml, provided that the difference between the errors at any two points is not more than 0.05 ml.

Table 16. Tolerance on capacity for burettes

Nominal capacity (ml)	Scale Subdivision (ml)	Tolerance on capacity (±ml)	
		Class A	Class B
1	0.01	0.006	0.01
2	0.02	0.01	0.02
5	0.02	0.01	0.02
5	0.05	0.02	0.04
10	0.02	0.01	0.02
10	0.1	—	0.05
25	0.05	0.03	0.05
25	0.1	0.05	0.1
50	0.1	0.05	0.1
100	0.2	0.1	0.2

Fig. 13.
Burette

Use of the Burette. See that the burette tap is lubricated with a thin film of grease. Before use rinse the burette twice with small portions of the solution (about 5 ml), allowing the burette to drain out well between the addition of each portion. Fill the burette with solution until the latter is slightly above the zero mark. Open the burette tap to fill the tip and expel all air bubbles, and, with the zero at eye level carefully and slowly run the liquid out until the bottom of the meniscus is level with or just below the zero mark. Remove the drop on the tip of the burette by momentarily touching it against a procelain tile or flask. Read the burette accurately after the film of solution adhering to the wall of the burette above the surface of the liquid has been allowed to drain for about 15 seconds. The burette must be read at eye level in order to avoid errors due to parallax; this is particularly important when using Class B apparatus. A piece of white paper or a white tile held behind the burette and just touching it at the appropriate level will assist observation of the meniscus and make for easy and accurate reading. Burettes (50 ml) are graduated in millilitres and tenths of a millilitre, but with a little experience, eye estimation is possible to a fifth of a division. Readings can be recorded to the nearest 0.02 ml. At the completion of a titration, 15 seconds must be allowed to elapse before the final reading is made, to allow for drainage.

Titration technique. The most suitable vessels for general work are conical flasks. Beakers are not recommended. If the latter are used, a stirring rod must also be employed to mix solutions during the titration. With a conical flask this can be done safely and easily by gently swirling the flask during the titration. The titration vessel should be kept well polished in order that the end point may be seen clearly. The solution being titrated is generally viewed against a white background (white tile).

Near the end point, it is advisable to split the drops of titrant. Partially open the tap so that a fraction of a drop flows out and remains attached to the burette tip. Touch the liquid against the inside of the flask, and wash the small volume of titrant into the bulk of the liquid with a few drops of Water. In any event, the upper internal portion of the flask should be washed down with a little Water just before the end point.

When it is considered that the end point has been reached, note the burette reading and then add a further drop of titrant, when a further distinct colour or other change should occur, unless the indicator changes to colourless. *Note.* This cannot be done if the solution is required for subsequent titration when excess titrant would interfere.

If the colour change at the end point is gradual, it is useful to have a comparison solution available. For example, if methyl orange is being used as indicator and the end point is gradual, two flasks containing the same volume of solution of approximately the same composition as the liquid being titrated can be prepared, one slightly acidic (red solution) and the other slightly alkaline (yellow solution). These comparison solutions will assist in deciding the colour change which indicates the end point.

Titrations should be carried out in duplicate and results should agree to within 0.05 ml (based on a 20 ml titration). If such agreement is not obtained, further titrations are necessary.

Any solution remaining in the burette after a series of titrations should be rejected and must not be returned to the stock solution. The burette is then washed out with distilled water after use and allowed to drain.

Measuring (graduated) Cylinders

These are vessels (Fig. 14) stoppered (B), or unstop-pered (A), of capacities varying from 5 ml up to about 2 litres. The smaller cylinders up to 100 ml are usually graduated in millilitres or fractions of a millilitre, whilst larger cylinders are graduated in units of 2, 5, 10, 20 or 50 ml, according to size. Cylinders are used for measuring out volumes of solution when only approximate volumes are re-quired.

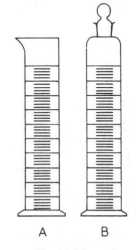

A   B

Fig. 14. Measuring cylinders

Effect of Temperature upon Volumetric Measurement

The effect of normal temperature changes on the volume of glass apparatus is negligible. However, graduated glassware must neither be heated nor filled with hot liquids because the glass will expand and may not contract to its initial volume on cooling to room temperature.

The effect of temperature variation on aqueous solutions is somewhat greater [cubical expansion of water and dilute aqueous solutions can be taken as approximately 0.00025 (0.025 per cent) per centigrade degree in the region 20° to 30°]. For general student work, corrections for the slight changes in temperatures in the laboratory can be ignored when aqueous solutions are used. However, when titrations in non-aqueous solvents (p. 175) are carried out, temperature corrections are improtant because the organic solvents used in volumetric work have much higher coefficients of expansion than that of water.

TECHNIQUE OF GRAVIMETRIC ANALYSIS

Much of the technique which has already been described is common to both gravimetric and volumetric analysis. In gravimetric work, certain additional techniques are necessary, namely, *precipitation, filtration, washing of the precipitate* and *drying* (or *ignition*) *of the precipitate;* these operations are considered in detail in the succeeding sections.

Precipitation

Gravimetric precipitations are usually made in beakers. Except during the actual precipitation, the beaker is covered with a clock-glass. A thin stirring rod, rounded at each end, is also required (see Fig. 15). The solution of the precipitating reagents is usually added slowly from a teat pipette or burette, with efficient stirring, to a suitably diluted solution of the sample.

Efficient stirring is necessary to avoid the possible high local concentrations of precipitating reagent which would tend to give contamination of the precipitate due to co-precipitation. The reagent is introduced down the side of the beaker to avoid splashing. Precipitation is usually made from hot dilute solutions, a procedure which tends to give an easily filterable precipitate and reduces the possibility of co-precipitation. The formation of coarse particles is favoured by *slow* addition of the precipitant, vigorous *stirring* of the solution during precipitation, and by carrying out the precipitation

Fig. 15

from hot solutions. Only a moderate excess of precipitating reagent is required. When the precipitate has settled somewhat, a few more drops of precipitant should be added to test for the completeness of precipitation; if further precipitation occurs, then the process of addition of precipitant, stirring and allowing the precipitate to settle and testing again must be carried out until there is no doubt about completeness of precipitation. The stirring rod should not come into contact with the sides or bottom of the beaker during stirring in order to avoid scratching particles of glass from the surfaces. Also a scratched surface is difficult to wash clean. Reagents should be examined for their clarity before use and filtered if necessary.

Before filtration, precipitates are digested by allowing them to stand over-night, or by heating the precipitate and its supernatant liquid nearly to boiling point for some time. Digestion in this manner increases the degree of coarseness of the precipitate. A further check for completeness of precipitation should be carried out when the supernatant liquid becomes clear during the period of digestion. No further precipitate should be produced when a few drops of the precipitant are added. However, should further precipitation occur, more precipitant must be added as described above.

If the precipitate is much more soluble in hot water than in cold, the solution is allowed to cool to room temperature before filtration. Otherwise, solutions are filtered hot to speed up the filtration.

Filtration

Apparatus used in Filtration

Various types of filtering apparatus are used in gravimetric analysis.

(a) Gooch crucible of either porcelain or silica in which filter-mats of purified asbestos are prepared prior to filtration. Ready made glass fibre filter mats may also be used for precipitates at temperatures below 200°.

(b) Sintered glass or sintered silica (Vitreosil) crucibles.

(c) Filter papers supported in filter funnels.

For many determinations, the use of Gooch or sintered crucibles is recommended because filtration and washing of the precipitates can be carried out rapidly and efficiently. The ignition of the precipitates can also be carried out without first burning off the filter paper. These advantages do not obtain when filter paper is used, but filter papers are used for gelatinous precipitates. Precipitates that require ignition are collected on filter paper, Gooch or sintered silica crucibles.

(a) *Gooch crucibles.* A Gooch crucible should be supported by a soft rubber collar in a glass adapter which passes through a rubber bung fitting into a Buchner-flask (Fig. 16). The rubber collar should not project below the bottom or above the top of the crucible. The tip of the adapter should project below the side arm of the Buchner flask to avoid the risk of the filtrate being sucked out of the flask. A trap is required to prevent any backflow from the pump contaminating the filtrate. Furthermore, it reduces the effects of changes in water-pressure which may cause the filter pad (see later) to lift from the base of the crucible.

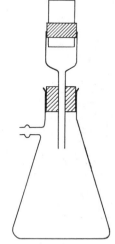

Fig. 16. Gooch crucible assembly

Disposable Glass Fibre Filter Discs

Standard glass filter discs are available (Whatman; 2.1 cm diameter) for use as filtration beds in Gooch crucibles. These filter discs are cheap, easy to handle, stable up to temperatures of at least 180°, and like the asbestos filter beds formerly used, are readily removed for cleansing and disposal at the end of the analysis.

Preparation of the Gooch crucible for filtration. Place the crucible in the suction filtration apparatus (Fig. 16). Sit a glass fibre filter disc gently on top of the porous base of the crucible. Gently wash the crucible and filter disc with water to remove any loose fibres, and apply gentle suction to drain off excess water. Suck the pad as dry as possible, covering the mouth of the crucible during the suction to prevent dust particles being drawn in. Dry (or ignite) the prepared Gooch crucible under the conditions specified for the drying (or ignition) of the precipitate, cool in a desiccator and weigh. Reheat, cool, and weigh; repeat until constant weight is attained.

(*b*) *Sintered glass or sintered silica crucibles.* The filtering beds are fused-in sintered discs of the same material as the crucible. The sintered glass type of crucible is most frequently used by students. They are available in various degrees of porosity numbered 1, 2, 3 and 4 indicating decreasing pore diameter with increasing number. A number 3 crucible is suitable for precipitates of medium particle size such as silver chloride while a number 4 is necessary for fine precipitates such as barium sulphate. Sintered glass crucibles should not be heated at temperatures above 400°. If the precipitate has to be ignited, or requires drying at temperatures above 400°, sintered silica crucibles (vitreosil) should be used.

Sintered crucibles are prepared for use in filtration by washing and passing Water through under suction and then drying (or igniting) to constant weight under the conditions which are specified for the drying (or ignition) of the filtered precipitate.

(*c*) *Filter paper supported in filter funnels.* In student quantitative work 'Whatman' No. 40, 41 or 42 filter papers are usually employed; these have very small ash values. No. 41 retains coarse particles and has a fast filtration speed, No. 40 retains medium size particles and has a medium speed, while No. 42 retains fine particles but has a slow filtration speed. The size of the filter paper to be used depends upon the bulk of the precipitate, the one of 11 cm diameter being more frequently used. The filter paper should extend to between 1 and 2 cm from the top of the funnel but not closer than 1 cm. Filtration with a filter paper is greatly facilitated by careful fitting of the paper to the funnel. A funnel with an angle as near as possible to 60° at the apex of the cone should be employed. The filter paper is folded in half and then in half again to form quarters, and the folds opened so that a 60° angle cone is formed with three thicknesses of paper on one side and a single thickness on the other. Because the angle of most filter funnels is rarely exactly 60° it is usually necessary to adjust the folds so that exact quarters are not made in order to make the paper fit the sides of the funnel closely. At least the upper portion of the paper must bed tightly against the funnel. Two funnels should be reserved for gravimetric work when their peculiarities are known, so that a filter paper can be fitted without repeated trials.

The paper is placed in the dry funnel, moistened with water and pressed down tightly to the sides of the funnel. No air space should be left between the upper half of the paper and the funnel. It is then filled with water; if the paper fits properly, the stem of the funnel will remain filled with liquid as the water passes through. The weight of the unbroken liquid column aids filtration.

Technique of Filtration and Washing the Precipitate

Using filter papers. Support the funnel containing the properly fitted paper in a filter stand and place a *clean* beaker so that the stem of the funnel just touches its inner side (Fig. 17). Splashing is then avoided and the filtrate which is sometimes required for subsequent operations, can be collected quantitatively. Pour the supernatant liquid above the precipitate down the glass rod into the filter without disturbing the precipitate more than is necessary. Keep the lower end of the glass rod close to, but not touching the filter paper on the side having three thicknesses of paper, so that the liquid is directed against the side of the filter and not toward the apex. Do not allow the liquid to rise above a level of 1 cm from the top edge of the paper. Keep the beaker inclined and a stream of liquid passing into the filter as long as the liquid filters freely. Wash the precipitate in the beaker by decantation as follows. Add about 10 ml of the wash liquid (usually heated) to the precipitate, stir the mixture, allow the precipitate to settle and pour off the supernatant liquid down the glass rod on to the filter paper. Repeat this procedure three or four times, then stir the precipitate with about 20 ml of wash liquid, and transfer as much of the solid as possible to the filter. Wash the traces of precipitate which adhere to the sides and bottom of the beaker into the filter with a jet of water from a wash-bottle as shown. Hold the wash-bottle* in the right hand, and the beaker in the left hand

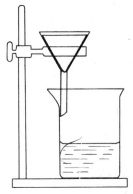

Fig. 17. Filtration

with the stirring rod pressed firmly against the lip of the beaker with the left thumb. A hot wash-bottle should have asbestos rope wrapped round its neck to protect the hand. Incline the beaker and direct a stream of wash liquid against the precipitate to wash it down the stirring rod into the filter. Small amounts of precipitate which adhere tenaciously to the walls of the beaker and the rod should be dislodged by rubbing with a 'policeman' (a glass rod covered at one end with a small piece of rubber tubing). When all the precipitate has been dislodged, rinse the 'policeman' with wash liquid and transfer the trace of precipitate down the rod to the filter. Finally hold the beaker and stirring rod up to the light, and examine carefully for traces of precipitate; repeat the treatment if necessary. (*Note*, a 'policeman' should not be used during precipitation or the early stages of filtration, and should be removed from the beaker immediately after use.)

When the precipitate has been transferred quantitatively to the filter, it should be washed immediately with wash liquid from a wash-bottle. Unwashed precipitate should not be allowed to stand for any length of time, because it will dry out and the mass will crack, and in this form it cannot be washed properly. The jet of liquid should be directed on to the filter in such a way that the precipitate is washed from the sides of the paper towards the apex of the cone. The clean margin of the paper above the precipitate always contains a certain amount of the mother liquor, and should also be well washed. The filter should be allowed to drain completely between each washing. From time to time, test for the completeness of washing by collecting small samples of the washings in a test-tube as they pass through the funnel, applying appropriate qualitative tests. When negative tests are obtained, the precipitate and paper are ready for drying and ignition.

Using a Gooch crucible. The prodedure for filtration and washing the precipitate is similar, in general, to the one already described for filter paper except that suction is

---

* A squeeze wash-bottle is very useful. Warm wash solution may be placed in these bottles but the bottle and contents must not be heated directly.

applied from a pump to draw the liquid through the crucible at a convenient rate. Liquid should not be poured into the crucible unless suction is being applied otherwise there is a danger of dislodging the filter pad.

First moisten the pad with a few drops of Water from a teat pipette. Apply suction, add more water, and then pour the liquid gently down a stirring rod on to the central portion of the filter mat. *Note*: a jet of liquid from a wash-bottle should *not* be directed into a prepared crucible. Transfer the precipitate to the crucible by the method described above. When washing the precipitate, apply only gentle suction. Collect test portions of filtrate in test-tubes inserted into the Buchner flask.

In some determinations, it is advantageous to wash the precipitate finally with ethanol to remove most of the water and thus reduce the time required for subsequent drying of the precipitate.

Using sintered glass or sintered silica crucibles. The technique is similar to that described for Gooch crucibles except that an asbestos pad is not used.

Drying and Ignition of the Precipitate

When the precipitate has been washed free from excess of the precipitant and other soluble impurities, it is converted to a substance of constant composition before it is weighed. The details of the actual temperature of drying or the temperature of ignition are given in the description of each determination. General techniques are outlined below.

Treatment of the Precipitate on Filter Paper

A precipitate which has been collected on filter paper is ignited to incinerate the paper before weighing. Two methods of igniting the precipitate and paper are commonly used; the method adopted depends upon whether or not the precipitate is reduced or changed upon ignition in presence of the filter paper.

Direct ignition of the precipitate and paper. It is preferable to use a silica crucible rather than a porcelain one because the former is much more resistant to thermal shocks. The crucible is ignited (perform the precipitation and ignition procedure simultaneously) to constant weight (±0.0002 g) under the conditions specified for the ignition of the precipitate.

Remove the *well drained* filter paper *carefully* from the funnel and fold into a small bundle to enclose the precipitate completely in several thicknesses of paper. Place in the ignited and weighed crucible supported on a pipeclay triangle on a vertical stand ring which can be lowered or raised at will as necessary during different stages of ignition. The filter paper in its funnel may be partially dried in a hot air oven before folding into a bundle and introducing into the crucible. Dry the paper and precipitate by heating the crucible about 10 cm above the tip of a luminous Bunsen flame about 2 cm high. The water must not be allowed to boil because portions of the precipitate may spatter about the crucible and some may even be ejected from the crucible. When the moisture has been expelled, increase the flame size slightly in order to char the paper *slowly*. The paper should not be allowed to burst into flame, or the gases allowed to ignite, because this leads to expulsion of fine particles of the precipitate. When the paper is completely blackened and gases are no longer evolved, heat the crucible with a large non-luminous flame to burn off the carbon and any tarry deposit; the precipitate and ash should remain in a compact mass at the base of the crucible. Now heat the crucible to the temperature stated in the assay of the substance in question. The flame (if a Bunsen is used) must not be allowed to enter the mouth of the crucible as fine particles of precipitate may be swept out; also reduction of the precipitate by the reducing gases of the Bunsen flame may occur. When ignition is complete, remove the flame and, after allowing the crucible to cool below dull redness in the atmosphere, transfer it while still hot to a desiccator by means of clean crucible tongs; allow to cool for at least 20 minutes and weigh. Ignite the crucible again at the specified temperature for 10 minutes, cool and reweigh. Repeat the heating, cooling and weighing until constant weight is attained.

If care is taken in the procedure described above, it is not necessary to use a crucible lid. If one is used it must be ignited with the empty crucible. During the removal of the moisture from the precipitate and filter paper in the crucible, the crucible should be only partially covered with the lid so that steam can easily escape. Keep the crucible covered during carbonisation of the paper but lift the lid at intervals in order that the progress of the charring can be followed. When charring is complete, the lid must be partially removed from the mouth of the crucible to allow free access of air.

Ignition with separate incineration of the paper. Some precipitates, e.g. barium sulphate, are extensively reduced when heated with charring filter paper. In these cases the bulk of the precipitate is sometimes removed from the paper before ignition and the paper and its trace of adhering precipitate ignited separately. Separation, however, can be avoided and the technique is therefore not described.

Alternative procedures include the use of (*a*) a sintered filter crucible or (*b*) chemical treatment of the partially reduced or changed material formed by direct ignition, to produce the desired material. Partially reduced barium sulphate is readily converted completely to barium sulphate by adding sulphuric acid and igniting in a fume cupboard.

Phase Separating Filtration

Special filter papers (Whatman No. 1PS), impregnated with a water-repellent, are available for effecting rapid small scale separations of immiscible mixtures of aqueous and organic liquids. The papers, which are used in a conventional filter funnel, are hydrophobic. Organic liquids readily pass through the paper, whilst aqueous liquids (and solids) are retained, so that rapid separation of two such liquids is readily attained.

5 Acidimetry and Alkalimetry

INTRODUCTION

Normality

A *normal solution* (designated N) is a solution which contains one gram equivalent of substance per litre of solution. Solutions of strengths other than normal are designated appropriately, e.g.

Twice normal	$2N$		
Half normal	$0.5N$	or	$N/2$
Deci-normal	$0.1N$	or	$N/10$
Centi-normal	$0.01N$	or	$N/100$

Equivalent Weight

The equivalent weight of a substance is determined by the actual reaction under consideration. The equivalent weight of an acid or base is that weight of it which contains 1.008 g of replaceable hydrogen (or its equivalent, such as the hydroxyl group). Thus the equivalent weights of hydrochloric acid (HCl) and sodium hydroxide (NaOH) are equal to their molecular weights, whilst that of sulphuric acid (H_2SO_4) or of oxalic acid $(COOH)_2$ which contain two replaceable hydrogens is equal to half the respective molecular weights.

In oxidation-reduction titrations, the equivalent weight is that weight which yields or combines with 1.008 g of 'available' hydrogen or 8.000 g of 'available' oxygen, 'available' being defined as available for use in the oxidation or reduction reaction under consideration.

The equivalent weight in precipitation reactions, as in argentimetry, is defined as that weight of substance which contains or combines with 1 g atom of a univalent metal.

The values taken for atomic and equivalent weights have, until recently, been based on the atomic weight $^{16}O = 16$. For the convenience of spectroscopists, atomic weights are now based on $^{12}C = 12$ but this involves only an insignificant change in the values hitherto accepted for the purposes of quantitative analytical chemistry.

Calculation of Equivalent Weight

Hydrochloric acid

$$HCl \equiv H$$
$$\therefore \quad 36.47 \text{ g HCl} \equiv 1000 \text{ ml } N$$

Sodium Carbonate

$$Na_2CO_3 + 2HCl \rightarrow 2NaCl + H_2O + CO_2$$
$$\therefore \quad Na_2CO_3 \equiv 2HCl \equiv 2H$$
$$\therefore \quad 106 \text{ g Na}_2CO_3 \equiv 2000 \text{ ml } N$$
$$\therefore \quad 53 \text{ g Na}_2CO_3 \equiv 1000 \text{ ml } N \text{ HCl}$$

Sodium Acid Oxalate. The equivalent varies with the reaction under consideration:

(i) *as an acid*

$$NaHC_2O_4 + NaOH \rightarrow Na_2C_2O_4 + H_2O$$

$$\therefore \quad NaHC_2O_4 \equiv NaOH \equiv H$$

$$\therefore \quad 112 \text{ g } NaHC_2O_4 \equiv 1000 \text{ ml } N \text{ NaOH}$$

(ii) *as a reducing agent*

$$NaHC_2O_4 + \bar{O} \rightarrow NaOH + 2CO_2$$

$$\therefore \quad NaHC_2O_4 \equiv \bar{O} \equiv 2H$$

$$\therefore \quad 112 \text{ g } NaHC_2O_4 \equiv 2000 \text{ ml } N$$

$$\therefore \quad 56 \text{ g } NaHC_2O_4 \equiv 1000 \text{ ml } N \text{ NaOH}$$

Molarity

A molar solution (designated M) is one which contains the gram-molecular weight of the substance in one litre of solution. The strengths of molar solutions are independent of the reaction under consideration. Molar solutions are sometimes used as a matter of convenience when a reagent, e.g. potassium iodate, has different equivalents in different reactions.

The Equivalence of Standard Solutions

If an acid A is titrated to neutrality against a base B, and the volume of acid and base used are v_A and v_B respectively, then, since the two solutions contain the same number of gram-equivalents,

$$v_A \times normality_A = v_B \times normality_B$$

Thus if the normality of one solution is known, and the equivalent volumes of the two solutions are determined by titration, the normality of the second solution can be found.

Preparation of Standard Solutions

If the chemical is available in a pure state, weigh out an exact quantity, dissolve it in water and make up to volume. Substances which are not usually obtainable in a pure state, e.g. mineral acids and caustic alkalis, are prepared as approximate solutions and standardised against a known pure standard, e.g. A.R. potassium hydrogen phthalate for alkalis and anhydrous sodium carbonate for acids. Solutions must be standardised under the same conditions as those to be used in any subsequent determination. If the unstandardised solution has to be used in the burette in the determinations, then it must be placed in the burette for standardisation and not in the titration vessel. The indicator used in standardisation must, in general, be that which has to be used in the subsequent determination.

Factors. In the preparation of standard solutions, it is often convenient to make solutions which, although of known strengths, are only approximately N, $0.5N$ or $0.1N$ etc. The relationship of the exact strength of such a solution to the nominal normality of the solution is then indicated by a factor. The factor is defined as the number of millilitres of exactly N, $0.5N$ etc. solution which are equivalent to 1 ml of the solution of the same nominal normality. It is the number by which the actual volume of the solution of approximate strength must be multiplied to obtain the equivalent volume of a standard solution of exact normality.

Calculation of the factor in the standardisation of N (approx.) H_2SO_4 by N Na_2CO_3.

Suppose 20 ml N Na_2CO_3 was neutralised by 19.40 ml of N (approx.) H_2SO_4.

Then, 19.40 ml N (approx.) $H_2SO_4 \equiv 20$ ml N Na_2CO_3

$$\therefore \quad 1 \text{ ml } N \text{ (approx.) } H_2SO_4 \equiv \frac{20}{19.40} \text{ ml } N$$

$$\equiv 1.031 \text{ ml } N$$

Therefore the strength of the H_2SO_4 solution is N (1.031). The *factor* of the solution is 1.031 and all volumes of this solution when multiplied by 1.031 will give the equivalent volume of normal solution.

It should be noted that, in equating the two solutions, the one of *unknown* strength is placed on the *left-hand side* of the equation, i.e.

$$19.40 \text{ ml of } N \text{ (approx.)} \equiv 20 \text{ ml } N \text{ } Na_2CO_3$$

Adherence to this rule will ensure correct calculation of the factor of an unknown solution and avoidance of the pitfall of reciprocal factors.

A tenfold quantitative dilution of an N solution gives an $0.1N$ solution with the same factor as the original N solution, e.g. n ml N H_2SO_4 (1.031) diluted with water to 10 n ml gives a solution which is $0.1N$ H_2SO_4 (1.031).

Method of Recording Quantitative Analysis in the Practical Note-book

All measurements such as weighings and burette readings should be recorded on the left-hand page of the note-book.

A statement of the determination to be carried out, the equations, the description of the method, the calculations and the final statement of the result should be recorded on the right-hand page of the note-book.

On p. 129 is a specimen pair of pages from a practical note-book.

Standardisation of N H_2SO_4

One of the methods of standardisation depends upon the use of exsiccated sodium carbonate:

$$H_2SO_4 + Na_2CO_3 \rightarrow Na_2SO_4 + CO_2 + H_2O$$

Method. Heat sufficient A.R. (AnalaR) exsiccated sodium carbonate in a nickel crucible at 260°–270° for half an hour to remove any trace of moisture. Cool in a desiccator and weigh an appropriate portion into a beaker containing Water (to avoid caking which otherwise occurs when it is wetted in the confines of a funnel). Dissolve with the aid of heat if necessary, cool, and transfer to an appropriate graduated flask. Dilute to volume and mix thoroughly.

Pipette 20 ml of this solution into a conical flask, add methyl orange indicator

Determination of the percentage of Na_2Co_3 in the sample of Exsiccated Sodium Carbonate

This determination depends upon the reactions expressed by the following equations:

$$Na_2CO_3 + 2HCl \rightarrow 2NaCl + CO_2 + H_2O$$
$$\therefore Na_2CO_3 \equiv 2HCl \equiv 2H$$
$\therefore$ 106.0 g $Na_2CO_3 \equiv 2000$ ml N solution
$\therefore$ 26.50 g $Na_2CO_3 \equiv 1000$ ml 0.5N solution
$\therefore$ 0.0265 g $Na_2CO_3 \equiv 1$ ml 0.5N solution

$\therefore$ 1 g of sample should give a burette reading of about 30 ml of 0.5N solution.

Method. Accurately weighed about 1 g sample into a conical flask containing a little water and titrated with 0.5N HCl solution.

Indicator. Methyl Orange—1 drop

End Point. The orange colour nearest to yellow

Calculations

I

Weight of sample = 0.9840 g
Burette reading = 34.03 ml 0.5N (1.065) HCl
$\therefore$ 0.9840 g sample = 34.03 × 1.065 ml 0.5N

$$\therefore 1 \text{ g sample} = \frac{34.03 \times 1.065}{0.9840} \text{ ml of } 0.5N$$

= 36.83 ml 0.5N HCl

II

Weight of sample = 0.9521 g
Burette reading = 32.90 ml 0.5N (1.065) HCl
$\therefore$ 0.9521 g sample = 32.90 × 1.065 ml 0.5N

$$\therefore 1 \text{ g sample} = \frac{32.90 \times 1.065}{0.9521} \text{ ml } 0.5N$$

= 36.81 ml 0.5N HCl

Average of I and II, 1 g sample = 36.82 ml 0.5N
= 36.82 ml × 0.02650 g Na_2CO_3
= 0.9757 g Na_2CO_3

Conclusion. The sample of Exsiccated Sodium Carbonate was found to contain 97.57 per cent Na_2CO_3.

Results

	1	2
1st weight of weighing bottle + sample =	21.2312 g	21.1901 g
2nd weight of weighing bottle + sample =	20.2472 g	20.2380 g
Weight of sample =	0.9840 g	0.9521 g
Burette readings: Final	34.03	32.90
Initial	0.00	0.00
Volume of 0.5N HCl (1.065) used	34.03 ml	32.90 ml

(1 drop) and titrate with the sulphuric acid solution until the first shade of orange colour is obtained. Repeat with another 20 ml portion. The two titrations should agree to within 0.05 ml, otherwise repeat until agreement is obtained.

The method can also be applied to the standardisation of other strengths of acid solutions.

Standard solutions of hydrochloric acid, sulphuric acid and sodium carbonate remain stable if properly stored.

Standardisation of N NaOH

Solutions of sodium hydroxide are always standardised by titration with standard acids of equivalent normality. They always contain a small but variable amount of carbon dioxide fixed as sodium carbonate so that the reactions expressed by equations (1–3) must be considered when standardising sodium (or potassium) hydroxide solutions:

$$NaOH + HCl \rightarrow NaCl + H_2O \tag{1}$$

$$Na_2CO_3 + HCl \rightarrow NaHCO_3 + NaCl \tag{2}$$

$$NaHCO_3 + HCl \rightarrow NaCl + H_2CO_3 \tag{3}$$
$$\downarrow$$
$$CO_2 + H_2O$$

As the hydrochloric acid is run into the sodium hydroxide solution the reactions expressed by equations (1), (2) and (3), occur in that order. The several end points of these reactions can be determined by suitable choice of indicators. Phenolphthalein, for example, is sensitive to carbon dioxide, i.e. a solution which is sufficiently alkaline to be *just pink* in colour will be changed to colourless by carbon dioxide. Immediately the reaction expressed by equation (2) is completed, the next drop of HCl will liberate CO_2 (equation 3), and the indicator will change from pink to colourless. Thus the phenolphthalein end point represents the point at which all hydroxide and half the carbonate have been neutralised. Methyl orange, on the other hand, is insensitive to carbon dioxide so that the methyl orange end point is obtained when all three reactions are complete. The methyl orange factor is therefore a measure of total alkali (hydroxide + carbonate). Therefore, the methyl orange and phenolphthalein factors are different.

Method. Pipette 20 ml of approximately N NaOH into a conical flask, add 2 drops of phenolphthalein solution, and titrate with N HCl until the solution is just colourless. Note the burette reading for calculation of the phenolphthalein factor, add 1 drop of methyl orange solution, and continue the titration until the first orange colour is obtained. Record the burette reading and calculate the methyl orange factor. The standardisation should be carried out in duplicate.

Note on the Preparation of Carbonate-free Sodium Hydroxide Solution

Place a small amount of paraffin wax in a glass stoppered bottle, and warm the bottle in a water bath to melt the wax. Shake to coat the whole of the inside surface of the bottle with paraffin wax, and cool rapidly to form a solid film. Prepare a 50 per cent w/w solution of sodium hydroxide in water and store in the bottle for 24 hours to allow sodium carbonate to crystallise out. Decant the clear supernatant liquid and dilute with CO_2-free water to the required strength.

Alternatively, carbonate-free sodium hydroxide can be prepared by ion exchange techniques (see p. 169).

DIRECT TITRATION OF STRONG ACIDS

Titration of strong acids with sodium hydroxide solution gives salts which are not hydrolysed in aqueous solution. The solutions are therefore neutral at the equivalence point. The pH changes extremely rapidly in the region of the equivalence point, and an indicator which changes colour anywhere within the range between pH 4 and 10 is suitable. Methyl orange is often used for such titrations.

Hydrochloric Acid. *Determination of the percentage w/w of* HCl

$$HCl + NaOH \rightarrow NaCl + H_2O$$

$$\therefore \quad 36.47 \text{ g HCl} \equiv H \equiv 1000 \text{ ml } N \text{ NaOH}$$

$$\therefore \quad 0.03647 \text{ g HCl} \equiv 1 \text{ ml } N \text{ NaOH}$$

Method. Weigh out accurately 4 g of sample (see page 113 for technique of weighing water-miscible liquids) using a stoppered weighing bottle to prevent loss of hydrochloric acid vapour (this would also attack the metal parts of the balance and weights). Transfer the acid into a stoppered flask by washing out with about 40 ml of Water. Titrate with N NaOH using methyl orange as indicator, taking the first definite orange colour as the end point.

Cognate Determinations

Hydriodic acid. Free iodine is first removed by titration with 0.1N sodium thiosulphate, and the acid then determined by titration with 0.1N NaOH.

Hydrobromic Acid
Dilute Hypophosphorous Acid
Nitric Acid
Perchloric Acid (72 per cent w/w)
Perchloric Acid (60 per cent w/w)
Potassium Hydrogen Sulphate
Sulphuric Acid
Thiamine Hydrochloride. Determination of Cl present as hydrochloride. This is a direct titration with 0.1N NaOH using bromothymol blue solution as indicator. The end point is given by the bluish-green colour of the indicator (pH 7.0).

Thiamine Hydrochloride Injection
Thiamine Hydrochloride Tablets

Determination of Aldehydes and Ketones in Essential Oils

The determination of aldehydes depends upon the reaction with hydroxylamine hydrochloride expressed by the following equation:

$$R.CHO + NH_2OH.HCl \rightarrow R.CH:N.OH + H_2O + HCl$$

The liberated hydrochloric acid can be titrated with standard alkali. The reaction, which is reversible, goes to completion if the hydrochloric acid is removed by titration with alkali. As the end point is flat and somewhat difficult to detect, the titration is continued to the full yellow colour of the methyl orange indicator. Hydroxylamine hydrochloride is hydrolysed in solution and is acidic to methyl orange. To compensate for the acid present due to hydrolysis of the reagent, the latter is first neutralised to the full yellow colour of methyl orange. Therefore, the reagent consists mainly of hydroxylamine hydrochloride but a small proportion of free hydroxylamine also is present; both base and salt may react with aldehydes. Thus the acid titrated during the actual assay is somewhat less than should theoretically have been formed, and an empirical factor ($\times 1.008$) is introduced into the calculation to allow for this error.

0.5N Potassium Hydroxide in Ethanol (60 per cent)

Preparation. Dissolve sufficient potassium hydroxide in ethanol (60 per cent) to produce a solution containing 28.05 g of KOH per litre. N.B. Potassium hydroxide contains only *ca.* 80 per cent of KOH.

Standardisation. Pipette 0.5N HCl (20 ml) into a conical flask; add methyl orange [2 drops of 0.2 w/v solution in ethanol (60 per cent)] and titrate with 0.5N KOH in ethanol (60 per cent). Take the full yellow colour of the indicator as the end point. N.B. An unusually concentrated solution of methyl orange is used in this determination. This is because in alcoholic solution the indicator is less sensitive to pH change than in water.

Hydroxylamine Hydrochloride Reagent in Ethanol (60 per cent)

Preparation. Dissolve hydroxylamine hydrochloride (34.75 g) in ethanol (950 ml of 60 per cent) and add methyl orange [5 ml of 0.2 w/v solution in ethanol (60 per cent)]. Add 0.5N KOH in ethanol (60 per cent) dropwise until the full yellow colour of the indicator is obtained. Adjust the volume to 1000 ml with ethanol (60 per cent). Check the pH of the solution by means of the following test: to 10 ml add one drop of 0.5N KOH in ethanol (60 per cent)—no change in colour is produced. To a further 10 ml add one drop of 0.5N HCl—the colour changes slightly towards orange.

Cinnamon Oil. *Determination of the percentage w/w of cinnamic aldehyde* C_9H_8O (This oil is not now official in the British Pharmacopoeia)

Method. By means of a wire loop, suspend a glass stoppered tube (about 25 mm in diameter and 150 mm long) from the stirrup of a balance. Allow five minutes for equilibration to occur and then weigh the tube. By means of a teat pipette, introduce the sample (about 1 g) and reweigh. Add from a measuring cylinder toluene (5 ml) and hydroxylamine hydrochloride reagent in 60 per cent ethanol, (15 ml), shake vigorously and titrate with 0.5N KOH in ethanol (60 per cent) until the red colour changes to full yellow. Continue the titration with vigorous shaking until the lower layer retains the full yellow of the indicator after shaking vigorously for two minutes and allowing to separate. The reaction is slow and a titration cannot be performed satisfactorily in less than fifteen minutes. This first titration indicates the approximate amount of aldehyde in the sample. Add a further 0.5 ml titrant to the above mixture to ensure that the full yellow colour of the indicator has been obtained, and use this as a standard colour for matching the end point of two further accurate determinations.

The reaction is not quantitative unless an adequate excess of hydroxylamine hydrochloride is used. To ensure that this is so, the volume of hydroxylamine hydrochloride reagent in ethanol (60 per cent) added must exceed the burette reading of $0.5N$ KOH in alcohol (60 per cent) by between 1 and 2 ml.

1 ml $0.5N$ KOH in ethanol (60 per cent)
$$\equiv 0.06661 \ (0.06608 \times 1.008) \text{ g of cinnamic aldehyde } C_9H_8O$$

Lemon Oil. *Determination of the percentage w/w of aldehydes calculated as citral* $C_{10}H_{16}O$ (This oil is not now official in the British Pharmacopoeia). It may be determined by the above method using 10 g of sample and 7 ml of hydroxylamine hydrochloride reagent in ethanol (60 per cent). Since Lemon Oil contains less aldehydes and more terpene than Cinnamon Oil a correspondingly larger weight of sample must be used in the assay. Addition of toluene is unnecessary since the proportion of terpene is sufficiently large to cause the titration liquid to separate into two layers. 1 ml $0.5N$ KOH in ethanol (60 per cent) $\equiv 0.07672 \ (0.07611 \times 1.008)$ g of citral $C_{10}H_{16}O$.

Cognate Determinations

Isobutyl Methyl Ketone
Ninhydrin

Sodium Dithionite (Sodium Hydrosulphite)

The determination depends upon the reaction of sodium dithionite with formaldehyde:

$$Na_2S_2O_4 + 2H.CHO + H_2O \longrightarrow \underset{\substack{\text{sodium} \\ \text{formaldehyde} \\ \text{sulphoxylate}}}{HC\overset{\displaystyle OH}{\underset{\displaystyle SO_2Na}{H}}} + \underset{\substack{\text{sodium} \\ \text{formaldehyde} \\ \text{bisulphite}}}{HC\overset{\displaystyle OH}{\underset{\displaystyle SO_3Na}{H}}}$$

The sample is added to a solution of formaldehyde which has been previously neutralised with alkali to phenolphthalein, and the mixture allowed to stand in a stoppered flask. The solution is then transferred to a graduated flask, diluted with previously boiled and cooled water, neutralised and adjusted to volume. A portion is pipetted into a flask, neutralised, and then titrated with $0.1N$ iodine using mucilage of starch as indicator. The reaction expressed by the following equation occurs:

$$H{-}C\overset{\displaystyle OH}{\underset{\displaystyle SO_2Na}{H}} + 2I_2 + 2H_2O \rightarrow HCHO + NaHSO_4 + 4HI$$

Under the conditions of this determination only the sodium formaldehyde sulphoxylate is oxidised. The trace of excess of iodine is removed by adding one drop of $0.1N$ sodium thiosulphate.

The acidic substances liberated by the iodine are then titrated with $0.1N$

NaOH using phenolphthalein as indicator:

$$NaHSO_4 + 4HI + 5NaOH \rightarrow Na_2SO_4 + 4NaI + 5H_2O$$

$$\therefore \quad Na_2S_2O_4 \equiv H-\overset{\displaystyle OH}{\underset{\displaystyle SO_2Na}{CH}} \quad \equiv (NaHSO_4 + 4HI) \equiv 5NaOH \equiv 5H$$

DIRECT TITRATION OF WEAK ACIDS

Titration of weak acids with sodium hydroxide solution gives salts which will be hydrolysed in aqueous solution to a greater or lesser extent depending upon the dissociation constant of the acid. The pH of the solution at the equivalence point will be above pH 7, and the indicator most frequently used in such titrations is phenolphthalein.

Acetic Acid. *Determination of the percentage w/w of* CH_3COOH

$$CH_3COOH + NaOH \rightarrow CH_3COONa + H_2O$$

$$\therefore \quad 60.05 \text{ g } CH_3COOH \equiv H \equiv 1000 \text{ ml } N \text{ NaOH}$$

$$\therefore \quad 0.06005 \text{ g } CH_3COOH \equiv 1 \text{ ml } N \text{ NaOH}$$

Acetic Acid B.P. contains 32.5–33.5 per cent w/w CH_3COOH

$\therefore$ 5 g of sample ($\equiv ca.$ 1.6 g CH_3COOH) is approximately equivalent 25 ml N.

Method. Weigh out accurately 5 g of sample (p. 113 for technique of weighing water-miscible liquids) using a stoppered weighing bottle. Wash out the acid into a stoppered flask using about 50 ml of Water. Titrate with N NaOH using phenolphthalein as indicator, and take the first shade of pink as the end point.

Dilute Phosphoric Acid. *Determination of the percentage w/w of* H_3PO_4

This determination depends upon the partial neutralisation of the phosphoric acid to give disodium hydrogen phosphate. The latter tends to hydrolyse, and this is suppressed in the assay by the addition of sodium chloride and limitation of the volume of water.

$$H_3PO_4 + 2NaOH \rightarrow Na_2HPO_4 + 2H_2O$$

$$\therefore \quad H_3PO_4 \equiv 2NaOH \equiv 2H \equiv 2000 \text{ ml } N \text{ NaOH}$$

$$\therefore \quad 98.00 \text{ g } H_3PO_4 \equiv 2000 \text{ ml } N \text{ NaOH}$$

$$\therefore \quad 0.04900 \text{ g } H_3PO_4 \equiv 1 \text{ ml } N \text{ NaOH}$$

$\therefore$ 10 g of 10 per cent solution is equivalent to approximately 20 ml N NaOH.

Method. Weigh a clean dry weighing bottle, add the sample (about 10 g) and reweigh. Transfer the contents of the weighing bottle to a conical flask by means of a glass rod and small funnel. Wash the weighing bottle, rod and funnel with Water and collect the washings in a conical flask. *Not more than 30 ml of water should be used.*

Add sodium chloride (10 g; weighed on a rough balance) and titrate with N NaOH using 2 drops of solution of phenolphthalein as indicator.

Potassium Hydrogen Tartrate. *Determination of the percentage of* $C_4H_5O_6K$

The molecule of potassium hydrogen tartrate has one replaceable hydrogen and the substance can, therefore, be assayed by titration with sodium hydroxide using phenolphthalein as indicator.

$$
\begin{array}{c}
\text{CH(OH).COOK} \\
| \\
\text{CH(OH).COOH}
\end{array}
+ \text{NaOH} \longrightarrow
\begin{array}{c}
\text{CH(OH).COOK} \\
| \\
\text{CH(OH).COONa}
\end{array}
+ \text{H}_2\text{O}
$$

$$\therefore \quad C_4H_5O_6K \equiv NaOH \equiv H \equiv 1000 \text{ ml } N \text{ NaOH}$$

$$\therefore \quad 188.2 \text{ g } C_4H_5O_6K \equiv 1000 \text{ ml } N \text{ NaOH}$$

$$\therefore \quad 0.03764 \text{ g } C_4H_5O_6K \equiv 1 \text{ ml } 0.2N \text{ NaOH}$$

Because potassium hydrogen tartrate is not very soluble in cold water, a modified titration procedure is used.

Method. Weigh out accurately about 1.5 g of sample, add Water (100 ml) and boil. Titrate the hot solution with $0.2N$ NaOH using phenolphthalein as indicator, boiling the solution after each addition. Near the end point, boil after the addition of each drop of titrant.

Note. Although phenolphthalein is used as indicator, the methyl orange factor of the $0.2N$ NaOH must be used, because in the boiling liquid any H_2CO_3 from the carbonate in the sodium hydroxide will be removed as CO_2, and so will not affect the phenolphthalein indicator.

Cognate Determinations

Benzoic Acid. Prepare 30 ml (approx.) of neutralised ethanol (90 per cent) by adding 10 drops of phenol red and titrating with $0.5N$ sodium hydroxide until a full red colour is obtained. Use 15 ml of this neutralised ethanol to dissolve about 2.5 g of sample, accurately weighed, add 20 ml of water to precipitate the acid in a finely divided state and titrate with $0.5N$ sodium hydroxide until a full red colour is obtained.

Citric Acid. Citric acid (pK_a 5.4 at 25°) is a weaker acid than acetic acid (pK_a 4.74 at 25°); phenolphthalein however is used as indicator in the titration.

Formic Acid

Maleic Acid

Mustine Injection

Nicotinic Acid. Since the titration is carried out with $0.1N$ solution, a two-colour indicator produces a more satisfactory end point than phenolphthalein. The indicator phenol red is, however, sensitive to carbon dioxide and titration is therefore carried out in CO_2-free water and with carbonate-free $0.1N$ sodium hydroxide.

Nicotinic Acid Tablets (p. 345 for general notes on the determination of tablet constituents). Nicotinic acid is extracted from the powdered tablets with hot 95 per cent ethanol (neutralised to phenolphthalein), the extract diluted with water and determined as described under Nicotinic Acid.

Oxalic Acid. The titration is carried out in CO_2-free water.

Potassium Dihydrogen Phosphate. The second dissociation constant of phosphoric acid (pK_a) is 7.25 at 25°, so that at this stage of its dissociation, phosphoric acid is a much weaker acid than acetic acid (pK_a 4.74 at 25°). The acid salt is titrated with sodium hydroxide to the grey-green colour of thymol blue which represents a pH of about 9.2. Addition of salt and limitation of volume is then not necessary as in the case of phosphoric acid.

Potassium Hydrogen Phthalate

Salicyclic Acid. Determined as for Benzoic Acid.

Salicyclic Acid Ointment

Sodium Acid Phosphate. Determined as for Phosphoric Acid.

Sulphanilic Acid

Trinitrophenol

Trichloroacetic Acid

Undecanoic Acid

Special Modifications in the Direct Titration of Weak Acids

Acid Value of Fixed Oils

Acid value is defined as the number of mg of potassium hydroxide required to neutralise the free acid in 1 g of substance. The method of determination is specified and is outlined below. The amount of free acid present in most of the fixed oils is generally low (Table 8), and the acid value is used to eliminate low grade and rancid oils which tend to have higher acid values.

Method. Weigh accurately a 250 ml conical flask. Introduce the sample (about 10 g) and reweigh. Mix ether (25 ml), 95 per cent ethanol (25 ml) and 1 ml of solution of phenolphthalein (1.0 per cent in 95 per cent ethanol) and if necessary neutralise by titration with a few drops of *0.1N* aqueous KOH. Add this solution to the flask and shake to dissolve the fatty acids which are present in the oil. The oil itself is usually insoluble. Heat is unnecessary except for solid fats and waxes, for which it is essential to ensure complete extraction of the free fatty acids. Cool to room temperature before commencing the titration with *0.1N* aqueous KOH. Shake continuously during the titration and continue the titration until the pink colour of the indicator persists for 15 seconds. Do not wash down with water during this titration. Strict adherence to this time factor is essential as the glycerides of the oil become hydrolysed in the presence of excess potassium hydroxide.

$$R.COOH + KOH \rightarrow R.COOK + H_2O$$

$$\therefore \quad KOH \equiv H$$

$$\therefore \quad 56 \cdot 1 \text{ g KOH} \equiv 10\ 000 \text{ ml } 0.1N$$

$$\therefore \quad 5 \cdot 61 \text{ mg KOH} \equiv 1 \text{ ml } 0.1N$$

$$\therefore \quad \text{Acid Value} = \frac{\text{Burette reading } (\times \text{factor of alkali}) \times 5 \cdot 61}{\text{Weight of sample (in g)}}$$

Cognate Determinations

Acid Value of Oleic Acid. This is determined as above, but using 1 g only. In this example, the acid value determination is really a determination of equivalent weight.

White Beeswax

Boric Acid. *Determination of the percentage of* H_3BO_3

Boric acid is too weak an acid to be titrated quantitatively in aqueous solution with sodium hydroxide solution using a visual indicator. However, it can be titrated with standard alkali in the presence of mannitol using phenolphthalein as indicator.

$$H_3BO_3 + NaOH \xrightarrow{\text{mannitol}} NaBO_2 + 2H_2O$$

$$\therefore \quad H_3BO_3 \equiv NaOH \equiv H \equiv 1000 \text{ ml } N \text{ NaOH}$$

$$\therefore \quad 61.84 \text{ g } H_3BO_3 \equiv 1000 \text{ ml } N \text{ NaOH}$$

$$\therefore \quad 0.06184 \text{ g } H_3BO_3 \equiv 1 \text{ ml } N \text{ NaOH}$$

Boric acid is esterified in the presence of polyhydric alcohols such as glycerol or mannitol forming a monobasic acid which is strong enough to give a satisfactory end point. It is stated that if glycerol is added a glycerylboric acid is formed. At least 30 per cent glycerol or mannitol is needed to prevent hydrolysis of the acid.

Method. Accurately weigh the sample (about 2 g), add Water (50 ml) and mannitol (20 g), and titrate with N NaOH using phenolphthalein as indicator.

Busulphan. This substance is hydrolysed by refluxing with water and the liberated methanesulphonic acid titrated with standard alkali using phenolphthalein as indicator.

$$CH_3SO_2.O(CH_2)_4O.SO_2CH_3 + 2H_2O \rightarrow HO(CH_2)_4OH + 2CH_3SO_2.OH$$

Phenylbutazone

Because Phenylbutazone is almost insoluble in water, aqueous acetone is used as the solvent. This reduces the apparent pK_a of the acid; bromothymol blue is therefore used as indicator. It is essential to carry out a blank titration on the same volume of solvent solution, the blank titration being subtracted from the first titration before calculating the result.

Cognate Determinations

Cycloserine. Isopropanol is used as the titration solvent, and carbonate-free 0.1N NaOH as titrant.

Frusemide. Ethanol (95 per cent) is used as the titration solvent.

Nicoumalone

Oxyphenbutazone

Sulphinpyrazone. Ethanol (95 per cent) is used as the titration solvent.

Aminocaproic Acid Injection

Aminocaproic acid (6-aminohexanoic acid) is amphoteric. It has a high melting point (204°, with decomposition), and is readily soluble in water (1 in 1.5 parts) to give solutions which are almost neutral (pH of a 20 per cent w/v solution is 7.5–8.0). It is, however, readily determined in aqueous solution by titration with sodium hydroxide in the presence of formaldehyde (formal titration). The latter forms a very weakly basic azomethine with the 6-amino group, so that the product is then capable of direct titration as a weak acid.

$$HCHO + H_2N.(CH_2)_5.COOH \rightarrow CH_2:N.(CH_2)_5.COOH$$

$$CH_2:N.(CH_2)_5.COOH + NaOH \rightarrow CH_2:N.(CH_2)_5.COONa + H_2O$$

Cognate Determination

Ammonium Chloride

DIRECT TITRATION OF STRONG BASES

Titration of strong bases with strong acids gives salts which are not hydrolysed in aqueous solution and the solution is therefore neutral. The pH changes in the region of the equivalence point are sufficiently large to permit a wide choice of indicator. Methyl orange is usually used.

Borax. *Determination of the percentage of* $Na_2B_4O_7,10H_2O$

Borax behaves in aqueous solution as a mixture of sodium metaborate and boric acid. The latter is too weak an acid to be titrated directly, but in the presence of glycerol or mannitol it forms complexes of considerably greater acidic strength. In consequence, borax can be titrated directly with standard NaOH in the presence of mannitol, using phenolphthalein as indicator.

$$Na_2B_4O_7, 10H_2O \xrightarrow{\text{Mannitol}} 2NaBO_2 + 2H_3BO_3 + 7H_2O$$

$$2H_3BO_3 + 2NaOH \xrightarrow{\text{Mannitol}} 2NaBO_2 + 4H_2O$$

$$\therefore \quad 381.4 \text{ g } Na_2B_4O_7,10H_2O \equiv 2000 \text{ ml } N \text{ NaOH}$$

$$\therefore \quad 0.1907 \text{ g } Na_2B_4O_7,10H_2O \equiv 1 \text{ ml } N \text{ NaOH}$$

Method. Dissolve the sample (about 0.4 g accurately weighed) in Water (25 ml). Add mannitol (2 g), heating if necessary until dissolved. Cool rapidly, and titrate the solution with *0.1N* NaOH using phenolphthalein as indicator.

Borax can also be titrated quantitatively with standard acid:

$$Na_2B_4O_7,10H_2O + 2HCl \rightarrow 4H_3BO_3 + 2NaCl + 5H_2O$$

$$\therefore \quad Na_2B_4O_7,10H_2O \equiv 2HCl \equiv 2H \equiv 2000 \text{ ml } N \text{ HCl}$$

$$\therefore \quad 381.4 \text{ g } Na_2B_4O_7,10H_2O \equiv 2000 \text{ ml } N \text{ HCl}$$

$$\therefore \quad 0.1907 \text{ g } Na_2B_4O_7,10H_2O \equiv 1 \text{ ml } N \text{ HCl}$$

To obtain the correct end point, the indicator must not be affected by the weak acid, H_3BO_3. This condition is met by the use of methyl red.

Method. Dissolve the sample (about 2.3 g accurately weighed) in Water (100 ml) with the aid of heat. Cool and titrate with *0.5N* HCl using methyl red as indicator.

The official assay serves also to detect admixture of either boric acid or alkali with the sample, as explained below.

Method. Accurately weigh the sample (3.0 g approx.), dissolve it in Water (60 ml) and titrate with *0.5N* hydrochloric acid using methyl red solution as indicator. Record the volume of $0.5N$ hydrochloric acid required. Reserve the titration liquid and repeat this portion of the assay. Calculate the volume of $0.5N$ HCl required to neutralise exactly 1 g of sample. These calculated volumes should *not* differ from one another by more than 0.05 ml. Boil the titration liquid to expel carbon dioxide. Cool the solution, add mannitol (20 g) and titrate with N sodium hydroxide using phenol-phthalein solution as indicator.

Mixtures of Sodium Carbonate or Boric Acid with Borax

Borax can be titrated quantitatively with standard acid thus:

$$Na_2B_4O_7,10H_2O + 2HCl \rightarrow 4H_3BO_3 + 2NaCl + 5H_2O \qquad (1)$$

provided that the indicator is insensitive to the weak acid H_3BO_3.

The liberated H_3BO_3 may then be titrated with standard alkali in the presence of glycerol or mannitol.

$$4H_3BO_3 + 4NaOH \xrightarrow{\text{mannitol}} 4NaBO_2 + 8H_2O \qquad (2)$$

Therefore from (1) and (2)

$$Na_2B_4O_7,10H_2O \equiv 2HCl \equiv 4H_3BO_3 \equiv 4NaOH \equiv 4000 \text{ ml } N \text{ NaOH}$$

From (1)

$$2000 \text{ ml } N \text{ HCl} \equiv 381.4 \text{ g } Na_2B_4O_7,10H_2O$$

From (2)

$$4000 \text{ ml } N \text{ NaOH} \equiv 381.4 \text{ g } Na_2B_4O_7,10H_2O$$

Thus the amount of H_3BO_3 liberated from *pure* borax by each ml of standard acid requires 2 ml of the same strength standard alkali for neutralisation in the reaction expressed by equation (2). In mixtures of borax with either sodium carbonate or boric acid, this is not the case.

Consider a mixture of borax and sodium carbonate. The acid reading will be the sum of the readings resulting from the reaction with the borax and the sodium carbonate. After titration with acid, the solution is boiled to remove CO_2 to which phenolphthalein indicator is sensitive. The sample will give an equivalent acid volume *larger* than half the volume of equivalent strength sodium hydroxide solution.

$$\text{Suppose 1 g Sample} \equiv 11.25 \text{ ml } 0.5N \text{ HCl } (1.000)$$

$$\equiv 20.10 \text{ ml } 0.5N \text{ NaOH } (1.000)$$

Since the sodium carbonate does not affect the sodium hydroxide titration, the H_3BO_3 produced from the borax in 1 g Sample $\equiv 20.10$ ml $0.5N$ NaOH. $\therefore$ $(20.10)/2 = 10.05$ ml $0.5N$ HCl must have been used to liberate this amount of H_3BO_3 from the $Na_2B_4O_7$ in 1 g sample.

But actually 11.25 ml $0.5N$ HCl were required,

$\therefore$ $11.25 - 10.05 = 1.20$ ml $0.5N$ HCl must have been used for the reaction with the sodium carbonate.

$$Na_2CO_3 + 2HCl \rightarrow 2NaCl + H_2O + CO_2$$

$$\therefore \quad 106.0 \text{ g } Na_2CO_3 \equiv 2000 \text{ ml } N$$

$$\therefore \quad 0.0265 \text{ g } Na_2CO_3 \equiv 1 \text{ ml } 0.5N \text{ HCl}$$

∴ Sodium carbonate in 1 g Sample $\equiv 1.20 \times 0.0265 \equiv 0.0318$ g Na_2CO_3 and Borax in 1 g Sample $\equiv 20.10 \times 0.04768 \equiv 0.9584$ g $Na_2B_4O_7,10H_2O$

∴   The mixture contains 3.18 per cent Na_2CO_3, and

95.84 per cent $Na_2B_4O_7,10H_2O$

Consider a mixture of borax and boric acid

Suppose 1 g Sample $\equiv 10.15$ ml *0.5N* HCl (1.000)

$\equiv 21.20$ ml *0.5N* NaOH (1.000)

With added H_3BO_3 present, the acid reading (calculated as exactly *0.5N*) will be *smaller* than the NaOH reading (calculated as exactly *0.5N*), because H_3BO_3 will give a reading with the NaOH solution but not with the HCl solution.

∴ 10.15 ml *0.5N* HCl are required to liberate H_3BO_3 from the $Na_2B_4O_7$ in the sample.

This amount of H_3BO_3 will require $10.15 \times 2 = 20.30$ ml *0.5N* NaOH in the second titration.

But 21.20 ml *0.5N* NaOH were actually required.

∴ $21.20 - 20.30 = 0.90$ ml extra *0.5N* NaOH is required above that needed to neutralise the H_3BO_3 liberated from the $Na_2B_4O_7$ in 1 g sample by the acid. From eq. (2)

$$H_3BO_3 \equiv NaOH \equiv H$$

∴   61.84 g $H_3BO_3 \equiv 1000$ ml N

∴   0.03092 g $H_3BO_3 \equiv 1$ ml *0.5N* NaOH

∴   0.90 ml *0.5N* NaOH $\equiv 0.03092 \times 0.90 = 0.02783$ g H_3BO_3

From eq. (1)

$$Na_2B_4O_7,10H_2O \equiv 2HCl \equiv 2H$$

∴   381.4 g $\equiv 2000$ ml N

∴   0.09535 g $\equiv 1$ ml *0.5N* HCl

∴   10.15 ml *0.5N* $\equiv 0.09535 \times 10.15 = 0.09679$ g $Na_2B_4O_7,10H_2O$

∴   The mixture contains 2.78 per cent H_3BO_3

96.79 per cent $Na_2B_4O_7,10H_2O$

Cognate Determinations

Ethylenediamine Hydrate
Potassium Carbonate
Potassium Hydroxide solution
Sodium Bicarbonate
Compound Sodium Bicarbonate Tablets
Sodium Bicarbonate Injection
Sodium Carbonate
Exsiccated Sodium Carbonate (p. 129)
Poldine Methylsulphate
Tetramethylammonium Hydroxide

Special Modifications in the Direct Titration of Strong Bases

Calcium Hydroxide Solution

Although this is a fairly strong alkali and is titrated with hydrochloric acid solution, methyl orange is not used as indicator. The solution absorbs carbon dioxide from the atmosphere to form carbonate but since the free $Ca(OH)_2$ and not the total alkali content is required, phenolphthalein is used as indicator.

Calcium Hydroxide. *Determination of the percentage of* $Ca(OH)_2$

This substance absorbs CO_2 from the atmosphere and some $CaCO_3$ is always present. The object of the determination is to find the percentage of $Ca(OH)_2$; this substance is separated from the $CaCO_3$ by means of the solubility of the former but not the latter in sucrose solution.

Method. Shake the appropriate weight gently in a litre bottle to give a fine suspension with 10 ml ethanol (90 per cent) previously neutralised to phenolphthalein. Add 10 per cent solution of sucrose (490 ml) previously neutralised to phenolphthalein, shake the mixture vigorously for about 5 minutes and then at frequent intervals for 4 hours. Filter an aliquot portion (250 ml) and titrate with N HCl using phenolphthalein as indicator. The burette reading is equivalent to the $Ca(OH)_2$ in half the weight of sample taken.

Sodium Hydroxide. *Determination of the total alkali calculated as* NaOH, *and carbonate calculated as* Na_2CO_3

This determination depends upon the reactions expressed by the following equations:

$$Na_2CO_3 + BaCl_2 \rightarrow 2NaCl + BaCO_3$$

$$NaOH + HCl \rightarrow NaCl + H_2O$$

$$BaCO_3 + 2HCl \rightarrow BaCl_2 + H_2O + CO_2$$

$$\therefore \quad NaOH \equiv HCl \equiv H$$

$$\therefore \quad 40.01 \text{ g NaOH} \equiv 1000 \text{ ml } N$$

$$\therefore \quad 0.04001 \text{ g} \equiv 1 \text{ ml } N \text{ HCl}$$

and

$$BaCO_3 \equiv Na_2CO_3 \equiv 2H$$

$$\therefore \quad 106.00 \text{ g } Na_2CO_3 \equiv 2000 \text{ ml } N$$

$$\therefore \quad 0.0530 \text{ g } Na_2CO_3 \equiv 1 \text{ ml } N \text{ HCl}$$

Barium chloride solution is added to precipitate the carbonate as $BaCO_3$. The free alkali can then be titrated with standard acid using phenolphthalein as indicator. Finally the carbonate (now as barium carbonate) can be titrated with standard acid using bromophenol blue as indicator. The end point of the first titration is equivalent to the conversion of the NaOH into sodium chloride. The carbonate does not react since it has been converted into barium carbonate which is insoluble as long as the solution is not acidic.

Method. Weigh accurately about 2 g of sample. Add water (about 25 ml) and barium chloride solution (5 ml) and titrate with N HCl using phenolphthalein as indicator.

Titrate *slowly* with *continuous shaking* to prevent the acid attacking the carbonate. Take the burette reading at the end point, i.e. when the solution becomes just colourless. Now add bromophenol blue solution (15 drops) (the solution will now be blue) and continue the titration with N HCl until a full green colour is obtained. The first full green colour will disappear on shaking as the acid attacks the undissolved carbonate. Therefore titrate until the full green colour remains permanent on shaking.

Calculate the total alkali from the total burette reading, and the carbonate from the difference in the readings at the two end points using phenolphthalein and bromophenol blue as indicators.

Cognate Determination

Potassium Hydroxide

Sodium Salicylate. *Determination of the percentage of* $C_6H_4(OH).COONa$

This determination depends upon the reaction expressed by the following equation:

$$C_6H_4(OH).COONa + HCl \rightarrow C_6H_4(OH).COOH + NaCl$$

Any usual indicator would be affected by the salicylic acid [$C_6H_4(OH).COOH$] liberated in the titration, i.e. salicylic acid is a sufficiently strong acid to give a pH which will be 'acid' to the usual indicators. Therefore a modification must be introduced in order to remove this acid.

Use is made of the fact that salicylic acid is more soluble in ether than it is in water. When a substance is shaken with two immiscible solvents in which it is soluble, the substance distributes itself so that the concentration in each liquid is in the ratio of the solubilities of the substance in each of the solvents separately, e.g.

Salicylic Acid soluble in water 1 in 500⎫ Ratio of solubilities
Salicylic Acid soluble in ether 1 in 2 ⎭   is $1:250$

Therefore very approximately (since ether is miscible with water to some extent) when salicyclic acid is shaken with ether and water, the concentration in the ether layer will be 250 times that in the water layer. Using equal volumes of ether and water and 1 g of acid, only 0.004 g $(\frac{1}{250})$ of the acid will remain in the aqueous layer. This fact is utilised in the determination.

$$C_6H_4(OH).COONa + HCl \rightarrow C_6H_4(OH).COOH + NaCl$$

$$\therefore \quad C_6H_4(OH).COONa \equiv H$$

$$\therefore \quad 160.1 \text{ g } C_6H_4(OH).COONa \equiv 1000 \text{ ml } N \text{ HCl}$$

$$\therefore \quad 0.08005 \text{ g } C_6H_4(OH).COONa \equiv 1 \text{ ml } 0.5N \text{ HCl}$$

Method. Dissolve the sample, about 3 g accurately weighed, in a little Water in a beaker and transfer quantitatively to a separator. Alternatively, introduce the sample directly into the separator via a wide-mouthed funnel. About 50 ml of Water in all should be used to effect solution and transference. Add ether (50 ml) and bromophenol blue indicator (20 drops) (yellow in strong acid, green in slightly acid, blue in alkali). Titrate with $0.5N$ HCl until a full green colour is obtained. In viewing the colour of the solution, place white paper half way round the separator and *examine in daylight.* The separating funnel must be stoppered and shaken at intervals during the titration. It does not matter if the end point is slightly over-run at this stage in the determination, since there is still sufficient salicylic acid in the aqueous layer to give a pH which is acid to the indicator. This acid must be further reduced in concentration if the correct end point is to be obtained. Run off the aqueous layer into a clean separating funnel and rinse the stem with about 2 ml of Water. Wash the ether layer remaining with two portions each of 5 ml of Water and add to the aqueous layer in the second separator. Again rinse the stem of the first separator. To this aqueous solution add 20 ml of ether and shake well. The salicylic acid still left in the aqueous layer becomes redistributed between the ether and aqueous layer and so, after shaking, very little salicylic acid remains in the aqueous solution; the aqueous layer therefore becomes blue again. Continue the titration, shaking well after each addition, until a full green colour is obtained. A bluish-green colour indicates under-titration and a yellowish-green colour indicates over-titration.

Cognate Determination

Sodium Benzoate. Determine as for Sodium Salicylate.

Thiopentone Sodium. *Determination of the percentage of* Na

Thiopentone sodium is a mixture of the monosodium derivative of 5-ethyl-5-(1-methylbutyl)-2-thiobarbituric acid and exsiccated sodium carbonate. It is determined by direct titration with *0.1N* HCl using methyl red as indicator, until the pink colour of the indicator is obtained. The solution is then boiled gently for 2 minutes to expel carbon dioxide, and the titration continued until the pink colour is restored.

Cognate Determinations

Methohexitone Injection
Thiopentone Injection

DIRECT TITRATION OF WEAK BASES

Titration of weak bases with standard mineral acid solutions gives salts which will be hydrolysed in aqueous solution to a greater or lesser extend depending upon the dissociation constant of the base. The pH of the solution at the equivalence point will be below pH 7, and therefore methyl red is the indicator used most frequently in such titrations.

Aminophylline. *Determination of the percentage of ethylenediamine*

The sample is dissolved in water and titrated with $0.1N$ H_2SO_4 to the green colour of bromocresol green.

Ammoniated Mercury. *Determination of the percentage of* NH_2HgCl

This determination depends upon the reactions expressed by the following equations:

$$NH_2HgCl + 2KI + 2H_2O \rightarrow NH_4OH + KOH + KCl + HgI_2$$

The HgI_2 dissolves in the excess KI:

$$HgI_2 + 2KI = K_2HgI_4 \text{ (soluble)}$$

The liberated alkali is titrated with standard acid.

$$\therefore \quad NH_2HgCl \equiv NH_4OH + KOH \equiv 2H$$
$$\therefore \quad 252 \text{ g } NH_2HgCl \equiv 2000 \text{ ml } N \text{ HCl}$$
$$\therefore \quad 0.01261 \text{ g } NH_2HgCl \equiv 1 \text{ ml } 0.1N \text{ HCl}$$
$$\therefore \quad \text{Approx. } 0.3 \text{ g } NH_2HgCl \equiv 25 \text{ ml } 0.1N \text{ HCl}$$

Method. Place *0.1N* HCl in the burette so that it is ready for titration immediately the potassium iodide is added. Weigh accurately the sample (about 0.3 g to 0.5 g) into a stoppered flask, add Water (50 ml). Calculate the approximate volume of acid which would be required in the titration and add 80 per cent of this volume; this step is taken to minimise the loss of ammonia which is released upon the addition of the potassium iodide. Add potassium iodide (4 g) and 5 drops of methyl orange-xylene cyanol FF solution and shake gently. Lumps of sample surrounded by red mercuric iodide may be present, but if the flask is gently shaken these lumps will dissolve and the solution will become yellow in colour. The solution can be kept *just* pink by adding *0.1N* HCl as long as lumps remain. As the lumps break up the solution will again turn yellow due to the liberation of more alkali as the sample reacts with potassium iodide. At the end point there must not be any undissolved particles remaining and if necessary a glass rod can be used to break up the particles.

Cognate Determination

Ammoniated Mercury Ointment

Pyridine. *Determination of the percentage* C_5H_5N

The sample is dissolved in water and titrated with N HCl to the full yellow colour of bromophenol blue (pH 2.8).

BACK TITRATIONS

Determinations involving back titrations consist in the addition of excess of a standard volumetric solution (from a pipette) to a weighed amount of sample, and determination of the excess not required by the sample. Hence the amount of volumetric solution used by the substance is determined.

In general this method is used for:

(1) volatile substances, e.g. ammonia, some of which would be lost during the titration,

(2) insoluble substances, e.g. Calcium Carbonate, which require excess volumetric solution to effect a quantitative reaction,

(3) substances for which a quantitative reaction proceeds rapidly only in presence of excess of the reagent, e.g. Lactic Acid,

(4) substances which require heating with a volumetric reagent during the determination in which decomposition or loss of the reactants or products would occur in the process, e.g. Formaldehyde.

Ammonia Solution. *Determination of the percentage w/w of* NH_3

This determination depends upon the reaction expressed by the following equation:

$$NH_3 + HCl \rightarrow NH_4Cl$$

$$\therefore \quad NH_3 \equiv H \equiv 1000 \text{ ml } N \text{ HCl}$$

$$\therefore \quad 17.03 \text{ g } NH_3 \equiv 1000 \text{ ml } N \text{ HCl}$$

$$\therefore \quad 0.01703 \text{ g } NH_3 \equiv 1 \text{ ml } N \text{ HCl}$$

A weight of sample which will consume between $\frac{2}{3}$ and $\frac{3}{4}$ of the volumetric solution should be used.

Method. Weigh the sample (about 1.8 to 2.0 g Strong Solution or 6 to 7 g Dilute Solution) in a stoppered weighing bottle. Introduce the bottle and contents into a large flask containing N HCl (50 ml). Open the bottle under the surface of the acid to avoid loss of ammonia. Mix thoroughly and back titrate the excess of acid with N NaOH using methyl red as indicator.

Method of Calculating the Result of an Experiment Involving a Back Titration
p. Suppose 50 ml N (1.100) H_2SO_4 was added and the burette reading $=$ 10.00 ml N (0.990) NaOH

Volume of N HCl available for the sample

$$= 50 \times 1.100 \qquad = 55.00 \text{ ml } N$$

Volume of N HCl not required by sample

$$= 10.00 \times 0.990 = 9.90 \text{ ml } N$$

Volume of N HCl required by sample

$$= 45.10 \text{ ml } N$$

Cognate Determination

Morpholine

Chalk. *Determination of the percentage of* $CaCO_3$, *calculated with reference to the dried substance*

There are two types of standard in the British Pharmacopoeia.

(1) A limit of the amount of pure chemical that the official substance shall contain. This may be a *minimum* only, e.g. Sulphuric Acid contains not less than 95 per cent w/w H_2SO_4 or a *minimum and maximum*, e.g. Citric Acid contains not less than 99.50 per cent and not more than the equivalent of 101 per cent $C_6H_5O_7,H_2O$. A maximum is included when, for example, a substance containing water of crystallisation can effloresce under certain conditions. Where no maximum is stated, however, and the standard is expressed in terms of the chemical formula for the substance of the mono-graph, an upper limit of 100.5 per cent is implied (British Pharmacopoeia general notice).

(2) A limit of the amount of pure chemical in the sample when *calculated with reference* to the dried or ignited substance, e.g. $CaCO_3$.

The dried substance is not titrated but the determination is carried out on the ordinary undried material; meanwhile a quantity of the substance is dried at the required temperature, and the percentage loss on drying ascertained. The percentage purity of the sample calculated with reference to the substance dried at the specified temperature, can then be calculated.

This method of determination is better than one involving the assay on the dried material because

(a) it is quicker since both parts of the determination can be carried out simultaneously and

(b) dried substances are difficult to weigh, since they absorb atmospheric moisture very quickly.

Standards of this type necessitate the inclusion of a limit of moisture under tests for purity in the Official Monograph, as otherwise a sample of the substance containing large percentages of water would still comply with the standard.

Loss on drying. Determine the loss on drying as described for sodium chloride (p. 211).

Assay. Chalk is insoluble in water, and the carbon dioxide released on titration with acid would interfere with the indicator unless removed. The sample, in water, is treated with excess N hydrochloric acid, the solution boiled to remove carbon dioxide, cooled, and the excess acid titrated with N sodium hydroxide using methyl orange as indicator.

$$CaCO_3 + 2HCl \rightarrow CaCl_2 + H_2O + CO_2$$

$$\therefore \quad CaCO_3 \equiv 2HCl \equiv 2H \equiv 2000 \text{ ml } N \text{ HCl}$$

$$\therefore \quad 100.1 \text{ g } CaCO_3 \equiv 2000 \text{ ml } N \text{ HCl}$$

$$\therefore \quad 0.05004 \text{ g } CaCO_3 \equiv 1 \text{ ml } N \text{ HCl}$$

Method. Weigh the sample (about 1.5 g) into a conical flask. Add N HCl (50 ml) from a pipette, heat gently until all the material has dissolved, and then boil for 1–2 min to remove CO_2. Cool, and back titrate the excess hydrochloric acid using methyl orange as indicator.

Calculation

A sample of chalk contained 98.90 per cent $CaCO_3$ (from titration) and on drying lost 0.7620 per cent of its weight.

$$\therefore \quad 100 \text{ g sample (undried)} \equiv 98.90 \text{ g } CaCO_3$$

$$\therefore \quad (100 - 0.762) = 99.238 \text{ g dried sample} \equiv 98.90 \text{ g } CaCO_3$$

$$\therefore \quad 100 \text{ g dried sample} \equiv \frac{98.90 \times 100}{99.238}$$

$$\equiv 99.66 \text{ g } CaCO_3$$

$\therefore$ *The sample of chalk contains 99.66 per cent $CaCO_3$ calculated with reference to the substance dried to constant weight.*

Benzoyl Chloride. *Determination of the percentage w/v of C_6H_5COCl*

$$C_6H_5COCl + 2NaOH \rightarrow C_6H_5COONa + NaCl + H_2O$$

$$\therefore \quad C_6H_5COCl \equiv 2NaOH \equiv 2H \equiv 2000 \text{ ml } N$$

Method. Dissolve the sample in excess standard NaOH and back titrate the excess alkali. Phenolphthalein is used as the indicator because C_6H_5COOH is a weak acid.

Alkali Metal Salts of Aliphatic Acids

When these substances are heated strongly the organic portion of the molecule is destroyed and the metals (Na or K) are converted quantitatively into Na_2CO_3 or K_2CO_3; these carbonates can be determined by titration.

Potassium Citrate. *Determination of the percentage of* $C_6H_5O_7K_3,H_2O$

$$2C_6H_5O_7K_3,H_2O \xrightarrow{\text{heat}} 3K_2CO_3$$

$$\therefore \quad 2C_6H_5O_7K_3,H_2O \equiv 6H$$

$$\therefore \quad 648.8 \text{ g } C_6H_5O_7K_3,H_2O \equiv 6000 \text{ ml } N$$

$$\therefore \quad 0.05407 \text{ g } C_6H_5O_7K_3,H_2O \equiv 1 \text{ ml } 0.5N \text{ HCl}$$

$$\therefore \quad 1.5 \text{ g} \equiv \text{approx. } 20 \text{ ml } 0.5N$$

Method. Weigh the sample (about 1.5–1.7 g) accurately, into a clean dry crucible. The weight must be kept within the limits stated, because if too much sample is weighed and 25 ml of *0.5N* acid is used there will be no back titration. If insufficient sample is weighed, difficulty will be experienced in washing the filter paper free from excess of acid.

Place this crucible containing the sample inside another crucible to serve as a jacket crucible. The initial heating should be gently carried out over a small Bunsen flame, since the molten mass tends to spit and bubble over the side of the crucible. Constant attention is needed at this stage, until the mass no longer swells up and charring commences. Do not allow the emitted gases to ignite as this may cause mechanical expulsion of particles of the residue. When the residue is all black, continue to heat for 10 minutes at a moderate temperature. Strong heat should be avoided to prevent fusion of the potassium carbonate with the silica of the crucible. Remove the jacket crucible and burn off any tar adhering to the inside of the inner crucible using a moderate Bunsen flame. *It is essential to burn off all the tarry material.* Allow the crucible to cool. Place the crucible with its contents on its side in a 250 ml beaker containing Water (about 10 ml). To the beaker partially covered with a clock-glass add *0.5N* HCl (25 ml) from a pipette and boil gently, breaking up the mass in the crucible with a glass rod, to ensure that all the carbonate reacts with the acid. Filter and collect the filtrate in a conical flask. Wash the beaker and contents with 5 ml portions of boiling water. Pass the washings through the filter and collect them in the conical flask. (If the filtrate is yellow or brown, due to incomplete removal of tarry matter it is useless to continue.) Allow the filter to drain completely between each washing. When washing the filter paper pay special attention to the top of the paper and the fold. *It is essential to wash efficiently and yet the volume of water used must be kept to the minimum.* After washing, test the beaker, crucible and filter for freedom from acidity with *one drop* of *diluted* methyl orange indicator.

Cool the flask and back titrate the excess of *0.5N* acid with *0.5N* NaOH. The end point is the last shade of orange before yellow, i.e. the point where the next drop of alkali will turn the solution yellow (it is useful to have a comparison flask containing indicator and about the same amount of water made alkaline with one drop of alkali).

If the correct amount of sample has been weighed out only a small back titration will result.

Cognate Determinations

Potassium Acetate. $2CH_3.COOK \rightarrow K_2CO_3$ ($\equiv 2H$)
Sodium Citrate. $2C_6H_5O_7Na_3 \rightarrow 3Na_2CO_3$ ($\equiv 6H$)
Sodium Citrate Tablets. $2C_6H_5O_7Na_3,2H_2O \rightarrow 3Na_2CO_3$ ($\equiv 6H$)
Sodium Acetic Citrate. $C_6H_6O_7Na_2,1\frac{1}{2}H_2O \rightarrow Na_2CO_3$ ($\equiv 2H$)
Sodium Lactate Injection. $2C_3H_5O_3Na \rightarrow Na_2CO_3$ ($\equiv 2H$)
Compound Sodium Lactate Injection

BACK TITRATIONS WITH BLANK DETERMINATIONS

In general blank determinations are used if the volumetric solution is unstable or if it alters in strength during the assay. It is necessary to perform blank determinations in assays which involve heating a liquid containing excess of standard alkali, cooling, and back titrating the excess. Heating and cooling an alkaline liquid results in an apparent change in strength if certain indicators are used. This may be due to interaction of the reagent with the glass or to the absorption of atmospheric CO_2. The amount of the change will be dependent upon the conditions used. In effect the alkali must be standardised under the conditions to be used in the determination; this is called a blank determination. Examples of this method of determination are given below.

Dry Acetone. *Determination of the percentage w/v of* H_2O (*Limit of* $\not> 0.3$ *per cent*)

This determination depends upon the reaction of water with acetyl chloride to yield acetic and hydrochloric acids:

$$H_2O + CH_3.COCl \rightarrow HCl + CH_3.COOH$$

The reaction is carried out in dry toluene, pyridine being used as a catalyst. Excess of acetyl chloride is destroyed at the completion of the reaction by the addition of absolute ethanol, with which it reacts to form ethyl acetate and hydrochloric acid.

$$C_2H_5OH + CH_3.CO.Cl \rightarrow CH_3.CO.O.C_2H_5 + HCl$$

Thus acetyl chloride releases two equivalents of acid in reaction with water, but only one in reaction with ethanol.

Method. Dilute acetyl chloride (12 ml; measuring cylinder) in a graduated flask with toluene to 100 ml. Pipette 10 ml of the solution (*Care*—use an automatic pipette) into a dry 250 ml glass-stoppered bottle. Add pyridine (2 ml) and shake, taking care not to wet the stopper. Pipette 50 ml of the acetone sample into the bottle, and again shake vigorously without wetting the stopper. Allow to stand at room temperature for five minutes, add from a pipette absolute ethanol (1.5 ml), shake vigorously and allow to stand for 10 minutes. Add absolute ethanol (25 ml) and titrate the liberated acid with N sodium hydroxide, using α-naphtholphthalein as indicator, overrunning the end point by about 0.5–1.0 ml. Back titrate the excess alkali with N HCl. Perform a blank determination omitting the sample of acetone. The difference between the titration obtained using sample and in the blank titration gives the volume of N NaOH equivalent to the water present.

Lactic Acid. *Determination of the percentage w/w of lactic acid and lactide together calculated as* $CH_3.CHOH.COOH$

Samples of lactic acid consist of a mixture of lactide and the free acid. The object of the official assay is to determine both the lactide and lactic acid calculated as $CH_3.CHOH.COOH$. Hydrolysis of the lactide with excess of sodium hydroxide solution produces lactic acid, and this, together with the free lactic acid originally present, is neutralised by the sodium hydroxide solution. The excess of sodium hydroxide is back titrated with standard hydrochloric acid.

$$CH_3\text{—}CH\text{—}C\text{=}O$$

$$\underset{O=C\text{——}CH\text{—}CH_3}{\overset{O \quad O}{|\quad\quad|}} \xrightarrow[2H_2O]{NaOH} 2CH_3.CHOH.COOH$$

$$CH_3.CHOH.COOH + NaOH \rightarrow CH_3.CHOH.COONa + H_2O$$

$$\therefore \quad C_3H_6O_3 \equiv NaOH \equiv H \equiv 1000 \text{ ml } N \text{ NaOH}$$

$$\therefore \quad 90.08 \text{ g } C_3H_6O_3 \equiv 1000 \text{ ml } N \text{ NaOH}$$

$$\therefore \quad 0.09008 \text{ g } C_3H_6O_3 \equiv 1 \text{ ml } N \text{ NaOH}$$

Method. Weigh the sample (3 to 4 g) in a weighing bottle. Wash out the sample via a funnel into a conical flask, using about 50 ml of Water. Lactic acid is viscous and so the weighing bottle must be washed out very carefully. Add N (approx.) NaOH (50 ml) from a pipette. Immerse the flask in a boiling water bath for 5 minutes. Cool by covering the mouth of the flask with a small inverted beaker and allowing a stream of cold water to flow over it. Back titrate the excess of alkali with N HCl using phenolphthalein as indicator. The pink colour of phenolphthalein gradually fades in contact with alkali. If this occurs, add a few more drops of indicator. Carry out a blank determination using N (approx.) NaOH (50 ml).

Note on the use of phenolphthalein in this determination. It would appear that methyl orange might be a suitable indicator in this determination since sodium hydroxide solution is being titrated with hydrochloric acid. But the choice of indicator is always governed by the weakest acid (or base) present in the system. In this case lactic acid is liberated at the end point and therefore phenolphthalein is used as indicator. Two blank titrations must also be performed in a similar manner to the actual determination except that now the sample is omitted. Hence it is not necessary to use standardised sodium hydroxide solution. If unstandardised approx. N sodium hydroxide is supplied, *do not attempt to standardise the solution.*

The volume of N HCl used in the blank titration measures the amount of alkali availabe to react with the lactic acid and lactide in the sample, and the volume of N HCl used, when the sample is present, measures the volume of excess alkali.

Method of writing up determinations involving back titrations and blank determinations

Weight of sample = 3.5210 g
Burette reading   = 12.00 ml N HCl (1.055)
Blank readings   = 50.01 ml N HCl (1.055)
(average)

Then

Volume of NaOH solution available for sample   = 50.01 ml N (1.055)

Volume of NaOH solution not required by sample = 12.00 ml N (1.055)

Volume of NaOH required by sample   = 38.01 ml N (1.055)

$$\therefore \quad 3.5210 \text{ g sample} \equiv 38.01 \times 1.055 \text{ ml } N$$

$$\therefore \quad 1 \text{ g sample} \equiv \frac{38.01 \times 1.055}{3.5210} \text{ ml } N$$

$$\equiv 11.39 \text{ ml } N \text{ NaOH}$$

Cognate Determination

Acetylsalicyclic Acid (Aspirin). *Determination of the percentage* $CH_3CO.OC_6H_4.COOH$

This determination depends upon the reactions expressed by the following equations:

$$CH_3.COOC_6H_4.COOH + 2NaOH \rightarrow CH_3.COONa + C_6H_4(OH)COONa$$

$$\therefore \quad CH_3.COOC_6H_4.COOH \equiv 2NaOH \equiv 2H \equiv 2000 \text{ ml } N$$

$$\therefore \quad 180.1 \text{ g } C_9H_8O_4 \equiv 2000 \text{ ml } N$$

$$\therefore \quad 0.04504 \text{ g } C_9H_8O_4 \equiv 1 \text{ ml } 0.5N \text{ NaOH}$$

The determination depends upon the alkaline hydrolysis of aspirin to acetic acid and salicylic acid (and immediate formation of their sodium salts), followed by back titration of the excess of alkali using phenol red as indicator.

The determination is similar to the lactic acid assay, except that the heating period in the boiling water bath is 10 minutes and $0.5N$ acid and alkali are used. A blank determination is essential because alkali is being heated and then cooled.

Cognate Determinations

Aspirin Tablets
Carbenicillin Sodium
Cloxacillin Injection
Cloxacillin Sodium
Methicillin Injection
Methicillin Sodium
Phenethicillin Capsules
Phenethicillin Potassium
Suxamethonium Chloride Injection

Formaldehyde Solution. *Determination of the percentage w/w* CH_2O

Formaldehyde can be oxidised by means of H_2O_2:

$$HCHO + \bar{O} \xrightarrow{\text{ } H_2O_2 \text{ }} H.COOH$$

The formic acid produced will react with standard alkali

$$H.COOH + NaOH \rightarrow H.COONa + H_2O$$

Formic acid is volatile but loss may be prevented by performing the oxidation in the presence of excess standard alkali which is subsequently back titrated with standard acid.

$$\therefore \quad H.CHO \equiv H.COOH \equiv H \equiv 1000 \text{ ml } N \text{ NaOH}$$

$$\therefore \quad 30.03 \text{ g } H.CHO \equiv 1000 \text{ ml } N \text{ NaOH}$$

$$\therefore \quad 0.03003 \text{ g } H.CHO \equiv 1 \text{ ml } N \text{ NaOH}$$

Method. Add the sample (about 3 g), accurately weighed, to a mixture of hydrogen peroxide (25 ml from a pipette) and N sodium hydroxide (50 ml) in a conical flask and warm on a water-bath until effervescence ceases (usually about 30 min). Cool (p. 149), and titrate the excess of alkali with N hydrochloric acid, using phenolphthalein solution as indicator.

A blank determination is required because alkali is heated and then cooled. Since a trace of mineral acid may be present in hydrogen peroxide this reagent is added from a pipette.

Saponification Value of Fixed Oils

Saponification Value is defined as the number of mg of potassium hydroxide required to neutralise the fatty acids resulting from the complete hydrolysis of 1 g of the substance when determined by the undermentioned method. The Acid Value of most edible oils is small relative to the Saponification Value, and is included in the latter. The Saponification Value is a measure of both free and combined acids.

Method. Accurately weigh the sample (approximately 2 g) in a small glass sample tube. Introduce the tube and sample into a 250 ml conical flask. Pipette ethanolic KOH (*0.5N* approx. 25 ml) into the flask. Fit a reflux condenser and heat with the flask immersed in a boiling water-bath for 1 hour. Remove the water-bath and wash down the condenser with not more than 5 ml of neutral alcohol and add solution of phenolphthalein (1 ml). (More indicator than usual is required because phenolphthalein does not function well as an indicator in strong ethanolic solution.) Remove the condenser, place a beaker over the neck of the flask and cool under a tap. When quite cool titrate the excess potassium hydroxide with *0.5N* HCl. Perform a blank determination at the same time. When performing the exercise in duplicate, it is preferable to carry out the determination and blank determination simultaneously.

$$R.COOR' + KOH \equiv R.COOK + R'OH$$

$$\therefore \quad 56.1 \text{ g } KOH \equiv H \equiv 2000 \text{ ml } 0.5N$$

$$\therefore \quad 28.05 \text{ mg } KOH \equiv 1 \text{ ml } 0.5N$$

Calculation

b = burette reading for blank

a = burette reading for sample

$$\text{then Saponification Value} = \frac{(b-a) \times (\text{factor}) \times 28.05}{\text{Weight of sample (in g)}}$$

The Saponification Value of the following oils is determined in this way:

Almond Oil
Arachis Oil
Castor Oil
Cod-liver Oil
Cottonseed Oil
Halibut-liver Oil
Olive Oil
Sesame Oil
Theobroma Oil
Wool Fat

All edible oils have Saponification Values lying between 188 and 196 and hence this test alone is of little value for identification purposes (see Table 8 on p. 56). The Saponification Value is a measure of the equivalent weight of the acids present, and is therefore useful as an indication of purity. Adulteration with mineral oils would be shown by low Saponification Values, whereas rancidity, which leads to the formation of low molecular weight acids, would be indicated by an abnormally high Saponification Value.

Cognate Determinations

Cetostearyl Alcohol; Emulsifying Wax; Wool Alcohols
These substances contain esters which are difficult to hydrolyse and the time of refluxing is extended to 2 hours. Ideally, these substances should be free from acids and esters, so the Saponification Value acts as a limit test for both.

Determination of Esters

The determination of esters is performed by hydrolysing the substance to an alcohol and an acid using excess of standard ethanolic KOH solution, and then back titrating the excess alkali. A blank determination is performed.

$$R.COOR' + KOH \rightarrow R.COOK + R'OH$$
$$\therefore \quad R.COOR' \equiv KOH \equiv H \equiv 2000 \text{ ml } 0.5N$$

Method. Accurately weigh the sample (approx. 2 g) in a glass sample tube. Transfer to a 250 ml flask, and add ethanol (5 ml; 95 per cent, previously boiled, cooled and neutralised to phenolphthalein). Neutralise the free acid in the solution by titrating with $0.1N$ ethanolic potassium hydroxide until just pink to phenolphthalein. Add $0.5N$ ethanolic KOH (20 ml) and reflux with the flask immersed in a boiling water bath for 1 hour. Cool, add Water (20 ml) and back titrate the excess alkali with $0.5N$ HCl, adding a further 0.2 ml of phenolphthalein solution. Carry out a blank determination, heating and cooling under the same conditions. The difference in readings gives the amount of alkali required to saponify the ester.

The following esters are determined in this way:
Amyl Acetate. N reagents are used. 1 ml N KOH $\equiv 0.1302$ g $C_7H_{14}O_2$
Benzyl Benzoate. 1 ml $0.5N$ KOH $\equiv 0.1061$ g $C_{14}H_{12}O_2$
Benzyl Benzoate application
n-Butyl Acetate. 1 ml $0.5N$ KOH $\equiv 0.05808$ g $C_6H_{12}O_2$

Dimethyl Phthalate. 1 ml $0.5N$ KOH $\equiv 0.04854$ g $C_{10}H_{10}O_4$
Ethyl Acetate. 1 ml $0.5N$ KOH $\equiv 0.04405$ g $C_4H_8O_2$
Ethyl Oleate. 1 ml $0.5N$ KOH $\equiv 0.1553$ g $C_{20}H_{38}O_2$
Menthyl Acetate (in Peppermint Oil). 1 ml $0.5N$ KOH $\equiv$ 0.09915 g $C_{12}H_{22}O_2$
Methyl Salicylate

The free acid in this substance is also determined. The substance is hydrolysed by heating with aqueous alkali, with preliminary neutralisation, and the excess alkali back titrated. A blank determination is essential.

$$1 \text{ ml } 0.5N \text{ KOH} \equiv 0.07608 \text{ g } C_7H_8O_3$$

Cognate Determinations

Glutethimide
Glutethimide Tablets

Determination of Ester Value

Ester Value is defined as the number of mg of potassium hydroxide required to neutralise the acids resulting from the complete hydrolysis of 1 g of material. Ester Value is a measure of the combined acids present in the substance, and is determined by the method described above,

$$\text{Ester Value} = \frac{m \times (\text{factor of acid}) \times 28.05}{w}$$

where w = weight (in g) of the substance taken and m = difference in burette readings for sample and blank determinations.

White beeswax. The Ester Value is determined by subtraction of the Acid Value from saponification value. The Saponification Value is determined on 5 g of sample, which is refluxed with N absolute ethanolic potassium hydroxide for 75 minutes.

The Ester Value here determines the quality of the beeswax, which contains cerotic acid (approx. 20 per cent) and melissyl palmitate (approx. 80 per cent).

Ratio Number.

$$\text{Ratio Number} = \frac{\text{Ester Value}}{\text{Acid Value}}$$

Ratio number is applied only to beeswax (3.3 to 4.2) and is a further check on the relative proportions of the acid and ester constituents.

Benzyl Alcohol. *Determination of the percentage w/w of $C_6H_5CH_2OH$*

This substance is determined by heating with acetic anhydride in presence of pyridine to produce an ester, benzyl acetate:

$$C_6H_5CH_2OH + (CH_3CO)_2O \rightarrow C_6H_5CH_2OCOCH_3 + CH_3COOH$$

Thus each molecule of acetic anhydride reacts with the alcohol to produce one molecule of acetic acid which is then titrated with standard alkali using phenolphthalein as indicator. The alkali also reacts with the excess of acetic anhydride:

$$(CH_3CO)_2O + 2NaOH \rightarrow 2CH_3COONa + H_2O$$

The pyridine does not interfere with the titration, because it is a weak base and phenolphthalein is used as indicator. A blank determination is performed.

$$\therefore \quad C_6H_5CH_2OH \equiv (CH_3CO)_2O \equiv CH_3COOH \equiv NaOH \equiv H$$

and

$$(CH_3CO)_2O \equiv 2NaOH \equiv 2H$$

$$\therefore \quad 108.1 \text{ g } C_6H_5CH_2OH \equiv 1000 \text{ ml } N \text{ NaOH}$$

$$\therefore \quad 0.1081 \text{ g } C_6H_5CH_2OH \equiv 1 \text{ ml } N \text{ NaOH}$$

Method. Reflux 1.5 g (accurately weighed) with a mixture of acetic anhydride (1 part by volume) and pyridine (1 part by volume) (25 ml) on a water-bath for 30 min. Cool, dilute with Water (25 ml) and titrate the excess acetic anhydride and acetic acid (formed in the reaction) with N sodium hydroxide to phenolphthalein. Perform a blank determination with the reagent (25 ml). The difference in titre between the blank and the test determinations is equivalent to the benzyl alcohol.

Cognate Determinations

Chlorphenesin
Dienoestrol. Determined as for Benzyl Alcohol but with the following modifications:

(a) Reflux time—2 hours.
(b) After refluxing, cool in ice water, and remove the dienoestrol acetate by filtration through a No. 4 sintered glass crucible.
(c) Titrate with 0.5N sodium hydroxide (carbonate free; see p. 130).

Determination of Free Menthol in Peppermint Oil

Alcohols are acetylated by treatment with acetic anhydride, and an Ester Value then determined on the acetylated oil. This latter value includes the Ester Value of esters present in the original sample, and an Ester Value must therefore also be determined on the original material. The method of calculation is complicated by the fact that the acetylated oil does not contain the same proportion of esters as the original oil. The following formula has been derived to allow for these differences:

$$\text{Percentage of free alcohols} = \frac{(b-a)y}{0.42(1335-b)}$$

Where a = Ester Value of original oil
   b = Ester Value of acetylated oil
   y = molecular weight of menthol 156.3

The above expression is derived as follows:

If b is the Ester Value of the acetylated oil, and x is the increase in weight of 1 g of original oil on acetylation, then $b(1+x)$ mg of KOH are required to neutralise the acetic acid from $(1+x)$ g of acetylated oil. (1)

Acetylation is expressed by the following equation:

$$ROH + (CH_3CO)_2O \rightarrow R.O.COCH_3 + CH_3COOH$$

Hence y g of alcohol yield $(y+42)$ g of acetate.
Since y g of alcohol increase in weight by 42 g, then

$$\left(y \times \frac{x}{42}\right) \text{ g of alcohol increase in weight by } x \text{ g.}$$

∴ Percentage alcohol in the oil is given by $\dfrac{xy}{0.42}$ per cent (2)

Saponification is expressed by the following equation:

$$R.O.COCH_3 + KOH \rightarrow R.OH + CH_3COOK$$
$$\therefore \quad CH_3CO— \equiv KOH$$
$$\therefore \quad 42 \text{ g of Acetyl} \equiv 56\,100 \text{ mg of KOH}$$
$$\therefore \quad x \text{ g of Acetyl} = \frac{56\,100x}{42}$$
$$= 1335x \text{ mg of KOH}$$

Esters originally present in $(1+x)$ g require a mg of KOH for saponification.

Acetylated alcohols present in $(1+x)$ g require $1335x$ g of KOH for saponification.

∴ $(a+1335x)$ mg of KOH are required to saponify total esters in $(1+x)$ g of acetylated oil.

$$\therefore \quad a + 1335x = b(1+x)$$
$$\therefore \quad x(1335-b) = (b-a)$$
$$\therefore \quad x = \frac{(b-a)}{(1335-b)}$$

Substituting for x in (2) above,

$$\text{Percentage of Free Alcohol in the original oil} = \frac{(b-a)y}{0.42(1335-b)}$$

Method. (i) Determine the Ester Value of the oil.

(ii) Acetylate the oil (10 ml) by mixing it with acetic anhydride (20 ml) and freshly fused sodium acetate (2 g) and refluxing for 2 hours in a 200 ml long-necked flask fitted with an air reflux condenser. Cool, add Water (50 ml) and heat on a boiling water-bath for 15 minutes, shaking thoroughly. Cool, transfer to a separator and reject the lower aqueous layer. Wash the acetylated oil in the separator successively with saturated brine solution (50 ml), saturated brine solution containing 1 g of anhydrous sodium carbonate (50 ml), and saturated brine solution (50 ml). Make good separations and reject the lower layer in each case. Dry the acetylated oil with anhydrous sodium sulphate (3 g) for at least 15 minutes, and until 1 drop of the oil ceases to give a cloudy solution with carbon disulphide (10 ml) in a dry tube. Filter the acetylated oil through a dry filter paper in a dry funnel and determine its Ester Value using 2 g.

Determination of Acetyl Value

Acetyl Value is defined as the number of mg of potassium hydroxide required to neutralise the acetic acid liberated by the hydrolysis of 1 g of the acetylated substance.

Acetyl Value is a measure of the free alcohols present in the substance and is calculated from the difference between the Saponification Values of the acetylated and unacetylated substances. Acetyl Value is used in the control of cholesterol waxes, and similar substances which contain a high proportion of alcohols.

Method. Determine the Saponification Value in the usual way. Acetylate 10 g of the substance by refluxing with acetic anhydride under specified conditions for 2 hours. Cool, pour into Water (600 ml) containing a small quantity (0.2 g) of pumice powder. Boil gently for 30 minutes. Cool, transfer to a separator and reject the lower aqueous layer. Wash the acetylated product with at least three successive portions of warm saturated aqueous sodium chloride solution (50 ml) until the washings are no longer acid to litmus. Wash with warm Water (20 ml). Remove as much water as possible, and finally dry with powdered anhydrous sodium sulphate (1 g) and filter. Determine the Saponification Value of the acetylated substance.

Calculate the Acetyl Value from the formula:

$$\text{Acetyl Value} = \frac{(b-a)\,1335}{1335-a}$$

where a = Saponification Value of the substance
b = Saponification Value of the acetylated substance.

The following substances are determined in this way:
Castor Oil. Acetyl Value 140

Hydroxyl Value

Hydroxyl Value is determined in Cetomacrogol 1000 by refluxing with stearic anhydride in xylene, and back titrating the excess in aqueous pyridine.

Acetic Anhydride

The assay is designed to determine the proportion of acetic anhydride and to serve as a limit test for acetic acid.

The sample is treated with excess of standard alkali, allowed to stand, and the excess of alkali titrated with standard acid. Both the anhydride and the free acid react with the alkali:

$$(CH_3CO)_2O + 2NaOH \rightarrow 2CH_3COONa + H_2O \qquad (1)$$
$$CH_3COOH\,(\text{free}) + NaOH \rightarrow CH_3COONa + H_2O \qquad (2)$$

The volume of N NaOH which reacts with 1 g of sample is calculated. Let this value be (a).

The sample is dissolved in dry benzene and treated with aniline which is acetylated by the anhydride. Excess standard alkali is added, the mixture shaken and the excess alkali titrated with standard acid:

$$(CH_3CO)_2O + C_6H_5NH_2 \rightarrow C_6H_5NHCOCH_3 + CH_3COOH \quad (3)$$

$$CH_3COOH \text{ [free in sample]} + NaOH \rightarrow CH_3COONa + H_2O \quad (4)$$

$$CH_3COOH \text{ [released in eq. (3)]} + NaOH \rightarrow CH_3COONa + H_2O \quad (5)$$

The volume of N NaOH required to neutralise the total acid from 1 g of sample is calculated. Let the value be (b).

The amount of NaOH required for the reaction expressed by (4) will be the same as that required for the reaction expressed by (2). From eq. (1)

$$(CH_3CO)_2O \equiv 2NaOH \text{ in determination } (a)$$

From eqs. (3) and (5) combined

$$(CH_3CO)_2O \equiv CH_3COOH \equiv NaOH \text{ in determination } (b)$$

$\therefore$ $(b - a) =$ volume of NaOH required by the $(CH_3CO)_2O$ for the following equivalence:

$$(CH_3CO)_2O \equiv NaOH \equiv H \equiv 1000 \text{ ml } N$$

$$\therefore \quad 102 \text{ g } (CH_3CO)_2O \equiv 1000 \text{ ml } N$$

$$\therefore \quad 0.102 \text{ g } (CH_3CO)_2O \equiv 1 \text{ ml } N$$

because the amount of NaOH required by the free CH_3COOH will be cancelled out by the subtraction.

$$\therefore \quad (b - a) \times 100 \times 0.102 = \text{percentage of } C_4H_6O_3$$

$$\therefore \quad (b - a) \times 10.2 = \text{percentage of } C_4H_6O_3$$

DETERMINATION OF ORGANICALLY COMBINED NITROGEN

Hydrolysis with Distillation of Liberated Volatile Bases

The liberated base may be determined either by absorption in excess standard mineral acid, and back titration with standard alkali, or by absorption into a solution of boric acid, followed by direct titration with standard acid.

Ammonium Sulphamate. *Determination of the percentage of* $N_2H_6O_3S$

$$NH_4.O.SO_2.NH_2 + H_2O \xrightarrow{H^+} (NH_4)_2SO_4$$

$$(NH_4)_2SO_4 + 2NaOH \rightarrow Na_2SO_4 + 2H_2O + 2NH_3$$

$$\therefore \quad NH_4.O.SO_2NH_2 \equiv 2NaOH \equiv 2H \equiv 2000 \text{ ml } N \text{ NaOH}$$

$$\therefore \quad 114.12 \text{ g } NH_4.O.SO_2NH_2 \equiv 2000 \text{ ml } N \text{ NaOH}$$

$$\therefore \quad 0.05706 \text{ g } NH_4.O.SO_2NH_2 \equiv 1 \text{ ml } N \text{ NaOH}$$

Method. Accurately weigh the sample (about 0.5 g) into a 250 ml flask. Add Water (50 ml) and concentrated sulphuric acid (10 ml) and heat under reflux for 2 hours. Transfer to an ammonia distillation apparatus, add excess 30 per cent w/v NaOH solution and distil. Collect the distillate in N H_2SO_4 (50 ml) and back titrate the excess acid with N NaOH, using methyl red as indicator.

Cognate Determinations

Amphetamine sulphate

$$(C_9H_{13}N)_2.H_2SO_4 + 2NaOH \rightarrow 2C_9H_{13}N + Na_2SO_4 + 2H_2O$$

$$\therefore \quad (C_9H_{13}N)_2.H_2SO_4 \equiv 2NaOH \equiv 2H \equiv 2000 \text{ ml } N$$

The sample (about 0.4 g), accurately weighed, is dissolved in water (120 ml), sodium hydroxide solution (2 ml) added and the volatile base distilled into excess of $0.1N$ hydrochloric acid. Distillation is continued until only 5 ml of liquid is left in the distillation flask. The excess of acid is titrated with $0.1N$ sodium hydroxide, using methyl red solution as indicator.

Dexamphetamine Sulphate. Determine as for Amphetamine Sulphate.

Ethyl Cyanoacetate. The determination depends upon hydrolysis with sulphuric acid solution, and titration of the ammonia evolved on subsequent addition of sodium hydroxide.

$$CN.CH_2.COOC_2H_5 \rightarrow CH_3COOH + CO_2 + NH_3 + C_2H_5OH$$

Formamide. The determination depends upon hydrolysis with caustic soda solution and titration of the ammonia evolved.

$$H.CONH_2 + NaOH \rightarrow H.COONa + NH_3$$

Meprobamate
Meprobamate Tablets
Methylamphetamine Sulphate
Methylamphetamine Injection
Nikethamide Injection. The substituted amide portion of this molecule can be hydrolysed to diethylamine by heating with 50 per cent v/v sulphuric acid, the diethylamine being fixed as the sulphate. The nitrogen atom in the ring is not hydrolysed by this treatment.

$$\therefore \quad C_{10}H_{14}ON_2 \equiv (C_2H_5)_2NH \equiv H$$

The substance is heated with 50 per cent v/v sulphuric acid in a long-necked flask for 2 hours, the liquid cooled, diluted with water, and transferred to an ammonia distillation apparatus. Excess of caustic soda solution is then added to liberate the diethylamine as the free base, which is distilled and collected in excess of standard $0.1N$ HCl. The excess acid is then back titrated with $0.1N$ NaOH using methyl red as indicator. A blank determination is performed.

Neostigmine Methylsulphate. This substance is a dimethylcarbamic ester which is hydrolysed on heating with caustic soda solution and the dimethylamine distilled into 4 per cent boric acid and titrated with $0.02N$ H_2SO_4. A blank determination is performed.

Neostigmine Tablets

$$\therefore \quad C_{12}H_{19}O_2N_2Br \equiv (CH_3)_2NH \equiv H$$

Pyrazinamide. Determine as for Nicotinamide.
Sulphamic Acid. Determine as for Ammonium Sulphamate.

Saccharin Sodium. *Determination of the percentage of* $C_7H_4O_3NSNa,2H_2O$

The nitrogen is obtained from the substance in the form of NH_3 by hydrolysis first with alkali and then with acid.

$$\therefore \quad C_7H_4O_3NS \equiv NH_3 \equiv H \equiv 1000 \text{ ml } N$$
$$\therefore \quad 205.2 \text{ g } C_7H_4O_3NSNa \equiv 1000 \text{ ml } N \text{ NaOH}$$
$$\therefore \quad 0.02052 \text{ g } C_7H_4O_3NSNa \equiv 1 \text{ ml } 0.1N \text{ NaOH}$$

Method. Weigh the sample accurately (approx. 0.7 g) into a Kjeldahl flask. Add sodium hydroxide solution (10 ml; 30 per cent w/v), and boil gently over a small flame for two minutes. Add hydrochloric acid (15 ml) and boil for 50 minutes under reflux. Cool, rinse the condenser with Water (50 ml) and pass a current of air through the flask to sweep out acid vapours. Transfer to an ammonia distillation apparatus, add sodium hydroxide solution (20 ml; 30 per cent w/v), and distil the ammonia into $0.1N$ H_2SO_4 (40 ml). Back titrate the excess acid with $0.1N$ NaOH, using methyl red as indicator.

Cognate Determinations

Paramethadione
Paramethadione Capsules
Saccharin. Determine as for Saccharin Sodium
Troxidone
Troxidone Capsules

Reduction with Distillation of Liberated Ammonia

Potassium Nitrate. *Determination of the percentage of* KNO_3

If KNO_3 is treated with Devarda's alloy and alkali the NO_3^- ion is reduced to ammonia which is distilled into excess of *0.1N* hydrochloric acid.

$$HNO_3 + 8H \rightarrow NH_3 + 3H_2O$$

$$\therefore \quad KNO_3 \equiv N \equiv NH_3 \equiv H \equiv 1000 \text{ ml } N \text{ HCl}$$

$$\therefore \quad 101.1 \text{ g } KNO_3 \equiv 1000 \text{ ml } N \text{ HCl}$$

$$\therefore \quad 0.01011 \text{ g } KNO_3 \equiv 1 \text{ ml } 0.1N \text{ HCl}$$

Method. Dissolve the sample (about 0.3 g), accurately weighed, in Water (300 ml) in an ammonia distillation apparatus, add Devarda's alloy (3 g) and sodium hydroxide solution (10 ml), and distil. Collect the distillate in *0.1N* hydrochloric acid (50 ml) and titrate the excess of acid with *0.1N* sodium hydroxide, using methyl red solution as indicator. Repeat the operation without the potassium nitrate. The difference between the titres is the acid required to neutralise the ammonia formed from the potassium nitrate.

Kjeldahl Type Determinations

In the following substances, the nitrogen is more firmly combined than in the foregoing examples. A much more severe treatment is required before the nitrogen is obtained quantitatively as ammonia. The organic portion of the molecule is destroyed, and the carbon oxidised until a clear liquid is obtained. The method consists in digestion with concentrated sulphuric acid using either sodium or potassium sulphate to raise the boiling point of the acid and catalysts such as mercury, selenium, and copper, to hasten the reaction (mercuric oxide is used in official determinations). The method is unsatisfactory for nitrogen present as nitro, azo, cyano or hydrazo groupings.

Isoprenaline Sulphate. *Determine the percentage of* $C_{11}H_{17}NO_3,\frac{1}{2}H_2SO_4$

$$C_{11}H_{17}NO_3,H_2SO_4 \equiv N \equiv NH_3 \equiv H$$

$$\therefore \quad 260.3 \text{ g } C_{11}H_{17}NO_3,H_2SO_4 \equiv 1000 \text{ ml } N$$

$$\therefore \quad 0.02603 \text{ g } C_{11}H_{17}NO_3,H_2SO_4 \equiv 1 \text{ ml } 0.1N$$

$$\therefore \quad 0.5 \text{ sample is approximately equivalent to 20 ml } 0.1N \text{ } H_2SO_4$$

Method. Accurately weigh the sample (0.5 g) into a long-necked flask (Kjeldahl flask) add anhydrous sodium sulphate (3 g), nitrogen-free mercuric oxide (0.3 g), and nitrogen-free sulphuric acid (8.5 ml). Ensure that all solids are washed into the bottom of the flask. Shake gently to mix the contents. Support the flask on a stand in a fume cupboard, inclining the neck of the flask at about 60°; place a small funnel in the neck of the flask. Heat the flask with a small flame until all frothing ceases, and then increase the size of the flame so that the liquid refluxes in the neck of the flask. Continue the heating until a clear colourless liquid is obtained (this may take several hours in some Kjeldahl determinations) and then boil gently for a further two hours. Cool the solution, and, when cold, dilute to 75–85 ml with Water, add a piece of granulated zinc (this prevents bumping during distillation) and a solution of sodium hydroxide (15 g) and sodium thiosulphate (2 g) in Water (25 ml). Immediately connect the flask to a distillation apparatus, mix the contents, and distil the liberated ammonia into $0.1N$ H_2SO_4 (50 ml). Titrate with $0.1N$ NaOH using methyl red as indicator. A blank determination is performed, and the difference between the two titres is equivalent to the ammonia liberated.

Cognate Determinations

The weight of sample will vary from compound to compound depending upon the percentage of nitrogen present. Similarly some sulphuric acid is consumed in the process, being reduced to sulphur dioxide, consequent upon the oxidation of carbon and hydrogen in the molecule being determined. The volume of sulphuric acid, therefore, varies from one determination to the next to compensate for this loss, and to ensure that the same volume of sulphuric acid is always present at the completion of the digestion period. The constancy of this volume, and of the weight of anhydrous sodium sulphate ensures a constant digestion temperature.

Aminoacetic Acid	Approximate weight of sample 0.6 g
	Nitrogen-free sulphuric acid 7.3 ml
Pentamidine Isethionate	Approximate weight of sample 0.4 g
	Nitrogen-free sulphuric acid 8.3 ml
Primidone	Approximate weight of sample 0.2 g
	Nitrogen-free sulphuric acid 7.5 ml
Tyrosine	Approximate weight of sample 0.5 g
	Nitrogen-free sulphuric acid 8.9 ml

Modifications of the Kjeldahl Method

A preliminary treatment without the catalyst (mercuric oxide) is usually applied when halogens are present in the sample, since mercuric chloride or bromide would be formed and these are volatile. Hydrogen halides are removed by heating with concentrated sulphuric acid before applying the normal digestion procedure.

N-(1-Naphthyl)ethylenediamine	Approximate weight of sample 0.25 g
Hydrochloride	Nitrogen-free sulphuric acid 7.5 ml

Blood products are standardised on their protein content. This is determined by a Kjeldahl determination of nitrogen, but since other nitrogenous substances such as urea and amino acids are present, some preliminary fractionation is essential. This is achieved by precipitation of the protein from solution as molybdic acid complex, and then submitting this fraction to a modified Kjeldahl determination of nitrogen.

Dried Human Plasma. *Determination of the percentage w/v of protein in an aqueous solution of the substance, equal in volume to the volume of plasma from which it was obtained*

Method. Transfer 0.2 ml (pipette) of a solution of the substance in Water, equal in volume to the volume of plasma from which it was obtained, to a round-bottomed centrifuge tube (capacity 15 ml). Add Water (5 ml) and mix. Add sodium molybdate solution (7.5 per cent w/v, 0.2 ml) and a mixture of nitrogen-free sulphuric acid and Water (1:30; 0.2 ml), shake and centrifuge for 5 minutes. Carefully decant the supernatant liquid and drain the tube by inverting on a filter paper. When thoroughly drained, add to the residue in the tube aqueous copper sulphate solution (30 per cent w/v; 3 drops) and nitrogen-free sulphuric acid (1 ml), and boil gently for 10 minutes. Cool, add anhydrous sodium sulphate (1 g) and selenium (10 mg), boil gently for 1 hour and again cool. Transfer the solution to an ammonia distillation apparatus, make alkaline with caustic soda solution (50 per cent w/v; 6 ml) and steam distil the liquid for seven minutes, into 4 per cent w/v boric acid solution (5 ml). Cool, add Water (5 ml), add one drop of a solution of methyl red in ethanol (95 per cent) containing methylene blue (0.1 per cent), and titrate with $0.0143N$ HCl. An empirical factor based on the percentage of nitrogen in plasma proteins is applied to convert the result to percentage of protein.

$$1 \text{ ml } 0.0143N \text{ HCl} \equiv 0.00135 \text{ g protein}$$

Cognate Determinations

Dried Human Serum

Human Fibrinogen. 2.0–5.0 ml of an approximately 0.67 per cent w/v solution in 0.9 per cent w/v sodium chloride solution is used for the determination

$$1 \text{ ml } 0.0143N \text{ HCl} \equiv 0.0002001 \text{ g of nitrogen (N)}$$

6 Ion Exchange and Gel Filtration

INTRODUCTION

Ion exchange material may be defined as an insoluble matrix containing labile ions capable of exchanging with ions in the surrounding medium without physical change taking place in its structure. Natural or synthetic zeolites (complex aluminium silicates), used to soften water, are examples of materials of this nature. If water containing calcium ions is passed through a bed of zeolite containing sodium ions an exchange of ions takes place, illustrated by the following equation:

$$Ca^{2+} + 2Na^{+}Z \rightarrow 2Na^{+} + Ca^{2+}Z$$

Z = zeolite or cation exchanger

When the sodium ions of the bed are exhausted, the filter may be regenerated by treatment with a strong solution of sodium chloride when the reverse exchange takes place.

On account of their instability towards acids and alkalis the use of zeolites is limited. The present wide application of ion exchange methods has only been made possible by the development of organic ion exchange materials. This extension of the field was initiated by the work of Adams and Holmes on insoluble phenolformaldehyde resins. They considered that the phenolic hydroxyl groups, not involved in the condensation of the phenol with formaldehyde, should be capable of ionisation, enabling cation exchange to take place; similarly by condensing such aromatic amines as m-phenylenediamine with formaldehyde, a resin capable of anion-exchange was produced. More recently non-phenolic resins have been developed with a cross linked polystyrene matrix which show superior stability to materials derived from phenols.

Cation and anion exchangers may be subdivided into weakly acidic, weakly basic, strongly acidic and strongly basic resins.

Strong Acid Resins

Nuclear sulphonic acid ($-SO_2OH$), and methylenesulphonic acid ($-CH_2.SO_2OH$) groups confer strong acid characters on the resin matrix. Cross linked polystyrene resins, containing such groups, are stable over a wide pH range and effective even in acid solutions on account of the high dissociation constants of these acid groupings. Strong acid exchangers derived from phenolic resins tend to be attacked by alkali and are thus unsuitable for use in solutions of pH greater than pH 8 to 8.5. In dilute solution the exchange potentials for cation absorption on to strong acid resins increase with increasing atomic number and increasing valency, as illustrated by the following series:

$$Fe^{3+} > Al^{3+} > Ca^{2+} > Mg^{2+} > Na^{+} > H^{+}$$

The reaction between a salt, e.g. sodium chloride, and a strong acid resin (R = resin matrix) may be represented by the equation:

$$R.SO_2OH + Na^+ + Cl^- \rightarrow R.SO_2.ONa + H^+ + Cl^-$$

Since the hydrogen ion is the least readily absorbed ion, regeneration of the resin to the acid form requires treatment with excess of strong acid.

It has been shown that certain divinylbenzene cross-linked polystyrene sulphonic acid resins undergo auto-degradation with the appearance of carboxyl groups (Armitage, Lyle and Nair, *Proc. Soc. Analyt. Chem.*, 1972, **9**, 204). As a consequence, the uptake of cations becomes pH-dependent at constant ionic strength despite the fact that the sulphonic acid ionises over the pH range examined.

Weak Acid Resins

Weak acid resins contain the carboxyl group (COOH) and exhibit the characters of insoluble weak acids. They are only effective in solutions of high pH, i.e. under conditions in which the acid groups are highly ionised, and may be buffered enabling absorption to be carried out at controlled pH. They have similar exchange potentials for cations as strong acid resins, with the important exception of showing a high affinity for hydrogen ions, demonstrated by the following series:

$$H^+ > Ca^{2+} > Mg^{2+} > Na^+$$

Thus the equilibrium between aqueous sodium chloride and a weak acid resin:

$$R.COOH + Na^+ + Cl^- \rightleftharpoons R.COONa + H^+ + Cl^-$$

lies largely to the left. With the salt of a weak acid, however, ion exchange takes place quantitatively as virtually no hydrogen ions are produced in the reaction. Thus the reaction between a weak acid resin and aqueous sodium bicarbonate will proceed from left to right:

$$R.COOH + NaHCO_3 \rightarrow R.COONa + CO_2 + H_2O$$

Exchangers of this nature are therefore of use in the selective exchange of cations when associated in solution with anions of both strong and weak acids, only those cations equivalent to the weak acid anions being exchanged (e.g. separation of basic and acidic amino acids).

Basic Resins

Basic characters are conferred upon a resin by the presence of amino nitrogen, basic strength being determined partly by the character of the group and partly by its position. Thus, a resin with a nuclear substituted amino group is weaker than one with the amino group in a side chain, while one with a quaternary ammonium group in a side chain produces a resin comparable in strength with the caustic alkalis. The earlier anion exchange resins were based on phenol but the more recent resins possess a crosslinked polystyrene matrix. The relation of strongly basic resins to weakly basic resins is comparable to that existing between the two types of cation exchangers. Strongly basic

Table 17. Cation exchangers

Commercial Name	Type	Functional Group(s)	Maximum Exchange Capacity mg-equiv. per ml (backwashed and drained) approximately
Amberlite IR-100	Phenolic	$-OH$; $-CH_2SO_2OH$	0.6
Amberlite IR-105	Phenolic	$-OH$; $-CH_2SO_2OH$	1.0
Amberlite IR-105 G	Phenolic	$-OH$; $-CH_2SO_2OH$	0.9
Amberlite IR-112	Crosslinked polystyrene	$-SO_2OH$	1.4
Amberlite IR-120	Crosslinked polystyrene	$-SO_2OH$	2.0
Zeo-Karb 215	Phenolic	$-OH$; $-CH_2SO_2OH$	0.55
Zeo-Karb 315	Phenolic	$-OH$; $-CH_2SO_2OH$	0.6
Zeo-Karb 225	Crosslinked polystyrene	$-SO_2OH$	1.95
Amberlite IRC-50	Crosslinked methacrylic acid	$-COOH$	2.4
Zeo-Karb 216	Phenolic	$-OH$; $-COOH$	total *ca.* 1.6 weak acid *ca.* 1.0
Zeo-Karb 226	Crosslinked methacrylic acid	$-COOH$	2.0

Table 18. Anion exchangers

Commercial Name	Type	Functional Group(s)	Maximum Exchange Capacity mg-equiv. per ml (backwashed and drained) approximately
Amberlite IRA-400	Crosslinked polysterene	Quaternary ammonium groups	0.8
Amberlite IRA-410	Crosslinked polystyrene	Quaternary ammonium groups	0.7
De Acidite FF	Crosslinked polystyrene	Quaternary ammonium groups	0.8
Amberlite IRA-4B	Phenolic	$-OH$; nuclear amino groups	2.2
De Acidite E	Phenolic	$-OH$; nuclear amino groups	2.3
Amberlite IR-45	Crosslinked polystyrene	$-N(C_3H_7)_2$	1.8
De Acidite G	Crosslinked polystyrene	$-N(C_2H_5)_2$	1.5
De Acidite H	Crosslinked polystyrene	$-N(CH_3)_2$	1.1

resins are effective over a wide range of pH, whereas weakly basic resins are active only in solutions of low pH. Anions of higher valency are preferentially absorbed by both types of exchanger, the great difference being in their behaviour towards hydroxyl ions, for which strongly basic resins have little affinity and weakly basic resins a high affinity. The following series illustrate this:

$$\text{Strongly basic resin} \quad PO_4^{3-} > SO_4^{2-} > Cl^- > OH^-$$
$$\text{Weakly basic resin} \quad OH^- > PO_4^{3-} > SO_4^{2-} > Cl^-$$

In general, the capacity of an anion exchange material for taking up an acid is less the lower the dissociation constant of the acid; and the capacity for a given acid is greater the stronger the basic character of the exchanger. Weakly basic materials have relatively low capacities for weak organic acids, but the capacity increases with more strongly basic exchangers. Very highly basic materials have good capacities even for such weak acids as phenol, carbon dioxide, hydrocyanic acid, hydrogen sulphide, and boric acid. In a mixture of two acids of different dissociation constants the stronger acid will be held on the material in preference to the weaker acid. This principle has been used for the removal of small amounts of mineral acids from lactic, citric and tartaric acids.

When ion exchangers are used for analytical purposes, the solution is percolated downwards through a fixed bed of ion exchange material packed in a column. The entering solution is termed the influent and the filtrate from the column the effluent. The total number of exchange groups in the column, usually expressed as milli-equivalents, is called the total capacity of the column. A more useful figure is the break-through capacity, defined as the amount of ions which can be taken up quantitatively by the column under any particular set of conditions. The break-through capacity is always lower than the total capacity. The former is dependent upon various factors such as particle size, filtration rate and composition of the solution. Therefore, it is usual to employ a quantity of resin having a capacity well in excess of the theoretical requirements (usually 20 : 1).

Sephadex Ion Exchangers

Cross-linked dextran polymers with acidic or basic functional groups attached by ether links to the polysaccharide chain are also available for use as ion exchangers. The types and grades are described in Table 20.

Apparatus and Technique

The apparatus required for column operation is very simple. The column may be made from any kind of glass tube and in many cases a burette may be used. A plug of glass wool at the bottom supports the resin bed but allows solution to pass through the column. The resin must be soaked overnight in the solvent as the dry material swells considerably when solvent is taken up. The resin suspension is poured into the column and the particles allowed to settle to form the bed. The bed must be covered with solvent to prevent the entrance of air which would result in channelling and considerably decrease the efficiency of the column. The solution to be analysed is passed through the

Table 19

	Cation exchangers		Anion exchangers	
	Strongly acidic	Weakly acidic	Strongly basic	Weakly basic
Functional Group	Sulphonic acid	Carboxylic acid	Quaternary ammonium	Amino
Effect of increasing pH value on capacity	No effect	Increases	No effect	Decreases
Stability of salts	Stable	Hydrolyse on washing	Stable	Hydrolyse on washing
Conversion of salts to free acid or free base	Requires excess of strong acid	Readily regenerated	Requires excess of sodium hydroxide	Readily regenerated with sodium carbonate or ammonia
Rate of exchange	Rapid	Slow unless ionised	Rapid	Slow unless ionised

column at a suitable rate followed by water, sufficient in quantity to displace completely the solution from the bed.

In addition to resin capacity several other factors need to be considered to ensure quantitative ion exchange. An increase in the rate of flow results in a lower break-through capacity; thus the time in which a determination may be completed cannot be reduced below a certain limit. When it is essential to use a high flow rate, e.g. in the case of unstable solutions, it is necessary to use a correspondingly large resin bed. As exchange capacity per unit volume is limited, ion exchange is applied principally to solutions of relatively low ionic concentrations. For most analytical purposes the concentration of the solution needs to be about $0.1N$. The presence of hydrogen ions in a solution decreases the break-through capacity for other cations. Thus the acidity of the solution to be analysed must be considered when assessing the size of resin bed necessary for quantitative retention of cations.

Table 20. Sephadex ion exchangers

Name and grade	Class	Functional group	Ionic form	Capacity (meq/g)	Bed volume (ml/g dry polymer)
DEAE A-25 A-50	Weakly basic anion exchangers	Diethylaminoethoxy- ($Et_2NCH_2CH_2O$—)	Cl^-	3.5 ± 0.5	5–9 25–33
CM C-25 C-50	Weakly acidic cation exchangers	Carboxymethoxy- (—OCH_2COO^-)	Na^+	4.5 ± 0.5	6–10 32–40
SE C-25 C-50	Strongly acidic cation exchangers	Sulphoethoxy- (—$OCH_2CH_2SO_2O^-$)	Na^+	2.3 ± 0.3	5–9 30–38

Analytical Applications

In the application of ion exchange methods to analysis two procedures may be followed. The retained ion may be determined, subsequent to elution from the column; alternatively the exchanged ion present in the effluent may be determined. Both procedures are illustrated in the following examples.

The total amount of ions in a salt solution may be determined by alkali titration after the cations have been exchanged for hydrogen ions by using a cation exchange column. Removal of interfering ions of opposite charge may be achieved by use of ion exchange methods, e.g. removal of heavy metals from phosphates prior to analysis. In certain cases a separation of ions from non-electrolytes is of interest, e.g. the uptake of vitamin B_1 by means of cation exchanger from solutions containing interfering substances present in yeast. Ion exchange methods have been reported for the determination of alkaloids in a number of drugs and galenicals. By means of weak acid exchangers it is possible to separate more basic substances from less basic substances, e.g. separation of basic from neutral and dicarboxylic amino acids. Carbonate-free sodium hydroxide may be prepared by passing the alkali through a strong base anion exchanger. Passage of aqueous sodium chloride through an anion exchanger is an elegant method for the preparation of standard sodium hydroxide solution. It is possible to separate ions of like charge by making use of differences in their absorption potentials on exchange materials. An example of this procedure, termed ion exchange chromatography by analogy to absorption chromatography, is the separation of rare-earth metals.

PRACTICAL EXERCISES

Column preparation. Use a burette plugged at its lower end with glass wool. Add solvent until the tube is one third filled, and then add a slurry of resin, which has previously been soaked in solvent, until the required length of column is obtained. Wash the resin thoroughly with solvent, maintaining a head of 1 cm of liquid above the resin to avoid drying out.

Determination of Quinine in Ethanolic Solution

A weak cation exchanger (H-form) is used so that the displacement can be carried out with as small a volume of liquid as possible. As absorption is rather slow, a fairly long resin column is needed if the solution is to be put through the column at a reasonable rate. To avoid the need of a correspondingly greater volume of eluting solution (ethanol saturated with ammonia) the column is split in two, the bulk of the quinine being retained in the upper column.

Method. Prepare two 10 cm columns using a weak cation exchange resin (H-form) in 96 per cent ethanol and mount one column directly above the other. Pipette 10 ml of an approximately 1 per cent solution of quinine in 96 per cent ethanol on to the upper column and allow the effluent from this to flow on to the lower column at a rate of one drop per second. Adjust the flow from the lower column to the same speed. When most of the solution has passed through the resin, add 10 ml ethanol (96 per cent) to the upper tube and allow this to pass through the columns in series at the same rate. Repeat twice and reject the effluents from the lower column. Separate the two columns and place them side by side so that their further effluents can be collected in a single vessel. Elute each at the same flow rate with 6×10 ml of ethanol (96 per cent) saturated with ammonia. Evaporate the combined effluents to dryness for determination by weighing, or to small volume (2–3 ml) for polarimetric determination. In the latter case make the residual solution with ethanol (96 per cent) washings of the container, up to 25 ml and determine the rotation in a 2 dm polarimeter tube.

Determination of Quinine in Quinine Salts (formed from strong acids)

Prior to treatment with a weak cation resin the salt solution is passed over a strong anion resin to remove the strong acid of the salt. This procedure is necessary as weak cation resins cannot compete successfully with strong acids for the base.

Method. Pipette 10 ml of an approximately 1 per cent solution of quinine sulphate in 96 per cent ethanol (some water can be added to the solvent to facilitate solution) on to a 10 cm column of strong anion exchange resin (OH-form), prepared with 96 per cent ethanol. Pass the effluent (flow rate 1 drop per second) on to the upper of two weak cation exchange columns. Wash (1 drop per second), the three columns in series with 10 successive quantities, each of 10 ml ethanol (96 per cent) reject the effluent from the bottom tube. Separate the two weak cation exchange columns and elute each with 60 ml of ethanol (96 per cent) saturated with ammonia. Complete the determination as in the previous experiment.

Determination of Ephedrine Hydrochloride

Direct determination of alkaloidal salts by passing the salt solution through an anion exchange resin and titrating the liberated alkaloid with acid has been applied to a number of organic bases. In order to apply this method to salts of strong bases such as ephedrine, a strong anion exchanger must be used, with the consequent difficulty that any other salts present, e.g. sodium chloride, will be converted to alkalis which will also titrate. This problem can be solved either by ashing the alkaloidal salt and finding the titre of the extracted ash after passage through the column or by carrying out the three column process described above for quinine salts.

Method. Prepare a 10 cm column using a strong anion exchange resin (OH-form) in Water. Wash the column with demineralised water and note the colour given by the final washings with Universal indicator. Pipette 10 ml of an approximately 1 per cent solution of ephedrine hydrochloride on to the column and allow it to flow through at a rate of 1 drop per second. Wash the column with the same solvent until the effluent pH, determined by spotting on a white tile with dilute Universal indicator solution, is the same as that determined in the initial washing of the column with water (80 to 100 ml of washing water required). Titrate the total effluent and washings with $0.1N$ sulphuric acid from a microburette, using bromophenol blue as indicator (end point, pale blue to pale green), the acid being standardised to the same indicator.

Preparation of Standard Carbonate-free Sodium Hydroxide Solution

Preparation of the column. Soak 40 g of a strong anion exchange resin in dilute hydrochloric acid to remove carbonate ions, and transfer to a column (18 mm × 60 cm). When the resin has settled, wash with Water and regenerate with 2 litres N sodium hydroxide made from $18N$ sodium hydroxide so as to be carbonate-free (sodium carbonate is almost insoluble in very strong sodium hydroxide solutions). Wash the resin bed with 2 litres of freshly boiled and cooled Water until the effluent gives a negative test for both hydroxyl and chloride ions.

Method. Accurately weigh about 2.9 g of AnalaR sodium chloride and dissolve in 50 to 100 ml of freshly boiled Water. Pass the solution through the exchange column at a rate of 4 ml per minute. Wash with freshly boiled and cooled Water. Collect the effluent in a 500 ml volumetric flask fitted with a soda-lime guard tube to prevent contamination with atmospheric carbon dioxide and adjust the volume with freshly boiled and cooled Water. The flow rate may safely be increased to about 8 ml per minute after 250 ml have been collected.

GEL FILTRATION

The use of cross-linked dextran gels (Sephadex) for the separation of molecules of different size by the method of gel filtration was first described by Porath and Flodin (*Nature*, 1959, **183**, 1657). The cross-linking of dextran is achieved by reaction with epichlorhydrin to give a polymer network as shown in Fig. 18. The polymers, so formed, are almost devoid of ionic groups, but exhibit polar character due to their high content of hydroxyl groups.

These dextran polymers, although water-insoluble, swell considerably in polar solvents such as water, but because of cross-linking, the gel grains remain particulate and possess a low degree of porosity. The extent of swelling is dependent on the degree of cross-linking of the polymer, and is usually expressed in terms of water regain (g water absorbed per g of dry gel). As a result of the limited porosity of the gel grains, they tend to be permeable only to

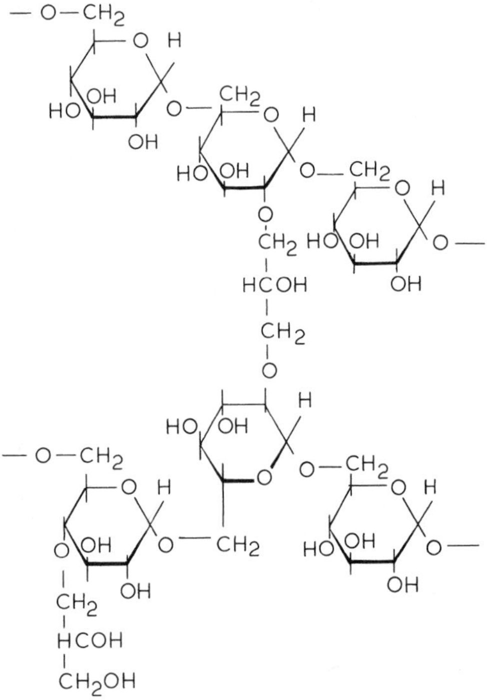

Fig. 18. Sephadex

Table 21. Properties of dextran polymer gels

Grade	Water regain (g/g)	Molecular weight excluded	Bed volume ml/g dry polymer	Swelling times hr
Sephadex G-10	1.0	700	2–3	3
Sephadex G-15	1.5	1 500	2.5–3.5	3
Sephadex G-25	2.5	5 000	5	12
Sephadex G-50	5.0	10 000	10	12
Sephadex G-75	7.5	50 000	13	24
Sephadex G-100	10	100 000	17	48
Sephadex G-200	20	200 000	30	72

the smaller molecules present in any system, and consequently high molecular weight molecules, such as proteins, are eluted preferentially from chromatographic columns of the gel.

Properties of Polymer Gels

The various grades of Sephadex are listed in Table 21, which shows that highly crossed-linked polymers such as Sephadex G-25, have a low water regain, and are capable of excluding compounds of molecular weight over 5000, whilst at the other extreme, Sephadex G-200 with a high water regain will only exclude molecules of molecular weight over 200 000.

Aqueous solutions filter rapidly through beds of the water-swollen gel, flow rates being highest where cross-linking is greatest. Solution of gel-material is negligible over a wide range of pH conditions.

A similar range of cross-linked polyacrylamide gels are marketed under the name of Bio-Gel P. These gels, which were discovered by Hjertén and Mosbach (*Anal. Biochem.*, 1962, **3,** 109), are formed by polymerisation of acrylamide in aqueous solution in the presence of N,N-methylene-bis-acrylamide. The gels are supplied in bead form in various particle sizes, porosities and molecular weight fractionating ranges, similar to those of Sephadex.

Gel Filtration

The liquid absorbed by the polymer granules is available in varying degree as solvent for solute molecules in contact with the gel. The distribution of solute between the inside and outside of the gel granules is a function of the space available and the distribution coefficient between granular and interstitial aqueous phases is independent of pH, ionic strength and concentration of the solvent (Gelotte, 1960).

The partition ratio between the granular and interstitial aqueous phases in a Sephadex column is defined by the distribution coefficient K_D, which can be calculated from the relationship $K_D = (V_e - V_o)/V_i$

where: V_e is the elution volume of the substance undergoing gel filtration

V_o is the void or interstitial volume

V_i is the inner volume, i.e. the volume of liquid taken up by the gel granules

Low molecular weight substances which can diffuse freely into the gel grains (i.e. $V_e = V_o + V_i$) have K_D values between 0.8–1.0. Values of K_D greater than 1 (i.e. $V_e > V_o$) indicate that there is absorption to the gel, and where there is no absorption (i.e. $V_e = V_o$), $K_D = 0$.

Table 22 gives a selection of K_D values obtained on gel filtration through grades of Sephadex. The highly cross-linked gels (G-25 and G-50) provide effective separation of proteins from amino acids, and for the de-salination of proteins and carbohydrate materials. Fractionation of amino acids by molecular weight is not possible, but preferential adsorption of aromatic and heterocyclic amino acids within the gel grains permits their separation from aliphatic amino acids which are eluted first. Peptide fractionation, and the separation of monosaccharides and oligosaccharides (e.g. in blood or urine) from low molecular weight polysaccharides can be accomplished with these polymer grades. whilst fractionation of proteins, and high molecular weight polysaccharides can be achieved with gel columns of Sephadex G-75, G-100 or G-200.

Gel Filtration in Non-aqueous Media

The dextran polymers already described also swell in polar solvents other than water including ethylene glycol, dimethyl sulphoxide and dimethyl formamide. They fail, however, to swell in methanol, ethanol or glacial acetic acid. Special dextran polymers in which the hydroxyl groups are protected by alkylation (Sephadex LH-20) are now available for work with polar organic solvents; strongly acidic ($> pH1$) or strongly oxidising systems must, however, be avoided. Table 23 lists some of the solvents which may be used with Sephadex LH-20, together with their solvent regain, and approximate bed volume per g of polymer.

Separations of comparatively low molecular weight substances can be achieved on Sephadex LH-20. Thus separation of fatty esters, e.g. tristearin (M.W. 891), tricaprin (M.W. 554) and triacetin (M.W. 218) can be readily

Table 22. K_D values obtained on gel filtration with Sephadex

Substance	K_D for various grades of Sephadex				
	G-25	G-50	G-75	G-100	G-200
Ammonium sulphate	0.9	—	—	—	—
Potassium chloride	1.0	—	—	—	—
Glycine	0.9	—	1.0	—	—
Phenylalanine	1.2	1.0	—	—	—
Tyrosine	1.4	1.1	—	—	—
Trytophan	2.2	1.6	1.2	—	—
Pepsin	0	0	0.3	—	—
Chymotrypsin	0	0	0.3	0.5	0.7
Trypsin	0	0	0.3	0.5	0.7
Serum albumin	0	0	0	0.2	0.4
γ-Globulin (19S type)	0	0	0	0	0
γ-Globulin (7S type)	0	0	0	0	0.2

Table 23. Solvent regain and bed volumes for Sephadex LH-20

Solvent	Solvent regain ml/g	Bed volume ml/g dry gel
Acetone	0.8	1.5
n-Butanol	1.6	3.0
Chloroform (+ 1 per cent ethanol)	1.8	3.0–3.5
Dimethylformamide	2.2	4.0
Dioxan	1.4	2.5–3.0
Ethanol	1.8	3.0–3.5
Ethyl acetate	0.4	0.5–1.0
Methanol	1.9	3.5–4.0
Tetrahydrofuran	1.4	2.5–3.0
Toluene	0.2	0.5
Water	2.1	4.0

effected in chloroform; cholesterol (M.W. 386) and dihydrocholic acid (M.W. 407) are separable using acid ethanol, and low molecular weight polymers of the polyethylene glycol series can be fractionated in either ethanol or chloroform.

Apparatus and Technique

Chromatographic columns are generally of glass with a diameter to height ratio of between 1:10 and 1:20. Separations involving substances with small differences in K_D may, however, require much longer columns with diameter to height ratios as high as 1:100. A plug of cotton or glass wool at the bottom of the tube supports the gel bed; the tube should be partly filled with water or buffer before insertion of the plug in order to avoid entrapping air bubbles.

The dry polymer must be allowed to swell completely in water or electrolyte solution before filling into the tube. The use of electrolyte solution prevents the beads of polymer from sticking together as may occur in water. Most grades of gel swell rapidly, but appreciable time is required for the gel to reach equilibrium. Fine particles, which if present in appreciable quantity can decrease the flow rate, should be removed by decantation, and the column filled by pouring in the gel suspension consisting of about one part of gel to two of fluid. Even packing is essential to avoid channelling of the column, and to assist this, the flow of eluate should be commenced slowly and carefully as soon as a few centimetres of gel have collected at the bottom of the tube. Continue the addition of gel to the column at a rate to provide a steady rise in the column bed, the surface of which should remain both even and horizontal throughout, until packing is complete. The upper surface of the column should be protected from disturbance by a filter paper and/or a plastic net. When completely filled, connect the column to a suitable reservoir of eluant, and allow the column to run freely overnight before use.

Remove most of the excess eluant from the top of the column by suction. Allow the remaining eluant to drain down to the surface of the gel, and just as the level of eluant reaches the surface of the gel, stop the flow of effluent and carefully run in the sample solution from a pipette. The sample should be applied in small volume and in as narrow a zone as possible if the maximum column efficiency is to be achieved. Open the stopcock to re-commence the effluent flow, and as the last of the sample solution reaches the surface of the gel, wash the last traces from the top surface with a small volume of eluant; add sufficient eluant to give roughly a 5 cm head of fluid, and maintain column flow and elution by re-connecting to the reservoir.

The effluent is best collected in fixed volumes by an automatic fraction collector, and may be monitored by chemical or preferably physico-chemical techniques, such as ultra-violet absorption.

7 Titration in non-aqueous Solvents

Substances which are either too weakly basic or too weakly acidic to give sharp end points in aqueous solution can often be titrated in non-aqueous solvents. The reactions which occur during many non-aqueous titrations can be explained by means of the concepts of the Lowry-Brönsted Theory. According to this theory an *acid* is a proton donor, i.e. a substance which tends to dissociate to yield a proton, and a *base* is a proton acceptor, i.e. a substance which tends to combine with a proton. When an acid HB dissociates it yields a proton together with the conjugate base B of the acid.

$$\underset{\text{acid}}{HB} \rightleftharpoons \underset{\text{proton}}{H^+} + \underset{\text{base}}{B^-}$$

Alternatively, the base B will combine with a proton to yield the conjugate acid HB of the base B, for every base has its conjugate acid and *vice versa*. It follows from these definitions that an acid may be either an electrically neutral molecule, e.g. HCl, or a positively charged cation, e.g. $C_6H_5NH_3^+$, or a negatively charged anion, e.g. HSO_4^-. A base may be either an electrically neutral molecule, e.g. $C_6H_5NH_2$, or an anion, e.g. Cl^-.

Some examples of acids and bases are set out below:

$$\text{Acids} \qquad \text{Bases}$$
$$HCl \rightleftharpoons H^+ + Cl^-$$
$$C_6H_5NH_3^+ \rightleftharpoons H^+ + C_6H_5NH_2$$
$$HSO_4^- \rightleftharpoons H^+ + SO_4^{2-}$$

Substances which are potentially acidic can function as acids only in the presence of a base to which they can donate a proton. Conversely basic properties do not become apparent unless an acid also is present.

Solvents

Aprotic solvents are neutral chemically inert substances such as benzene and chloroform. They have a low dielectric constant, do not react with either acids or bases and therefore do not favour ionisation. The fact that picric acid gives a colourless solution in benzene which becomes yellow on adding aniline shows that picric acid is not dissociated in benzene solution and also that in the presence of the base aniline it functions as an acid, the development of yellow colour being due to formation of the picrate ion.

undissociated   picrate ion
colourless   yellow

Since dissociation is not an essential preliminary to neutralisation, aprotic solvents are often added to 'ionising' solvents to depress solvolysis (which is comparable with hydrolysis) of the neutralisation product and so sharpen the end point.

Protophilic solvents are basic in character and react with acids to form solvated protons.

$$\underset{\text{acid}}{HB} + \underset{\substack{\text{basic} \\ \text{solvent}}}{Sol.} \rightleftharpoons \underset{\substack{\text{solvated} \\ \text{proton}}}{Sol.\,H^+} + \underset{\substack{\text{conjugate base} \\ \text{of acid}}}{B^-}$$

A weakly basic solvent has less tendency than a strongly basic one to accept a proton. Similarly a weak acid has less tendency to donate protons than a strong acid. As a result a strong acid such as perchloric acid exhibits more strongly acidic properties than a weak acid such as acetic acid when dissolved in a weakly basic solvent. On the other hand, all acids tend to become indistinguishable in strength when dissolved in strongly basic solvents owing to the greater affinity of strong bases for protons. This is called the levelling effect. Strong bases are levelling solvents for acids, weak bases are differentiating solvents for acids.

Protogenic solvents are acidic substances, e.g. sulphuric acid. They exert a levelling effect on bases.

Amphiprotic solvents have both protophilic and protogenic properties. Examples are water, acetic acid and the alcohols. They are dissociated to a slight extent. The dissociation of acetic acid, which is frequently used as a solvent for titration of basic substances, is shown in the equation below.

$$CH_3COOH \rightleftharpoons H^+ + CH_3COO^-$$

Here the acetic acid is functioning as an acid. If a very strong acid such as perchloric acid is dissolved in acetic acid, the latter can function as a base and combine with protons donated by the perchloric acid to form an 'onium' ion.

$$HClO_4 \rightleftharpoons H^+ + ClO_4^-$$
$$CH_3COOH + H^+ \rightleftharpoons \underset{\text{onium ion}}{CH_3COOH_2^+}$$

Since the $CH_3COOH_2^+$ ion readily donates its proton to a base a solution of perchloric acid in glacial acetic acid functions as a strongly acidic solution.

When a weak base, such as pyridine, is dissolved in acetic acid, the acetic acid exerts its levelling effect and enhances the basic properties of the pyridine. It is possible, therefore, to titrate a solution of a weak base in acetic acid with perchloric acid in acetic acid, and obtain a sharp end point when attempts to carry out the titration in aqueous solution are unsuccessful.

$$
\begin{aligned}
HClO_4 &+ CH_3COOH \rightleftharpoons CH_3COOH_2^+ + ClO_4^- \\
C_5H_5N &+ CH_3COOH \rightleftharpoons C_5H_5NH^+ + CH_3COO^- \\
CH_3COOH_2^+ &+ CH_3COO^- \rightleftharpoons 2CH_3COOH
\end{aligned}
$$

Adding $\quad HClO_4 \qquad + C_5H_5N \qquad \rightleftharpoons C_5H_5NH^+ \quad + ClO_4^-$

TITRATION OF ALKALI-METAL AND ALKALINE EARTH-METAL SALTS OF ORGANIC ACIDS

Preparation of *0.1N* Perchloric Acid

Slowly add perchloric acid (72 per cent; 8.5 ml) to glacial acetic acid (900 ml) with continuous and efficient mixing. Similarly add acetic anhydride (30 ml); adjust the volume to 1 litre with glacial acetic acid and allow the solution to stand for 24 hours before use. The acetic anhydride reacts with the water in the perchloric acid and acetic acid and renders the mixture virtually anhydrous. Although excess acetic anhydride is not always disadvantageous, care must be taken to avoid an excess when primary and secondary amines (which acetylate readily to give non-basic products) are to be titrated. The perchloric acid must be well diluted with acetic acid before adding the acetic anhydride. Failure to observe this precaution leads to formation of the *explosive* acetyl-perchlorate.

Standardisation of *0.1N* Perchloric Acid

Alkali and alkaline earth salts of organic acids function as bases in acetic acid solution.

$$R.COOM \rightleftharpoons RCOO^- + M^+$$
$$CH_3COOH_2^+ + RCOO^- \rightleftharpoons R.COOH + CH_3COOH$$

Potassium hydrogen phthalate may be used as a standardising agent for acetous perchloric acid. The reaction is expressed by the following equation:

$$\therefore \quad 204.14 \text{ g } C_8H_5O_4K \equiv HClO_4 \equiv H \equiv 1000 \text{ ml } N$$
$$\therefore \quad 0.02041 \text{ g } C_8H_5O_4K \equiv 1 \text{ ml } 0.1N \text{ HClO}_4$$

Method. Accurately weigh the potassium hydrogen phthalate (0.5 g approx.) into a 100 ml conical flask. Attach a reflux condenser fitted with a silica gel drying tube and add glacial acetic acid (25 ml). Warm until the salt has dissolved. Cool and titrate with 0.1N perchloric acid.

Indicator. Use 2 drops of either 0.5 per cent w/v acetous crystal violet (end point blue to blue green) or 0.5 per cent w/v acetous oracet blue B (end point blue to pink).

Indicators

The following indicators are in common use:

Indicator	Colour change		
	basic	*neutral*	*acidic*
Crystal Violet (0.5 per cent in glacial acetic acid)	violet	blue-green	yellowish-green
α-Naphtholbenzein (0.2 per cent in glacial acetic acid)	blue or blue-green	orange	dark-green
Oracet Blue B (0.5 per cent in glacial acetic acid)	blue	purple	pink
Quinaldine Red (0.1 per cent in methanol)	magenta		almost colourless

The same indicator must be used throughout for standardisation, titration and neutralisation of mercuric acetate solution, if used.

Potentiometric Titration (see Part 2)

The end point may be determined by titrating potentiometrically and plotting dE/dV against V.

Temperature

Non-aqueous solvents in general have greater coefficients of expansion than water, so that small temperature differences can cause significant errors unless suitable correction factors are used. Standardisation and titration should be carried out as far as possible at the same temperature. If this is not possible however, the volume of titrant may be corrected by applying the following formula:

$$V_c = V[1 + 0.001(t_1 - t_2)]$$

where V_c = corrected volume of titrant
$\qquad V$ = volume of titrant measured
$\qquad t_1$ = temperature at which titrant was standardised
$\qquad t_2$ = temperature at which titration was carried out

Sodium Chromoglycate. *Determination of the percentage* $C_{23}H_{14}Na_2O_{11}$
Titrate approximately 0.4 g of sample by the above method.

TITRATION OF PRIMARY, SECONDARY AND TERTIARY AMINES

The reaction between a primary amine and perchloric acid is expressed by the following equation.

$$R.NH_2 + HClO_4 \rightarrow [R.NH_3]^+ + ClO_4^-$$

Adrenaline. *Determination of the percentage of* $C_9H_{13}O_3N$

$$HO-\langle\!\!\!\!\bigcirc\!\!\!\!\rangle-CH(OH).CH_2.NH.CH_3 + HClO_4 \longrightarrow$$
$$HO$$

$$\left[HO-\langle\!\!\!\!\bigcirc\!\!\!\!\rangle-CH(OH).CH_2.\overset{+}{N}H_2.CH_3 \right] + ClO_4^-$$
$$HO$$

$$\therefore \quad 183.2 \text{ g } C_9H_{13}O_3N \equiv HClO_4 \equiv H \equiv 1000 \text{ ml } N$$
$$\therefore \quad 0.01832 \text{ g } C_9H_{13}O_3N \equiv 1 \text{ ml } 0.1N \text{ HClO}_4$$

Method. Accurately weigh the sample (about 0.3 g) into a 250 ml conical flask. Dissolve in glacial acetic acid (50 ml), gently warming the solution if necessary. Cool and titrate with *0.1N* acetous perchloric acid, using crystal violet or oracet blue B as indicator.

Cognate Determinations

Adrenaline Acid Tartrate
Bisacodyl. α-Naphtholbenzein is used as indicator.
Bisacodyl Suppositories
Erythromycin Stearate. The sample is extracted with chloroform, the solution evaporated, and the residue titrated.
Ethionamide Tablets
Metaraminol Tartrate
Methyldopa
Metronidazole. α-Naphtholbenzein is used as indicator.
Noradrenaline Acid Tartrate
Orciprenaline Sulphate
Orphenadrine Citrate
Prochlorperazine Maleate
Pyrimethamine. Quinaldine Red is used as indicator.
Titration of Amino Acids. Amino acids may be titrated by the above procedure. Those which are insoluble in glacial acetic acid may be dissolved in an excess of *0.1N* acetous perchloric acid, and the solution back titrated with *0.1N* sodium acetate in glacial acetic acid.

TITRATION OF HALOGEN ACID SALTS OF BASES

The halide ions chloride, bromide and iodide are too weakly basic to react quantitatively with acetous perchloric acid. Addition of mercuric acetate (which is undissociated in acetic acid solution) to a halide salt replaces the halide ion by an equivalent quantity of acetate ion, which is a strong base in acetic acid.

$$2R.NH_2.HCl \rightleftharpoons 2RNH_3^+ + 2Cl^-$$
$$(CH_3COO)_2Hg \text{ (undissociated)} + 2Cl^- \rightarrow HgCl_2 \text{ (undissociated)} + 2CH_3COO^-$$
$$2CH_3COOH_2^+ + 2CH_3COO^- \rightleftharpoons 4CH_3COOH$$

Benzhexol Hydrochloride. *Determination of the percentage of* $C_{20}H_{31}ON,HCl$
The determination depends upon the reactions expressed by the following equations:

$$2C_{20}H_{31}ON.HCl \rightleftharpoons 2C_{20}H_{31}ON,H^+ + 2Cl^-$$
$$(CH_3COO)_2Hg + 2Cl^- \rightarrow HgCl_2 + 2CH_3.COO^-$$
$$2CH_3COOH_2^+ + 2CH_3COO^- \rightleftharpoons 4CH_3COOH$$
$$\therefore \quad C_{20}H_{31}ON.HCl \equiv Cl^- \equiv CH_3COO^- \equiv HClO_4 \equiv H \equiv 1000 \text{ ml } N$$
$$\therefore \quad 337.9 \text{ g } C_{20}H_{31}ON,HCl \equiv 1000 \text{ ml } N \text{ HClO}_4$$
$$\therefore \quad 0.03379 \text{ g } C_{20}H_{31}ON,HCl \equiv 1 \text{ ml } 0.1N \text{ HClO}_4$$

Method. Accurately weigh the sample (about 0.7 g) and dissolve in warm glacial acetic acid (25 ml). Add mercuric acetate solution (15 ml; 5 per cent w/v in glacial acetic acid), and 0.2 ml of crystal violet solution (0.5 per cent in acetic acid), and titrate with *0.1N* perchloric acid.

Cognate Determinations

Amitriptyline Hydrochloride
Chlordiazepoxide Hydrochloride. Potentiometric titration.
Chlorproguanil Hydrochloride
Chlorpromazine Hydrochloride
Cyclopentolate Hydrochloride. Potentiometric titration.
Cyproheptadine Hydrochloride
Dextropropoxyphene Hydrochloride and Napsylate
Dextropropoxyphene Capsules
Dicyclomine Hydrochloride
Gallamine Triethiodide
Lignocaine Hydrochloride
Nortryptiline Hydrochloride
Nortryptiline Capsules
Orphenadrine Hydrochloride
Oxyphencyclimine Hydrochloride. Oracet blue B is used as indicator.
Oxyphencyclimine Tablets. Extract the base hydrochloride with chloroform, evaporate the solution to dryness and titrate as for Oxyphencyclimine Hydrochloride.
Phenmetrazine Hydrochloride. α-Naphtholbenzein is used as indicator.
Phenmetrazine Tablets. Extract the base hydrochloride with acid, basify with sodium hydroxide, extract with chloroform, and evaporate the solution to dryness. Titrate as for Phenmetrazine Hydrochloride.
Proguanil Hydrochloride
Promazine Hydrochloride
Propranalol Hydrochloride

Modifications

A number of hydrochlorides and quaternary salts give insoluble perchlorates during the titration under the above conditions. To avoid this, solutions in acetic acid are treated with acetous mercuric acetate and then titrated in dioxan or other suitable solvent with *0.1N* perchloric acid. The following substances are determined in this way:

Chlorcyclizine Hydrochloride. A potentiometric end point is used.
Desipramine Hydrochloride. Mentanil Yellow is used as indicator.
Suxamethonium Bromide
Suxamethonium Chloride

TITRATION OF ACIDIC SUBSTANCES

Preparation of *0.1N* Potassium Methoxide in Toluene-methanol

Method. To a mixture of methanol (40 ml) and dry toluene (50 ml) in a loosely covered flask, add freshly cut potassium (4 g) a little at a time. When the potassium has dissolved, add sufficient methanol to produce a clear solution. Add toluene with constant shaking until the mixture becomes cloudy. Repeat the alternate addition of methanol and benzene until 1 litre of solution is obtained, always using the minimum volume of methanol necessary to produce a clear solution.

0.1N Lithium Methoxide. Prepared as for *0.1N* Potassium Methoxide, using 0.7 g of lithium in place of potassium.

0.1N Sodium Methoxide. Prepared as for *0.1N* Potassium Methoxide, using 2.3 g of sodium in place of potassium.

Standardisation of *0.1N* Methoxide Solutions

Method. Use the apparatus shown in Fig. 19. Place dimethylformamide (10 ml) in the conical flask. Add 3 drops of thymol blue (0.3 per cent in methanol) and neutralise the acidic impurities in the dimethylformamide by titration with *0.1N* lithium methoxide in toluene-methanol. Avoid subsequent contamination of the liquid with atmospheric carbon dioxide. Immediately introduce benzoic acid (about 0.06 g) and titrate at once with methoxide in toluene-methanol.

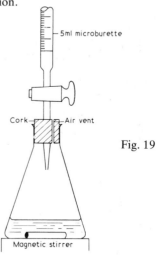

Fig. 19

* **Phenobarbitone.** *Determination of the percentage of* $C_{12}H_{12}O_3N_2$

$$\therefore \quad 232.2 \text{ g } C_{12}H_{12}O_3N_2 \equiv I \text{ iOMe} \equiv H \equiv 1000 \text{ ml } N$$
$$\therefore \quad 0.02322 \text{ g } C_{12}H_{12}O_3N_2 \equiv 1 \text{ ml } 0.1N \text{ LiOMe}$$

Method. Place dimethylformamide (40 ml) in the titration flask. Add 3 drops of quinaldine red solution (0.1 per cent w/v in ethanol) and neutralise by titration with *0.1N* lithium methoxide. Accurately weigh the sample (about 0.2 g) directly into the flask, taking care to avoid contamination of the liquid with atmospheric carbon dioxide. Titrate with *0.1N* lithium methoxide until the solution changes from pink to colourless. Alternatively the titration may be carried out with *0.1N* tetrabutylammonium hydroxide in toluene-methanol using thymol blue (0.5 per cent in methanol; 2 drops as indicator).

* **β-Naphthol.** *Determination of the percentage of* $C_{10}H_8O$

The determination depends upon the reaction expressed by the following equation:

* *Note:* this is not an official assay procedure.

$$\therefore \quad C_{10}H_8O \equiv CH_3OK \equiv H \equiv 1000 \text{ ml } N$$
$$\therefore \quad 144 \text{ g } C_{10}H_8O \equiv 1000 \text{ ml } N$$
$$\therefore \quad 0.0144 \text{ g } C_{10}H_8O \equiv 1 \text{ ml } 0.1N \text{ CH}_3OK$$

Method. Accurately weigh the sample (about 0.04 g) into previously neutralised ethylenediamine (25 ml). Add 2 drops of *o*-nitroaniline in toluene (0.15 per cent w/v) as indicator and titrate with *0.1N* potassium methoxide until the solution changes from clear yellow to orange-red in colour.

Preparation of *0.1N* Tetrabutylammonium Hydroxide in Toluene-methanol

Method. Dissolve tetrabutylammonium iodide (40 g) in absolute methanol (90 ml), add finely powdered purified silver oxide (20 g) and shake vigorously for 1 hour. Centrifuge a few ml of the mixture and test the supernatant liquid for iodide. If a positive reaction is obtained, add an additional 2 g of silver oxide and shake for a further 30 minutes. Repeat this procedure until the liquid is free from iodide, filter the mixture through a fine sintered glass filter and rinse the reaction vessel with 3 portions, each of 50 ml, of dry toluene. Add the washings to the filtrate and dilute to 1 litre with dry toluene. Flush the solution with carbon dioxide-free nitrogen for five minutes and protect from carbon dioxide and moisture during storage.

Standardisation of *0.1N* Tetrabutylammonium Hydroxide

Method. Accurately weigh about 60 mg of benzoic acid into dimethylformamide (10 ml), previously neutralised to the full blue colour of thymol blue (0.3 per cent w/v in methanol; 3 drops) by titration with *0.1N* tetrabutylammonium hydroxide. Allow the benzoic acid to dissolve, and titrate in an atmosphere of carbon dioxide-free nitrogen with *0.1N* tetrabutylammonium hydroxide.

$$C_6H_5COOH \equiv H \equiv 1000 \text{ ml } N$$
$$\therefore \quad 0.01221 \text{ g } C_7H_6O_2 \equiv 1 \text{ ml } 0.1N$$

Ethosuximide *Determination of the percentage of* $C_7H_{11}NO_2$

The determination depends upon the reaction expressed by the following equation:

$$\therefore \quad C_7H_{11}NO_2 \equiv Bu_4\overset{+}{N}OH^- \equiv H \equiv 1000 \text{ ml } N$$
$$\therefore \quad 141.2 \text{ g } C_7H_{11}NO_2 \equiv 1000 \text{ ml } N$$
$$\therefore \quad 0.01412 \text{ g } C_7H_{11}NO_2 \equiv 1 \text{ ml } 0.1N$$

The titration is carried out in dimethylformamide using magneson solution as indicator.

Cognate Determinations

Bendrofluazide
Cyclopenthiazide

Phenobarbitone. *Determination of the percentage of* $C_{12}H_{12}N_2O_3$

Phenobarbitone functions as a weak acid in aqueous media, but is relatively insoluble. It is much more soluble in pyridine, from which it is precipitated quantitatively as its silver salt on addition of silver nitrate. The supernatant pyridine, which contains the displaced HNO_3, can be titrated with NaOH solution.

$$\therefore \quad 232.2 \text{ g } C_{12}H_{12}O_3N_2 \equiv 2HNO_3 \equiv 2H \equiv 2000 \text{ ml } N$$
$$\therefore \quad 0.01161 \text{ g } C_{12}H_{12}O_3N_2 \equiv 1 \text{ ml } 0.1N \text{ AgNO}_3$$

8 Oxidation-Reduction Titrations

Oxidation is (*a*) addition of oxygen, e.g. $SO_2 + O \rightarrow SO_3$
    or (*b*) removal of hydrogen, e.g. $H_2S + O \rightarrow S + H_2O$
or in general
      (*c*) increase in the ratio of electronegative to the electropositive
       portion of the molecule.
$$Sn^{2+} + 2Cl^- + 2HgCl_2 \rightarrow Sn^{4+} + 4Cl^- + Hg_2Cl_2$$
$$2Fe^{2+} + 4Cl^- + Cl_2 \rightarrow 2Fe^{3+} + 6Cl^-$$

Reduction is (*a*) addition of hydrogen, e.g. $C_2H_2 + 2H \rightarrow C_2H_4$
    or   removal of oxygen, e.g. $CuO + 2H \rightarrow Cu + H_2O$
or in general (*c*) increase in the ratio of electropositive to electronegative
       portion of the molecule.

 Oxidation and reduction usually occur simultaneously in a reaction, one substance becoming reduced in the process of oxidising the other.
 Many oxidations which are quantitative can be used as a basis for volumetric determinations, a standard solution of the oxidising agent being used for the titration. The chief oxidising agents used in volumetric work are potassium permanganate, potassium dichromate, iodine, potassium iodate and ceric sulphate.

DETERMINATIONS INVOLVING THE USE OF POTASSIUM PERMANGANATE SOLUTION

The use of potassium permanganate as an oxidising substance in acid solution depends upon the reactions expressed by the following equations:

$$MnO_4^- + 8H^+ + 5e \rightarrow Mn^{2+} + 4H_2O$$
$$\therefore \quad KMnO_4 \equiv 5e$$
$$\therefore \quad 158.0 \text{ g } KMnO_4 \equiv 5000 \text{ ml } N$$
$$\therefore \quad 3.160 \text{ g } KMnO_4 \equiv 1000 \text{ ml } 0.1N \text{ } KMnO_4$$

The Preparation of Volumetric Solutions of Potassium Permanganate

Since potassium permanganate often contains a small proportion of manganese dioxide, volumetric solutions must be made up approximately and then standardised. The intense colour of the solution makes difficult the detection of undissolved solid. The use of heat in the preparation of potassium permanganate solutions is also undesirable, since traces of grease or other contaminants on the glass vessels used can catalyse its decomposition. The following procedure is therefore recommended.

Method. Weigh out approximately the appropriate amount of potassium permanganate (about 3.2 g for 1 litre) on a watch-glass. Transfer to a 250 ml beaker containing cold Water and stir thoroughly breaking up the crystals with a glass rod, to effect solution. Decant the solution, through a small plug of glass wool supported in a funnel, into the appropriate graduated flask, leaving the undissolved residues in the beaker. Add more Water to the beaker and repeat the process. Continue the process until all the potassium permanganate has dissolved. Make the solution up to the graduation mark, and shake well to ensure thorough mixing.

Standardisation of *0.1N* Potassium Permanganate Solution by means of pure Oxalic Acid

This standardisation depends upon the reactions expressed by the following equations:

$$\begin{array}{c} COOH \\ | \\ COOH \end{array} \rightarrow 2CO_2 + 2H^+ + 2e$$

$$\therefore \quad H_2C_2O_4,2H_2O \equiv 2e$$
$$\therefore \quad 6.302 \text{ g } H_2C_2O_4,2H_2O \equiv 1000 \text{ ml } 0.1N \text{ KMnO}_4$$

Method. Weigh out pure oxalic acid (about 6.3 g) accurately into a litre graduated flask, dissolve in Water and make up to volume. Pipette out 20 ml of this solution, add concentrated sulphuric acid (about 5 ml) and warm to about 70° (the flask can only just be held in the hand). Add the potassium permanganate solution from the burette. The first few drops result in a pink colour persisting for about 20 secs. Wait until the colour disappears and then continue the titration in the usual manner. Formation of a brown colour during the titration is caused by insufficient acid, by using too high a temperature or by the use of a dirty flask. Clean the flask with solution of hydrogen peroxide and dilute sulphuric acid before repeating the titration. The end point is reached when a faint pink colour persists for about 30 seconds upon shaking the flask.

Cognate Determinations

Oxalic Acid. Determination of the percentage of $C_2H_2O_4,2H_2O$

Ammonium Vanadate. *Determination of the percentage of* NH_4VO_3

This determination depends upon the quantitative reduction in acid solution of the vanadium from the pentavalent state to the divalent vanadyl (VO^{2+}) ion by means of SO_2, followed by quantitative reoxidation of the vanadyl ion to pentavalent vanadium by means of potassium permanganate solution.
Reduction:

$$VO_3^- + 4H^+ + e \rightarrow (VO)^{2+} + 2H_2O$$
$$SO_3^{2-} + 2H_2O \rightarrow SO_4^{2-} + 4H^+ + 2e$$

then oxidation:

$$(VO)^{2+} + 2H_2O \rightarrow VO_3^- + 4H^+ + e$$
$$\therefore \quad NH_4VO_3 \equiv (VO)^{2+} \equiv e$$
$$\therefore \quad 117.0 \text{ g } NH_4VO_3 \equiv 1000 \text{ ml } N$$
$$\therefore \quad 0.0117 \text{ g } NH_4VO_3 \equiv 1 \text{ ml } 0.1N \text{ KMnO}_4$$

Method. Accurately weigh the sample (about 0.5 g) into a conical flask containing Water (30 ml) and concentrated sulphuric acid (1 ml). Fit the flask with suitable inlet and outlet tubes passing through the cork. Heat, and pass SO_2 through the solution for 5-10 minutes until reduction is complete. The solution acquires the blue colour of vanadyl sulphate which is stable in sulphuric acid solution. *Remove the excess SO_2 by boiling gently and passing a rapid current of CO_2.* Test for complete removal of SO_2 by passing the issuing gas through a very dilute acidified solution of potassium permanganate which should not be decolorised. Wash the inlet tube with Water and remove the cork and tubes, cool to about 60° and titrate with *0.1N* KMnO₄. As titration proceeds the colour changes from blue through bluish green to green and then yellow. The end point is the first permanent orange colour, i.e. a combination of the yellow colour of the solution and the pink colour of the permanganate.

Ferrous Sulphate. *Determination of the percentage of* $FeSO_4,7H_2O$

This determination (no longer official) depends upon the reactions expressed by the following equations:

$$Fe^{2+} \rightarrow Fe^{3+} + e$$
$$\therefore \quad 278.0 \text{ g } FeSO_4,7H_2O \equiv e \equiv 10\ 000 \text{ ml } 0.1N$$
$$\therefore \quad 0.02780 \text{ g } FeSO_4,7H_2O \equiv 1 \text{ ml } 0.1N \text{ KMnO}_4$$

Method. Accurately weigh the sample (about 1 g) into a conical flask containing dilute sulphuric acid (20 ml). Titrate with *0.1N* KMnO₄ until one drop gives a permanent pink colour.

Reduced Iron. *Determination of the percentage of* Fe

Reduced iron is prepared by the reduction of ferric oxide by hydrogen. It contains not less than 80 per cent of metallic iron. The remainder of the sample consists of oxides of iron. The free iron content only is determined by a process which separates free iron from its oxides.

The determination depends upon the reactions expressed by the following equations:

$$Fe + CuSO_4 \rightarrow FeSO_4 + Cu \downarrow$$
$$Fe^{2+} \rightarrow Fe^{3+} + e$$
$$\therefore \quad Fe \equiv Fe^{2+} \equiv e$$
$$\therefore \quad 55.85 \text{ g Fe} \equiv 1000 \text{ ml } N$$
$$\therefore \quad 0.005585 \text{ g Fe} \equiv 1 \text{ ml } 0.1N \text{ KMnO}_4$$

Only the *metallic* iron displaces the copper from the copper sulphate to give ferrous sulphate. The theoretical quantity of $CuSO_4,5H_2O$ required is five times the weight of Fe taken, but in practice eight times as much is used. The ferrous sulphate solution is separated by filtration from iron oxides and copper and titrated with *0.1N* KMnO₄ solution in presence of dilute sulphuric acid. Note that sulphuric acid can only be added *after* filtration because if it were added before it would dissolve the iron oxides.

Method. The following directions must be followed implicitly if correct results are to be obtained:
(1) Use a clean stoppered bottle containing a little water to prevent the sample from caking.
(2) Dry the neck of the bottle to prevent the sample from adhering to the sides of the neck.
(3) Weigh about 0.25 g of sample into the bottle. Do not weigh more than 0.27 g or less than 0.23 g.

(4) Place the stoppered bottle, with stopper removed, in a boiling water-bath for about 5 min. Meanwhile prepare a boiling solution of copper sulphate, containing eight times as much copper sulphate (weighed to 2 decimal places) as the weight of sample taken in 30 ml of water.

(5) Add the boiling copper sulphate solution to the bottle in the water-bath, grease the stopper *slightly*, insert, and shake vigorously for 15 min, placing the bottle in the boiling water-bath frequently to keep it hot. Do not remove the stopper until the 15 min is completed. If the supernatant liquid turns brown in colour it is useless to continue the determination.

(6) Filter quantitatively and rapidly into a conical flask, containing about 20 ml recently boiled and cooled Water, collecting the washings in a conical flask. Keep the filter full during the filtration to minimise oxidation. Acidify the filtrate and washings with dilute sulphuric acid and titrate with *0.1N* KMnO$_4$. The end point is a greyish-purple colour, i.e. a combination of the blue of the excess copper sulphate solution and the pink of the permanganate.

Potassium Bromide. *Determination of the percentage of* KBr

This determination depends upon the reactions expressed by the following equations, in which the bromide ion is oxidised to bromine by acidified potassium permanganate:

$$2Br^- \rightarrow Br_2 + 2e$$
$$\therefore \quad 2KBr \equiv 2e$$
$$\therefore \quad 119 \text{ g KBr} \equiv 1000 \text{ ml } N \text{ KMnO}_4$$
$$\therefore \quad 0.0119 \text{ g KBr} \equiv 1 \text{ ml } 0.1N \text{ KMnO}_4$$

Method. Accurately weigh the sample (about 1.2 g) into a graduated flask (100 ml), dissolve in water, and dilute to volume. Pipette 10 ml into a conical flask, add dilute sulphuric acid (10 ml) and a few glass beads. Heat to boiling, and titrate (fume cupboard) with *0.1N* KMnO$_4$, added dropwise, until the pink colour just persists.

Potassium Ferrocyanide. *Determination of the percentage of* K$_4$Fe(CN)$_6$,3H$_2$O
This is determined by oxidation of the ferrous iron complex to the ferric state (ferricyanide) by titration with potassium permanganate.

$$[Fe(CN)_6]^{4-} \rightarrow [Fe(CN)_6]^{3-} + e$$

The sample is dissolved in recently boiled and cooled water, acidified with dilute sulphuric acid and titrated with *0.1N* KMnO$_4$ in the cold.

Hydrogen Peroxide Solution. *Determination of the percentage w/v of* H$_2$O$_2$

This determination depends upon mutual oxidation-reduction as expressed by the following equations:

$$2MnO_4^- + 6H^+ + 5H_2O_2 \rightarrow 2Mn^{2+} + 8H_2O + 5O_2$$
$$\therefore \quad 5H_2O_2 \equiv 2MnO_4^- \equiv 10e$$
$$\therefore \quad 34.02 \text{ g H}_2O_2 \equiv 2000 \text{ ml } N$$
$$\therefore \quad 0.001701 \text{ g H}_2O_2 \equiv 1 \text{ ml } 0.1N \text{ KMnO}_4$$

Method. Pipette the sample (10 ml) into a 100 ml graduated flask. Adjust to volume with Water and titrate 20 ml portions of the solution with *0.1N* KMnO$_4$ in the cold, after adding sulphuric acid (50 per cent v/v, 5 ml).

Calculation of volume strength of the solution

The 'volume strength' of the solution is the number of ml of oxygen at N.T.P. that can be obtained by complete thermal decomposition of 1 ml of solution. Under these conditions decomposition occurs according to the equation:

$$2H_2O_2 \rightarrow 2H_2O + O_2$$
$$\therefore \quad 68.04 \text{ g } H_2O_2 \equiv 22\,400 \text{ ml } O_2$$
$$\therefore \quad 1 \text{ g } H_2O_2 \equiv 329.2 \text{ ml } O_2$$

Consider a sample found to contain 5.70 per cent w/v H_2O_2
$$\text{then 100 ml sample} \equiv 5.70 \text{ g } H_2O_2$$
$$\text{and 1 ml sample} \equiv 0.0570 \text{ g } H_2O_2$$
$$\equiv 0.0570 \times 329.2 \text{ ml } O_2$$
$$\equiv 18.76 \text{ ml } O_2$$
$$\therefore \quad \text{Volume strength of the sample is 18.76}$$

Sodium Nitrite. *Determination of the percentage of* $NaNO_2$

This determination depends upon the reactions expressed by the following equations:

$$2NaNO_2 + H_2SO_4 \rightarrow 2HNO_2 + Na_2SO_4$$
$$NO_2^- + H_2O \rightarrow NO_3^- + 2H^+ + 2e$$
$$\therefore \quad NaNO_2 \equiv NO_2^- \equiv 2e$$
$$69.0 \text{ g } NaNO_2 \equiv 2000 \text{ ml } N$$
$$\therefore \quad 0.003450 \text{ g} \equiv 1 \text{ ml } 0.1N \text{ KMnO}_4$$

Method. Weigh the sample (0.6-0.7 g) (to give weaker than *0.1N* solution so that a bigger burette reading results) via a funnel into a 250 ml graduated flask. Dissolve in recently boiled and cooled Water and adjust to volume. This solution cannot be acidified and titrated with permanganate because HNO_2 would be lost as oxides of nitrogen. Therefore the sodium nitrite solution is placed in the burette.

Into a conical flask (exceptionally clean flask required) pipette *0.1N* KMnO$_4$ solution (25 ml) and add Water (about 30 ml). Add concentrated sulphuric acid (4-5 ml) rotating the flask as the addition is made. The temperature should now be about 40° and if this is not so, heat the flask by rotating it *above* a heated gauze. *Do not stand the flask upon the gauze*, because this causes local overheating and decomposition of the permanganate solution. Titrate with sodium nitrite solution *in a continuous stream of drops, shaking and rotating the flask vigorously and continuously.* Rotate the flask from time to time above the heated gauze to keep the temperature at 40°. The constant agitation during titration is essential because the nitrous acid reacts only slowly with the permanganate solution at 40° and otherwise there is the possibility of loss of nitrous acid. On the other hand the reaction cannot be speeded up by increasing the temperature because decomposition of the permanganate solution results.

When the solution in the flask becomes a paler pink, raise the temperature and add the nitrite solution dropwise. When the permanganate is less concentrated it is less liable to decompose, but the temperature must be raised because the permanganate reacts more slowly with the nitrous acid. The solution in the flask should be almost boiling when adding the last few drops of nitrite. Allow $\frac{1}{2}$ minute between the addition of each drop for the last few drops near the end point.

Note. If the permanganate solution becomes brown during the titration, it is useless to continue. The cause of the decomposition of the permanganate may be overheating of the flask on the gauze, lack of sufficient sulphuric acid or a greasy flask. If decomposition does occur, pipette more $0.1N$ KMnO$_4$ into a different flask and repeat.

Titanium Dioxide. *Determination of the percentage of* TiO$_2$

The determination depends upon:

(*a*) Conversion of the titanium dioxide (TiO$_2$) to titanic sulphate by fusion with potassium hydrogen sulphate.

(*b*) Quantitative reduction of the titanic sulphate in sulphuric acid solution by amalgamated zinc in a Jones reductor to titanous sulphate:

$$Zn \rightarrow Zn^{2+} + 2e$$
$$Ti^{4+} + e \rightarrow Ti^{3+}$$

(*c*) Treatment of the resulting titanous sulphate solution with excess acidified ferric ammonium sulphate, which is reduced to ferrous ammonium sulphate:

$$Ti^{3+} + Fe^{3+} \rightarrow Ti^{4+} + Fe^{2+}$$

(*d*) Titration of the ferrous iron solution with potassium permanganate:

$$Fe^{2+} \rightarrow Fe^{3+} + e$$
$$\therefore \quad TiO_2 \equiv Ti^{4+} \equiv Ti^{3+} \equiv Fe^{2+} \equiv e$$
$$\therefore \quad 79.9 \text{ g } TiO_2 \equiv 1000 \text{ ml } N$$
$$\therefore \quad 0.00799 \text{ g } TiO_2 \equiv 1 \text{ ml } 0.1N \text{ KMnO}_4$$

The Jones Reductor. The reductor, shown in Fig. 20 consists essentially of a column fitted with a sintered or porcelain disc near the base and a tap to control the outflow. The outlet tube passes through a bung into a Buchner flask which can be attached to a source of vacuum, so that the flow-rate can be increased if necessary. The column which should be almost 2 cm in diameter, is packed with amalgamated zinc to a height of about 36 cm. Amalgamated zinc is prepared from pure granulated zinc (A.R.) by boiling with mercuric chloride solution, and then washing with water.

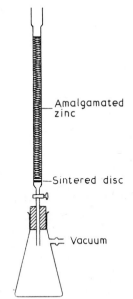

Amalgamated zinc

Sintered disc

Vacuum

Method. Accurately weigh the sample (about 0.3 g) into a platinum crucible. Add potassium hydrogen sulphate (about 3 g) and fuse gently. Cool, and extract the residue with dilute sulphuric acid (150 ml) used in small portions.

Activate the Jones reductor as follows. Pass 200 ml of N sulphuric acid through the column, and wash the column with Water (100 ml), collecting both together in the receiving flask containing ferric ammonium sulphate solution (15 per cent w/v in a 25 per cent v/v solution of dilute sulphuric acid; 50 ml). Titrate the resulting solution with $0.1N$ KMnO$_4$.

Fig. 20. Jones reductor

Reactivate the column with N H$_2$SO$_4$ (100 ml). Pass the titanium solution through the column, followed by N H$_2$SO$_4$ (100 ml) and Water (100 ml) collecting as before in 50 ml of acidified ferric alum solution. Titrate with $0.1N$ KMnO$_4$. The difference between the two titrations gives the amount of potassium permanganate required by the titanium dioxide.

Zinc Powder. *Determination of the percentage of* Zn

This determination also depends upon the quantitative reduction of ferric iron to the ferrous state by the zinc present and then titration with $KMnO_4$ solution of the resultant ferrous ion.

$$Zn + 2Fe^{3+} \rightarrow Zn^{2+} + 2Fe^{2+}$$
$$2[Fe^{2+} \rightarrow Fe^{3+} + e]$$
$$\therefore \quad Zn \equiv 2Fe^{2+} \equiv 2e$$
$$\therefore \quad 65.38 \text{ g Zn} \equiv 2000 \text{ ml } N$$
$$\therefore \quad 0.003269 \text{ g Zn} \equiv 1 \text{ ml } 0.1N \text{ } KMnO_4$$

Method. Accurately weigh the sample (about 1 g) into ferric ammonium sulphate (2 g) dissolved in Water (100 ml) in a 500 ml graduated flask. Shake until dissolved. Add dilute H_2SO_4 (100 ml), dilute with Water to the 500 ml mark, and titrate 50 ml portions of this solution with $0.1N$ $KMnO_4$.

This method determines the free Zn only and not the zinc oxide, because the oxide will not reduce the ferric salt.

DIRECT TITRATION WITH IODINE

Determinations involving direct titration with iodine make use of the oxidising power of iodine in aqueous solution.

$$I_2 + 2e \rightarrow 2I^-$$
$$\therefore \quad I_2 \equiv 2e$$
$$\therefore \quad 126.9 \text{ g } I_2 \equiv 1000 \text{ ml } N$$
$$\therefore \quad 12.69 \text{ g } I_2 \equiv 1000 \text{ ml } 0.1N \text{ iodine}$$

Because iodine is practically insoluble in water, use is made of the fact that it dissolves in solutions of potassium iodide to form KI_3, which behaves in solution as free iodine.

Preparation of Approximately *0.1N* Iodine Solution

Method. Weigh the iodine (3.2 g approx.) on a dispensing balance. Transfer to a beaker containing potassium iodide (7.5 g) and Water (10 ml). Dissolve (I_2 dissolves more readily in this concentrated solution of KI than in weaker solutions), transfer to a 250 ml graduated flask and adjust to volume.

Note. Iodine is volatile and standard solutions should be stored in tightly stoppered (glass) bottles. Standard solutions should not be collected from bulk containers in beakers or other open vessels, and when measured from suitable containers in pipettes (or burettes) should be titrated without delay. These precautions must be observed at all times, if satisfactory results are to be obtained.

Standardisation of Approximately *0.1N* Iodine Solution by means of Arsenic Trioxide

This standardisation depends upon the reactions expressed by the following equations:

$$As_2O_3 + 2H_2O \rightarrow As_2O_5 + 4H^+ + 4e$$
$$\therefore \quad As_2O_3 \equiv 4e$$
$$\therefore \quad 197.8 \text{ g } As_2O_3 \equiv 4000 \text{ ml } N$$
$$\therefore \quad 0.4946 \text{ g } As_2O_3 \equiv 100 \text{ ml } 0.1N \text{ iodine}$$

Because of the strong reducing properties of hydriodic acid, the oxidation with iodine is a reversible reaction

$$As_2O_3 + 2I_2 + 2H_2O \rightleftharpoons As_2O_5 + 4H^+ + 4I^-$$

The reaction is made to go completely to the right (i.e. to As_2O_5) by removal of the HI with sodium bicarbonate. Sodium hydroxide or sodium carbonate cannot be used to remove the HI because they react with iodine as follows:

$$6NaOH + 3I_2 \rightarrow 5NaI + NaIO_3 + 3H_2O$$
$$3Na_2CO_3 + 3I_2 \rightarrow 5NaI + NaIO_3 + 3CO_2$$

Method. Accurately weigh the arsenic trioxide (0.5 g approx.) into a beaker, add sodium hydroxide solution (20 per cent; about 2 ml), and warm to dissolve. Cool, transfer quantitatively to a 100 ml graduated flask and adjust to volume. Pipette 25 ml, acidify with dilute hydrochloric acid, testing for acidity with a little sodium bicarbonate when effervescence should occur. This acidification is necessary to remove the free NaOH which would react with the iodine. Add sodium bicarbonate to remove the excess acid, followed by a further 2 g (to remove the HI in the subsequent titration). Titrate with the approx. *0.1N* iodine solution. The end point is the first permanent pale straw colour.

Note. There is no need to add starch mucilage in this determination.

Cognate Determinations

Arsenic Trioxide. Determination of the percentage of As_2O_3
Sodium Periodate. Determination of the percentage of $NaIO_4$

This determination depends upon the reactions expressed by the following equations:

$$IO_4^- + 2I^- + H_2O \rightarrow IO_3^- + I_2 + 2OH^-$$
$$(2KOH + 2NaHCO_3 \rightarrow Na_2CO_3 + K_2CO_3 + 2H_2O)$$
$$AsO_3^{3-} + H_2O \rightarrow AsO_4^{3-} + 2H^+ + 2e$$
$$I_2 + 2e \rightarrow 2I^-$$

The sample, dissolved in water, is treated with sodium bicarbonate and potassium iodide and the liberated iodine titrated with *0.1N* sodium arsenite solution.

Antimony Sodium Tartrate. *Determination of the percentage of* $C_4H_4O_7NaSb$

This substance behaves in solution as a mixture of sodium acid tartrate and antimony trioxide, i.e.

$$2C_4H_4O_7NaSb \equiv Sb_2O_3$$
$$Sb_2O_3 + 2H_2O \rightarrow Sb_2O_5 + 4H^+ + 4e$$
$$2[I_2 + 2e \rightarrow 2I^-]$$
$$\therefore \quad C_4H_4O_7NaSb \equiv Sb_2O_3 \equiv 4e$$
$$\therefore \quad 617.6 \text{ g } C_4H_4O_7NaSb \equiv 4000 \text{ ml } N$$
$$\therefore \quad 0.01544 \text{ g } C_4H_4O_7NaSb \equiv 1 \text{ ml } 0.1N \text{ iodine}$$

Method. Dissolve the sample (about 0.5 g accurately weighed) in Water (50 ml). Add sodium potassium tartrate (5 g) and borax (2 g) to remove hydriodic acid as formed, and titrate with *0.1N* iodine solution. The end point is rather slow, therefore shake well, to make sure that the end point persists. There is a possibility of under-titration unless this provision is observed.

Cognate Determinations

Antimony Potassium Tartrate Injection
Antimony Trichloride. Determination of the percentage of $SbCl_3$
This determination depends upon the reactions expressed by the following equations:

$$2SbCl_3 + 3H_2O \rightarrow Sb_2O_3 + 6HCl$$
$$Sb_2O_3 + 2H_2O \rightarrow Sb_2O_5 + 4H^+ + 4e$$
$$2[I_2 + 2e \rightarrow 2I^-]$$
$$\therefore \quad 2SbCl_3 \equiv Sb_2O_3 \equiv 2I_2 \equiv 4e \equiv 4000 \text{ ml } N$$

The sample is dissolved in water containing sodium potassium tartrate, to increase the solubility of the Sb_2O_3, and titrated with *0.1N* iodine in the presence of $NaHCO_3$.

Potassium Antimonate. Determination of the percentage of $KSbO_3, 3H_2O$
The antimony is precipitated as Sb_2S_3 by passing H_2S through a solution of the sample in dilute hydrochloric acid. The precipitate is dissolved in concentrated hydrochloric acid, the solution boiled to remove hydrogen sulphide, and neutralised with sodium carbonate in the presence of sodium potassium tartrate. The solution is then titrated with *0.1N* iodine in the presence of sodium bicarbonate.

Acetarsol. *Determination of the percentage of* $C_8H_{10}AsNO_5$

$$HO\text{—}\langle\;\rangle\text{—}AsO(OH)_2$$
$$AcHN$$

Arsenic in organic arsenicals such as Acetarsol is determined, after preliminary treatment, by oxidation of arsenic in the trivalent state to the pentavalent state by means of *0.1N* iodine solution.
Before the determination of As can be carried out, the As must be removed from its organic combination. The following is a summary of the steps involved.
(1) The organic matter is destroyed by heating strongly with concentrated sulphuric and fuming nitric acids.

(2) The nitric acid is removed by adding ammonium sulphate and heating until all N_2O has been removed:

$$2HNO_3 + (NH_4)_2SO_4 \rightarrow 2NH_4NO_3 + H_2SO_4$$
$$\downarrow \text{heat}$$
$$2N_2O + 4H_2O$$

(3) Potassium iodide is added to the acid solution and the liberated HI reduces the pentavalent arsenic to the trivalent state. The solution is boiled during the reduction and most of the liberated iodine is expelled.

$$As_2O_5 + 4HI \rightarrow As_2O_3 + 2I_2 + 2H_2O$$

(4) The liquid is cooled and any remaining iodine removed by titration with 0.1N sodium sulphite solution.

(5) The solution is made just alkaline with sodium hydroxide solution and then acidified with dilute sulphuric acid to remove the free NaOH. Sodium bicarbonate is added to ensure that the reaction proceeds quantitatively in the correct direction and the liquid is titrated with 0.1N iodine solution:

$$As_2O_3 + 2I_2 + 2H_2O \rightleftharpoons As_2O_5 + 4HI$$
$$\therefore \quad 2C_8H_{10}AsNO_5 \equiv As_2O_5 \equiv As_2O_3 \equiv 2I_2 \equiv 4e$$
$$\therefore \quad 2 \times 275.1 \text{ g } C_8H_{10}AsNO_5 \equiv 4000 \text{ ml } N$$
$$\therefore \quad 0.01375 \text{ g } C_8H_{10}AsNO_5 \equiv 1 \text{ ml } 0.1N \text{ iodine}$$

Method. Accurately weigh the sample (about 0.25 g) into a 500 ml conical flask. Add sulphuric acid (7.5 ml) and fuming nitric acid (SG 1.5; 1.5 ml) and boil gently (fume cupboard) for 45 minutes. Cool, add fuming nitric acid (0.5 ml), and heat until brown fumes cease to be evolved. Cool, add ammonium sulphate (5 g) in small quantities at a time, and heat again (fume cupboard) until there is no further evolution of gas and the resulting liquid is colourless. Cool, dilute with water (100 ml), add KI (1 g), and boil gently (fume cupboard) until the volume is reduced to 50 ml. Cool, declorise by the addition of a few drops of 0.1N sodium sulphite. Dilute with Water (60 ml), add phenolphthalein solution (1 drop) and make just alkaline with sodium hydroxide solution. Acidify slightly with dilute sulphuric acid, neutralise with sodium bicarbonate, and then add 4 g in excess, and titrate with 0.1N iodine.

Cognate Determinations

Melarsoprol
Sodium Stibogluconate
Sodium Stibogluconate Injection

Ascorbic Acid. *Determination of the percentage of* $C_6H_8O_6$

The determination depends upon the quantitative oxidation of ascorbic acid to dehydroascorbic acid with iodine acid solution.

$$
6\;
\begin{array}{l}
\text{CO}\!-\!\rceil\\
|\\
\text{C.OH}\\
\|\\
\text{C.OH}\\
|\\
\text{CH}\!-\!\rfloor\\
|\\
\text{CHOH}\\
|\\
\text{CH}_2\text{OH}
\end{array}\Bigg\}\,O
\;+\,I_2 \longrightarrow\;
6\;
\begin{array}{l}
\text{CO}\!-\!\rceil\\
|\\
\text{C}\!=\!\text{O}\\
|\\
\text{C}\!=\!\text{O}\\
|\\
\text{CH}\!-\!\rfloor\\
|\\
\text{CHOH}\\
|\\
\text{CH}_2\text{OH}
\end{array}\Bigg\}\,O \;+2HI
$$

$$C_6H_8O_6 + I_2 \rightarrow C_6H_6O_6 + 2HI$$
$$\therefore \quad C_6H_8O_6 \equiv I_2 \equiv 2e$$
$$\therefore \quad 176.1 \text{ g } C_6H_8O_6 \equiv 2000 \text{ ml } N \text{ iodine}$$
$$\therefore \quad 0.0881 \text{ g } C_6H_8O_6 \equiv 1 \text{ ml } N \text{ iodine}$$

Dimercaprol. *Determination of the percentage w/w of* $C_3H_8OS_2$

This determination depends upon the oxidation of the thiol groups with iodine. The reaction is not reversible.

$$
\begin{array}{l}
\text{CH}_2\text{SH}\\
|\\
\text{CH.SH}\\
|\\
\text{CH}_2\text{OH}
\end{array}
+I_2 \longrightarrow
\begin{array}{l}
\text{CH}_2\!-\!\text{S}\!-\!\text{S}\!-\!\text{CH}_2\\
|\qquad\qquad\qquad|\\
\text{CH}\!-\!\text{S}\!-\!\text{S}\!-\!\text{CH}\\
|\qquad\qquad\qquad|\\
\text{CH}_2\text{OH}\qquad\ \text{CH}_2\text{OH}
\end{array}
+4HI
$$

$$\therefore \quad 2\times124.2 \text{ g } C_3H_8OS_2 \equiv 2I_2 \equiv 4000 \text{ ml } N$$
$$\therefore \quad 0.006211 \text{ g } C_3H_8OS_2 \equiv 1 \text{ ml } 0.1N \text{ iodine}$$

Method. Weigh accurately 0.2 g in a small sample tube. Transfer the tube and contents to a 250 ml conical flask, add *0.1N* hydrochloric acid (40 ml) and titrate with *0.1N* iodine.

Cognate Determinations

Dimercaprol Injection

Thioglycollic Acid. The assay is divided into two parts and the product is required to contain not less than 89 per cent w/w $C_2H_4O_2S$ as determined by both parts of the assay.

(1) The sample is dissolved in water and titrated with N NaOH using cresol red as indicator:

$$
\begin{array}{l}
\text{CH}_2\text{SH}\\
|\\
\text{COOH}
\end{array}
+\text{NaOH} \longrightarrow
\begin{array}{l}
\text{CH}_2\text{SH}\\
|\\
\text{COONa}
\end{array}
+H_2O
$$

(2) Sodium bicarbonate is added to the above neutralised liquid, and the solution titrated with *0.1N* iodine solution:

$$
2\;
\begin{array}{l}
\text{CH}_2\text{SH}\\
|\\
\text{COONa}
\end{array}
+I_2 \rightleftharpoons
\begin{array}{l}
\text{CH}_2\!-\!\text{S}\!-\!\text{S}\!-\!\text{CH}_2\\
|\qquad\qquad\qquad\ \ |\\
\text{COONa}\qquad\ \text{COONa}
\end{array}
+2HI
$$

The sodium bicarbonate is necessary to remove the HI thus enabling the reaction to proceed to completion from left to right.

IODINE-SODIUM THIOSULPHATE TITRATIONS

Standardisation of Iodine Solutions by Sodium Thiosulphate

AnalaR sodium thiosulphate $(Na_2S_2O_3,5H_2O)$ can be used to standardise iodine solution, which oxidises the former to sodium tetrathionate.

$$2S_2O_3^{2-} \rightarrow S_4O_6^{2-} + 2e$$
$$I_2 + 2e \rightarrow 2I^-$$
$$\therefore \quad 2S_2O_3^{2-} \equiv I_2 \equiv 2e$$
$$\therefore \quad 2 \times 248.2 \text{ g } Na_2S_2O_3,5H_2O \equiv 2000 \text{ ml } N$$
$$\therefore \quad 24.82 \text{ g } Na_2S_2O_3,5H_2O \equiv 1000 \text{ ml } 0.1N \text{ iodine}$$

Method. Transfer 6.025 g (accurately weighed) of A.R. sodium thiosulphate, to a 250 ml graduated flask. Dissolve in Water, adjust to volume and shake. If large volumes of sodium thiosulphate solutions are being prepared for storage, the solution should be stabilised against decomposition by the addition of one or two drops of sodium hydroxide solution (20 per cent w/v).

Pipette 25 ml of *0.1N* iodine into a conical flask, and titrate with the standard sodium thiosulphate solution until just colourless.

Cognate Determinations

Iodine

Sodium Thiosulphate

Sulphur Ointment. Sulphur is converted to sodium thiosulphate by refluxing with excess sodium sulphite solution:

$$Na_2SO_3 + S \rightarrow Na_2S_2O_3$$

The excess sodium sulphite, which would react with iodine, is converted by addition of formaldehyde into formaldehyde sodium bisulphite and the sodium thiosulphate then titrated with *0.1N* iodine.

Back Titration of Excess Iodine with Sodium Thiosulphate

Benzylpenicillin

After preliminary hydrolysis with sodium hydroxide solution to convert the antibiotic to the corresponding penicilloic acid, treatment with acid yields D-penicillamine (and benzylpenillic acid) which is oxidised almost quantitatively by iodine to the corresponding disulphide. Excess iodine is back-titrated with *0.02N* sodium thiosulphate solution.

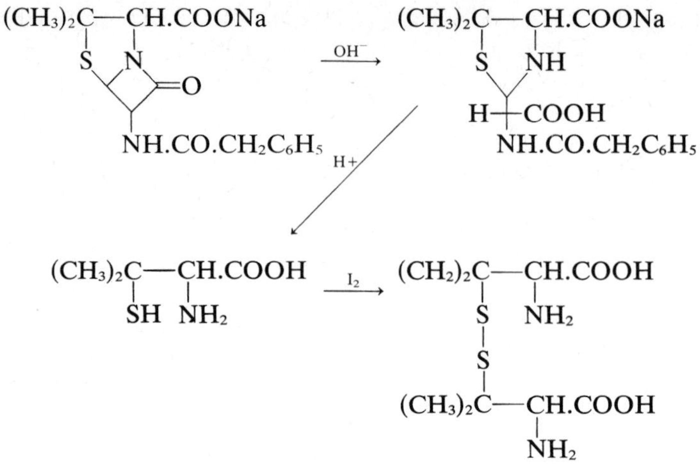

$$(CH_3)_2C-CH.COONa$$
$$S \quad N$$
$$=O$$
$$NH.CO.CH_2C_6H_5$$

$\xrightarrow{OH^-}$

$$(CH_3)_2C-CH.COONa$$
$$S \quad NH$$
$$H+COOH$$
$$NH.CO.CH_2C_6H_5$$

$H+$

$$(CH_3)_2C-CH.COOH$$
$$SH \quad NH_2$$

$\xrightarrow{I_2}$

$$(CH_2)_2C-CH.COOH$$
$$S \quad NH_2$$
$$S$$
$$(CH_3)_2C-CH.COOH$$
$$NH_2$$

$$C_{16}H_{17}N_2NaO_4S \equiv I \equiv e$$

Benzylpenicillin Sodium is standardised against a chemical reference substance of known potency

Cognate Determination

Ecothiopate Iodide

Dichloralphenazone

The sample in sodium acetate buffer is treated with standard iodine solution to yield a precipitate of phenazone-iodine complex. This is dissolved in chloroform, and excess iodine in the aqueous phase is titrated with $0.1N$ sodium thiosulphate.

A separate assay for chloral is also required to ensure that the composition of the chloral-phenazone complex is consistent from batch to batch.

Cognate Determination

Dichloralphenazone Tablets

Glyoxal Sodium Bisulphite. $[CH(OH).SO_3Na]_2, H_2O$

The combined sodium bisulphite is oxidised by the addition of a known excess of iodine, the excess being back titrated with sodium thiosulphate.

$$[CH(OH).SO_3Na]_2, H_2O + I_2 + H_2O \rightarrow 2CH_2O + 2NaHSO_4 + 2HI$$

Mercuric Chloride. *Determination of the percentage of* $HgCl_2$

This substance is reduced to metallic mercury by formaldehyde in alkaline solution. Oxidation of the mercury to mercuric iodide by excess iodine in the presence of acetic acid and potassium iodide is followed by back titration of the excess of iodine.

$$HgCl_2 + HCHO + 3NaOH \rightarrow Hg + 2NaCl + H.COONa + 2H_2O$$
$$Hg + I_2 \rightarrow HgI_2$$
$$HgI_2 + 2KI \rightarrow K_2HgI_4 \text{ (soluble)}$$
$$\therefore \quad HgCl_2 \equiv Hg \equiv I_2 \equiv 2e$$
$$\therefore \quad 271.5 \text{ g } HgCl_2 \equiv 2000 \text{ ml } N$$
$$\therefore \quad 0.01358 \text{ g } HgCl_2 \equiv 1 \text{ ml } 0.1N \text{ iodine}$$

Method. Dissolve the sample, accurately weighed (between 0.25 and 0.35 g), in Water (80-90 ml) in a 250 ml stoppered bottle. Add calcium chloride solution (10 ml), potassium iodide (1 g) dissolved in a little Water, formaldehyde solution (3 ml) and caustic soda solution (15 ml). On the addition of alkali the calcium chloride reacts to form a fine precipitate of calcium hydroxide which keeps the mercury in a fine state of subdivision. This hastens the reaction with the iodine. Shake continuously for 2 minutes—the mercuric chloride is reduced to mercury in this time. Add acetic acid (about 20 ml), which should produce an acid solution—mix thoroughly and test with litmus. If necessary add more acid to produce an acid solution. Then add 0.1N iodine solution (50 ml) (pipette) and shake vigorously until all the mercury has dissolved (10 to 15 minutes vigorous shaking is required). Ensure that no very fine globules of mercury remain undissolved. Back titrate the excess of iodine with 0.1N sodium thiosulphate solution. The end point is almost colourless (actually a very faint greenish-yellow due to the K_2HgI_4 in solution).

Sodium Metabisulphite. *Determination of the percentage of $Na_2S_2O_5$*

This substance reacts as follows:

$$Na_2S_2O_5 + 2HCl \rightarrow 2NaCl + H_2O + 2SO_2$$
$$2[SO_2 + H_2O \rightarrow H_2SO_3]$$
$$2[SO_3^{2-} + H_2O \rightarrow SO_4^{2-} + 2H^+ + 2e]$$
$$2[I_2 + 2e \rightarrow 2I^-]$$
$$\therefore \quad Na_2S_2O_5 \equiv 2I_2 \equiv 4e$$

Sodium Sulphide. *Determination of the percentage of $Na_2S,9H_2O$*

This substance is decomposed by acid liberating hydrogen sulphide, which reacts quantitatively with iodine:

$$Na_2S + 2HCl \rightarrow 2NaCl + H_2S$$
$$S^{2-} \rightarrow S + 2e$$
$$I_2 + 2e \rightarrow 2I^-$$
$$\therefore \quad Na_2S,9H_2O \equiv I_2 \equiv 2e$$

Sulphurous Acid. *Determination of the percentage w/w of SO_2*

This substance is oxidised by iodine to sulphuric acid, and the excess iodine back titrated with sodium thiosulphate.

$$SO_3^{2-} + H_2O \rightarrow SO_4^{2-} + 2H^+ + 2e$$
$$I_2 + 2e \rightarrow 2I^-$$
$$\therefore \quad SO_2 \equiv SO_3^{2-} \equiv I_2 \equiv 2e$$

Sodium Sulphite. *Determination of the percentage of* $Na_2SO_3,7H_2O$

This is determined as for sulphurous acid, which is liberated by treatment with hydrochloric acid:

$$Na_2SO_3 + 2HCl \rightarrow 2NaCl + H_2SO_3$$
$$SO_3^{2-} + H_2O \rightarrow SO_4^{2-} + 2H^+ + 2e$$
$$I_2 + 2e \rightarrow 2I^-$$
$$\therefore \quad Na_2SO_3,7H_2O \equiv I_2 \equiv 2e$$

Release of Iodine and Titration with Sodium Thiosulphate

Chloramine. *Determination of the percentage of* $C_7H_7O_2NClSNa,3H_2O$

In acid solution this substance behaves as though it were a hypochlorite and yields active chlorine.

$$2NaOCl + H_2SO_4 \rightarrow Na_2SO_4 + 2HOCl$$
$$2[HO\frown Cl + I^- \rightarrow HO^- + I\text{---}Cl]$$
$$2[I^\frown + I\frown Cl \rightarrow I_2 + Cl^-]$$

$$C_7H_7O_2NClSNa \equiv I_2 \equiv 2e$$
$$\therefore \quad 281.7 \text{ g } C_7H_7O_2NClSNa,3H_2O \equiv 2000 \text{ ml } N$$
$$\therefore \quad 0.01409 \text{ g } C_7H_7O_2NClSNa,3H_2O \equiv 1 \text{ ml } 0.1N \text{ } Na_2S_2O_3$$

Method. Dissolve the sample (about 0.4 g), accurately weighed, in Water (about 50 ml) in a glass stoppered bottle. Add potassium iodide (1 g) and dilute sulphuric acid (5 ml). Shake and allow to stand for 10 minutes. Titrate the liberated iodine with $0.1N$ sodium thiosulphate solution. If a white precipitate forms during the titration, dilute with a little water so that all substances are kept in solution.

Chlorinated Lime. *Determination of the percentage w/w of available* Cl_2

$$CaCl(OCl) + 2CH_3.COOH \rightarrow Ca(CH_3.COO)_2 + HCl + HClO$$
$$HCl + HClO \rightarrow H_2O + Cl_2$$
$$2HI + Cl_2 \rightarrow 2HCl + I_2$$
$$\therefore \quad 35.46 \text{ g } Cl \equiv 1000 \text{ ml } N$$
$$\therefore \quad 0.003546 \text{ g} \equiv 1 \text{ ml } 0.1N \text{ } Na_2S_2O_3$$

An official sample contains about 30 per cent Cl_2 and therefore about 0.4 g sample $\equiv$ about 40 ml $0.1N$ thiosulphate.

Method. Triturate the sample (about 4 g), accurately weighed, with Water in a mortar to a smooth cream. This must be performed rapidly to avoid loss of chlorine.

Transfer the paste to a 1 litre graduated flask by means of a glass rod and funnel, washing out the mortar and funnel several times with Water. Dilute the suspension to the 1 litre mark and shake well. There must not be any large particles in the suspension. Rinse out a 100 ml graduated flask with a portion of the suspension, shaking the litre flask well before pouring out the suspension.

Rinse out a conical flask containing a little dilute acid with some of the suspension from the litre graduated flask to oxidise any organic material which may be present. Then *wash out well* with Water. Transfer the 100 ml of suspension from the graduated flask to this conical flask, washing out the suspension completely from the graduated flask by means of Water. Add KI (3 g), acetic acid (5 ml), and titrate the liberated iodine with *0.1N* sodium thiosulphate solution.

Chromium Trioxide. *Determination of the percentage of* CrO_3

Chromium trioxide is an oxidising agent:

$$2CrO_3 + 12H^+ + 6e \rightarrow 2Cr^{3+} + 6H_2O$$
$$3[2I^- \rightarrow I_2 + 2e]$$
$$\therefore \quad CrO_3 \equiv 3e$$
$$\therefore \quad 100.0 \text{ g } CrO_3 \equiv 3000 \text{ ml } N$$
$$\therefore \quad 0.003334 \text{ g } CrO_3 \equiv 1 \text{ ml } 0.1N \text{ } Na_2S_2O_3$$

Prepare an aqueous solution of chromium trioxide and determine as follows:

Method. Pipette the required volume of solution into a litre flask, taking care not to get any of the solution on the sides of the flask. Add dilute sulphuric acid (10 ml) and potassium iodide (3 g). Cork and allow to stand for 10 minutes. Dilute to about 500 ml with recently boiled and cooled Water, and titrate the liberated iodine with *0.1N* sodium thiosulphate solution, shaking well during the titration. As the titration proceeds the colour changes from a brown to a yellowish green. Add starch indicator—a dark green colour is obtained. Titrate slowly with continuous shaking, and allow $\frac{1}{2}$ minute between each drop near the end point. The colour changes from dark green to dark blue $\rightarrow$ pale blue $\rightarrow$ pale green. The permanent pale green colour is the end point.

Note. A poor end point is obtained if the starch indicator is added too soon, or if too much sulphuric acid is added, or if the dilution is insufficient. The sulphuric acid must be present to keep the chromium in solution as $Cr_2(SO_4)_3$. However, sulphuric acid interferes with the end point when starch is present. Hence the solution is diluted as much as possible to minimise this interference.

Cognate Determinations

Potassium Chromate
Potassium Dichromate

Copper Sulphate. *Determination of the percentage of* $CuSO_4,5H_2O$

The determination of copper compounds depends on the instability of cupric iodide formed by the reaction between the copper salt and potassium iodide. The cupric iodide breaks up into cuprous iodide and iodine, and this iodine can be titrated with standard sodium thiosulphate solution. The reactions are expressed by the following equations:

$$2CuSO_4 + 4KI \rightarrow 2CuI_2 + 2K_2SO_4$$
$$2CuI_2 \rightarrow Cu_2I_2 + I_2$$
$$\therefore \quad 2CuSO_4,5H_2O \equiv I_2 \equiv 2e$$
$$\therefore \quad 2 \times 249.7 \text{ g } CuSO_4,5H_2O \equiv 2000 \text{ ml } N$$
$$\therefore \quad 0.02497 \text{ g} \equiv 1 \text{ ml } 0.1 \text{ Na}_2S_2O_3$$

Method. Accurately weigh the sample (about 1 g) into a flask. Dissolve in Water (about 50 ml). Add a little sodium carbonate until turbid to remove any free mineral acid. Add acetic acid to clear the solution and then 5 ml in excess. Add potassium iodide (3 g), dilute to about 200 ml with Water, and titrate the liberated iodine with *0.1N* sodium thiosulphate solution. During the titration the brown colour of the iodine becomes less intense. When a pale yellowish colour has been obtained add starch indicator (1 ml). The solution should now be blue. Continue the titration drop by drop, and when near the end point add 2 g of potassium thiocyanate and shake well. Complete the titration shaking well and allowing 10 seconds between each drop, because the end point is rather slow. The end point is white or flesh colour and one drop of *0.1N* sodium thiosulphate should result in a change from the blue colour to this end point.

Note. The starch indicator must not be added until near the end of the titration.

Cognate Determinations

Copper Acetate
Copper Nitrate

2,6-Dichlorophenolindophenol Sodium Salt. *Determination of the percentage of* $O:C_6H_2Cl_2:N.C_6H_4ONa,2H_2O$

This substance oxidises hydriodic acid to iodine in acid solution and the iodine is then titrated with standard sodium thiosulphate solution.

$$O:C_6H_2Cl_2:N.C_6H_4ONa + 2HI \rightarrow HO.C_6H_2Cl_2.NH.C_6H_4ONa + I_2$$

Ferric Ammonium Citrate. *Determination of the percentage of* Fe

All the iron is oxidised to the ferric state. Actually very little ferrous ion is present. The ferric ion then liberates an equivalent amount of iodine from acidified potassium iodide solution.

$$2[Fe^{3+} + e \rightarrow Fe^{2+}]$$
$$2I^- \rightarrow I_2 + 2e$$
$$\therefore \quad 2Fe \equiv 2Fe^{3+} \equiv 2e$$
$$\therefore \quad 2 \times 55.85 \text{ g } Fe \equiv 2000 \text{ ml } N$$
$$\therefore \quad 0.005585 \text{ g } Fe \equiv 1 \text{ ml } 0.1N \text{ Na}_2S_2O_3$$

Method. Dissolve the sample (0.5 g) in Water (15 ml—measuring cylinder). Add sulphuric acid (1 ml), and warm *gently*, until yellow, to decompose the iron and ammonium citrate complex. Cool, and add *0.1N* $KMnO_4$ drop by drop from a burette until the pink colour persists for 5 seconds. This oxidises Fe^{2+} to Fe^{3+}. Care must be exercised not to add too much $KMnO_4$ solution, since although the pink colour will gradually disappear an excess will nevertheless lead to high results. Add hydrochloric acid (15 ml) and potassium iodide (2 g), and set aside for three minutes to complete the liberation of the iodine. Add Water (about 60 ml) and titrate with $0.1N$ sodium thiosulphate shaking well during the titration. Avoid washing down during the titration until almost at the end point so that the concentrations of the various substances are kept correct. The end point is almost colourless. Starch indicator is not required.

Cognate Determinations

Ferric Ammonium Sulphate
Ferric Chloride Solution. Determination of the percentage w/v of $FeCl_3$

Standardisation of $0.1N$ Sodium Thiosulphate with Potassium Iodate

As there is always some doubt as to the exact water content of sodium thiosulphate crystals, it is usual to standardise thiosulphate solutions by such substances as potassium iodate, potassium bromate or potassium dichromate, all of which can be obtained in a high state of purity.

The standardisation with potassium iodate depends upon the reactions expressed by the following equation:

$$IO_3^- + 5I^- + 6H^+ \rightarrow 3I_2 + 3H_2O$$
$$3[I_2 + 2e \rightarrow 2I^-]$$
$$3[2S_2O_3^{2-} \rightarrow S_4O_6^{2-} + 2e]$$
$$\therefore \quad KIO_3 \equiv 3I_2 \equiv 6e$$
$$\therefore \quad 214.0 \text{ g } KIO_3 \equiv 6000 \text{ ml } N$$
$$\therefore \quad 3.567 \text{ g } KIO_3 \equiv 1000 \text{ ml } 0.1N \text{ } Na_2S_2O_3$$

Reaction between KIO_3 and excess KI in acid solution results in liberation of iodine which can be titrated with sodium thiosulphate solution.

Method. Accurately weigh about 1.3 g of AnalaR potassium iodate into a 250 ml graduated flask. Dissolve in Water and adjust to volume. The substance is only slowly soluble. Pipette the solution (25 ml) into a conical flask, add potassium iodide (2 g) and dilute sulphuric acid (4 ml) and titrate the liberated iodine with the approx. $0.1N$ sodium thiosulphate solution.

Cognate Determinations

Potassium Iodate
Iodic Acid
Iodine Pentoxide
Iodine Trichloride
Potassium Bromate. This can be determined in the same way as potassium iodate by the addition of potassium iodide and dilute hydrochloric acid:

$$KBrO_3 + HI \rightarrow HIO_3 + KBr$$
$$IO_3^- + 5I^- + 6H^+ \rightarrow 3I_2 + 3H_2O$$
$$\therefore \quad KBrO_3 \equiv IO_3^- \equiv 3I_2 \equiv 6e$$
$$\therefore \quad 167.02 \text{ g } KBrO_3 \equiv 6000 \text{ ml } N$$
$$\therefore \quad 0.002784 \text{ g } KBrO_3 \equiv 1 \text{ ml } 0.1N \text{ } Na_2S_2O_3$$

Back Titration of Excess iodine with Sodium Thiosulphate

Phenol. *Determination of the percentage w/w of* C_6H_5OH
The phenol is treated with excess *0.1N* bromine:

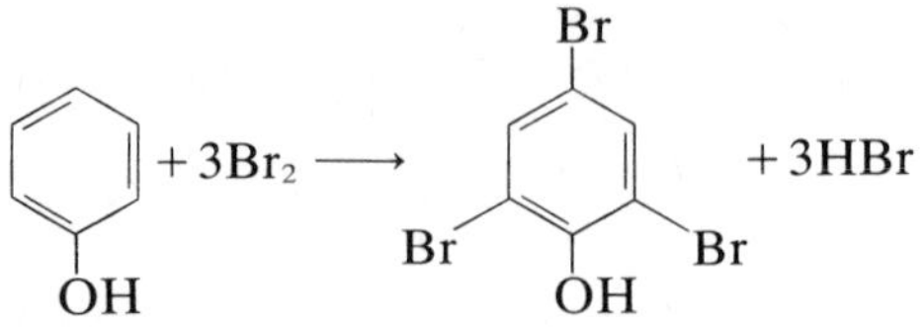

The excess of bromine not required for the formation of tribromophenol is determined by adding potassium iodide and titrating the liberated iodine with sodium thiosulphate solution.

$$Br_2 + 2I^- \rightarrow 2Br^- + I_2$$
$$\therefore \quad C_6H_5OH \equiv 3Br_2 \equiv 3I_2 \equiv 6e$$
$$\therefore \quad 94.11 \text{ g } C_6H_5OH \equiv 6000 \text{ ml } N$$
$$\therefore \quad 0.001569 \text{ g } C_6H_5OH \equiv 1 \text{ ml } 0.1N \text{ Bromine}$$

Solution of bromine is not stable and therefore a solution containing potassium bromide and potassium bromate is used. This solution on acidification liberates bromine:

$$BrO_3^- + 5Br^- + 6H^+ \rightarrow 3Br_2 + 3H_2O$$

The quantities of potassium bromide and potassium bromate used in the solution are such that on acidification an approximately *0.1N* bromine solution is produced. The exact strength of the solution is not required because a blank determination is performed.

Method. Dissolve the sample (about 2 g accurately weighed) in Water in a litre graduated flask and adjust to volume. Pipette 25 ml into a glass-stoppered bottle. Add bromine solution (50 ml) from a pipette and concentrated hydrochloric acid (5 ml). Shake repeatedly for half an hour and then set aside for 15 minutes. The shaking is to complete the bromination of the phenol to tribromophenol. Add potassium iodide (1 g), shake well, wash the stopper and titrate the liberated iodine with standard sodium thiosulphate solution. When the solution is pale yellow in colour, add chloroform (10 ml) and a few drops of starch mucilage, and titrate with vigorous shaking to a colourless end point. Carry out a blank determination on the bromide-bromate solution (50 ml).

Cognate Determinations

Liquefied Phenol
Isoniazid. Determined by direct titration with potassium bromate in presence of bromide and acid.

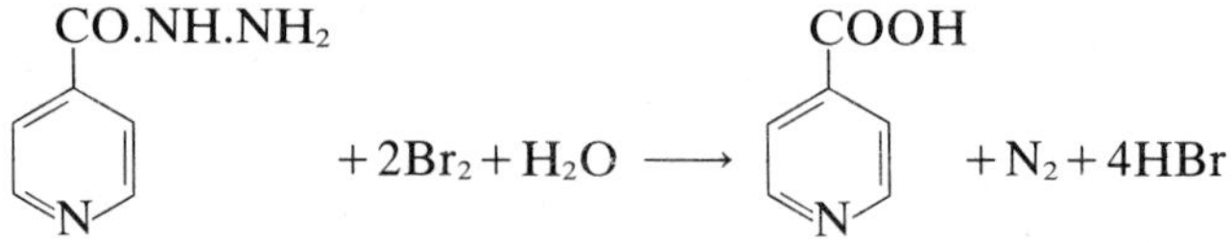

$$\therefore \quad C_6H_7ON_3 \equiv 4Br \equiv 40\ 000 \text{ ml } 0.1N \text{ Bromine}$$

Methyl Hydroxybenzoate
Propyl Hydroxybenzoate
Phenindione. This substance is treated with bromine in ethanol to give the dibromo addition compound. Excess bromine is removed by the addition of β-naphthol, and traces of bromine vapour by means of a current of air. The bromine atom in 2-bromo-2-phenylindane-1,3-dione is labile, and treatment with potassium iodide releases iodine, which can be titrated with sodium thiosulphate.

$$2KBr + I_2$$

Potassium Ferricyanide. *Determination of the percentage of* $K_3Fe(CN)_6$

The determination depends upon the reactions expressed by the following equations:

$$2[Fe(CN)_6]^{3-} + 2I^- \rightarrow 2[Fe(CN)_6]^{4-} + I_2$$
$$2Zn^{2+} + [Fe(CN)_6]^{4-} \rightarrow Zn_2Fe(CN)_6$$
$$\therefore \quad 2 \times 329.3 \text{ g } K_3Fe(CN)_2 \equiv I_2 \equiv 2e \equiv 2000 \text{ ml } N \text{ iodine}$$
$$\therefore \quad 0.03293 \text{ g } K_3Fe(CN)_6 \equiv 1 \text{ ml } 0.1N$$

Standardisation of Potassium Permanganate with Sodium Thiosulphate

The standardisation depends upon the reactions expressed by the following equations:

$$2[MnO_4^- + 8H^+ + 5e \rightarrow Mn^{2+} + 4H_2O]$$
$$5[2I^- \rightarrow I_2 + 2e]$$

The iodine so released is titrated with $0.1N$ sodium thiosulphate.

$$\therefore \quad 2KMnO_4 \equiv 5I_2 \equiv 10e \equiv 100\,000 \text{ ml } 0.1N \text{ Na}_2S_2O_3$$

Method. Pipette 25 ml of the approx. $0.1N$ KMnO$_4$ solution into a conical flask. Add dilute sulphuric acid (10 ml) and potassium iodide (3 g). Quantitative liberation of iodine occurs immediately. Titrate the liberated iodine with $0.1N$ sodium thiosulphate solution, shaking continuously during the titration to prevent the acid attacking the thiosulphate. Starch indicator is unnecessary because the end point is colourless.

Cognate Determinations

Manganese Dioxide
Sodium Peroxide. Iodine is liberated from acidified potassium iodide solution by the peroxide, and titrated with sodium thiosulphate.

$$Na_2O_2 + 2HI \rightarrow 2NaI + H_2O_2$$
$$H_2O_2 + 2H^+ + 2I^- \rightarrow 2H_2O + I_2$$

204 OXIDATION-REDUCTION TITRATIONS

IODINE VALUE OF FIXED OILS

The iodine value is defined as the weight of iodine absorbed by 100 parts by weight of the substance. It is a measure of the unsaturated compounds present in the substance, and is based upon the addition of halogen across a carbon-carbon double bond. Iodine itself reacts too slowly; instead either iodine monochloride or bromine in pyridine may be employed as reagents in official determinations. A quantitative measure of the unsaturation present is obtained provided that certain experimental conditions are complied with, since there is a tendency with both reagents for substitution reactions as well as addition to take place. This is especially liable to occur where appreciable concentrations of steroid material are present, and in these instances the so-called pyridine-bromide method is obligatory.

Pure glyceryl trioleate has a theoretical iodine value of 87.62 and most edible oils have iodine values in this region, e.g. Olive Oil (79–88) (Table 8, p. 56). Such oils are often described as non-drying oils. The so-called drying-oils and semi-drying oils, which have more centres of unsaturation, exhibit much higher iodine values, e.g. Linseed Oil (170–200); Cod-liver Oil (155–177). On the other hand the solid fats, which contain a high proportion of esters of saturated fatty acids, have relatively low iodine values, e.g. Theobroma Oil (35–40).

Iodine Monochloride Method

This determination depends upon reactions expressed by the following equations:

Excess iodine monochloride is converted to iodine by the addition of the potassium iodide, and titrated with sodium thiosulphate in the usual way.

$$I^- + I\!-\!Cl \longrightarrow I_2 + Cl^-$$

Method. Weigh the oil (see below for calculation of the required amount) in a small specimen tube and place it in an iodine flask or a 250 ml clean dry glass-stoppered bottle. Add 10 ml of carbon tetrachloride to dissolve the oil. Run in 20 ml of iodine monochloride solution from an automatic pipette (see Fig. 12, p. 117), insert the stopper previously moistened with potassium iodide solution and set aside in the dark for 30 minutes (exactly), at a temperature of about 17°. The reaction is performed in the dark to avoid side reactions such as substitution which are light-catalysed. After 30 minutes, add 10 per cent aqueous potassium iodide solution (15 ml), and back titrate the excess iodine with $0.1N$ $Na_2S_2O_3$ using starch mucilage as indicator. Carry out a blank determination using all the reagents, but omitting the sample. Calculate the iodine value from the formula:

$$\text{Iodine value} = \frac{(b-a) \times 0.01269 \times 100 \times \text{factor}}{\text{Weight of sample (in g)}}$$

where b = blank burette reading; a = test burette reading; factor = factor of standard sodium thiosulphate solution used.

Solution of Iodine Monochloride contains:

Iodine trichloride	8 g
Iodine	9 g
Carbon tetrachloride	300 ml
Glacial acetic acid to	1000 ml

Dissolve the iodine trichloride in about 200 ml of glacial acetic acid and dissolve the iodine in the carbon tetrachloride. Mix the two solutions and adjust to 1000 ml with glacial acetic acid. This operation should be carried out in a fume cupboard.

Calculation of the Weight of Oil Required

Each 20 ml of iodine monochloride solution is approximately equivalent to 40 ml of $0.1N$ $Na_2S_2O_3$. Allowing for at least 100 per cent excess, this leaves approximately 10 ml iodine monochloride ($\equiv$ 20 ml $0.1N$ $Na_2S_2O_3$) available for reaction with the oil.

$$\text{Sample of oil} \equiv 20 \text{ ml } 0.1N \text{ } Na_2S_2O_3$$
$$\equiv 20 \times 0.01269 \text{ g Iodine}$$
$$\equiv 0.2538 \text{ g Iodine}$$

For calculation purposes the oil is regarded as equivalent to 0.2 g of iodine.

Let the expected iodine value of the sample be x.

Then x g of iodine will be absorbed by 100 g of sample

$$\therefore \quad 0.2 \text{ g of iodine will be absorbed by } 100 \times \frac{0.2}{x} \text{ g of sample}$$
$$= \frac{20}{x} \text{ g of sample.}$$

Hence the weight of oil (in grammes) required for each determination is calculated by dividing the highest expected iodine value into 20.

If more than half the available halogen is absorbed, the test must be repeated on a smaller quantity of substance.

Pyridine Bromide Method

This method depends upon the reactions expressed by the following equations:

$$\text{C}{=}\text{C} + Br_2 \text{ (pyridine)} = -\overset{+}{\text{C}}-\overset{|}{\underset{Br}{\text{C}}}- + Br^- + \text{pyridine}$$

$$Br^- + -\overset{+}{\underset{Br}{\text{C}}}-\overset{|}{\underset{Br}{\text{C}}}- \longrightarrow \overset{|}{\underset{Br}{\text{C}}}-\overset{|}{\underset{Br}{\text{C}}}-$$

Excess bromine is destroyed by addition of potassium iodide, and the iodine so released titrated with sodium thiosulphate.

$$I^- + Br_2 \quad \rightarrow \quad IBr + Br^-$$
$$I^- + IBr \quad \rightarrow \quad I_2 + Br^-$$

$$\therefore \quad \text{C}{=}\text{C} \equiv Br_2 \equiv I_2 \equiv 2e \equiv 20\,000 \text{ ml } 0.1N$$

Method. Weigh the sample in a small specimen tube and place in a dry glass-stoppered bottle with 10 ml of carbon tetrachloride as in the iodine monochloride method. Add 25 ml of pyridine bromide solution, stopper and allow to stand in the dark for 10 minutes. Add aqueous potassium iodide solution and proceed as in the iodine monochloride method.

   Pyridine Bromide Reagent contains:

Pyridine	8 g
Sulphuric acid	10 g
Bromine	8 g
Glacial acetic acid to	1000 ml

Dissolve the pyridine and sulphuric acid in glacial acetic acid (20 ml) keeping the solution cool. Add bromine disolved in glacial acetic acid (20 ml) and dilute to 1000 ml with acetic acid.

Calculation of the Weight of Oil Required

Each 25 ml of pyridine bromide solution contains 0.20 g of bromine. Allowing for rather more than 100 per cent excess, this leaves 0.08 g of bromine (0.125 g iodine) for reaction with the oil.

Let the expected iodine value of the sample be x.
Then

$$x \text{ g of iodine will be absorbed by } 100 \text{ g of sample}$$

$$\therefore \quad 0.125 \text{ g of iodine will be absorbed by } \frac{100 \times 0.125}{x} \text{ g of sample}$$

$$= \frac{12.5}{x} \text{ g of sample}$$

Hence the weight of oil (in g) for each determination is calculated by dividing the highest expected Iodine Value into 12.5.

If more than half the available halogen is absorbed, the test must be repeated, using a smaller quantity of substance.

Halibut-liver Oil contains a high proportion of unsaponifiable matter, and two Iodine Values are determined, both by the pyridine bromide method. The first determination is on the oil itself and the second on the residue left in the determination of unsaponifiable matter using 1 g of oil. The iodine value of the glycerides is obtained by calculation from the two values using the formula

$$\text{Iodine Value} = \frac{100x - Sy}{100 - S}$$

where

x = Iodine Value of the oil
S = Percentage of unsaponifiable matter in the oil
y = Iodine Value of unsaponifiable matter

The above expression is derived as follows:
By definition,

   y g of Iodine are absorbed by 100 g of unsaponifiable matter

$$\therefore \quad \frac{Sy}{100} \text{ of Iodine are absorbed by } S \text{ g of unsaponifiable matter}$$

   (present in 100 g of oil).

But x g of Iodine are absorbed by 100 g of oil.

$\therefore$ Iodine absorbed by the glycerides in 100 g of oil $= x - \dfrac{Sy}{100}$

$$= \frac{100x - Sy}{100}$$

But weight of glycerides in 100 g of oil $= 100 - S$

$\therefore$ Iodine absorbed by 100 g of glycerides $= \dfrac{100x - Sy}{100} \times \dfrac{100}{100 - S}$

$\therefore$ Iodine Value of glycerides $= \dfrac{100x - Sy}{100 - S}$

POTASSIUM IODATE TITRATIONS

Potassium iodate is a fairly strong oxidising agent. Under suitable conditions it reacts quantitatively with both iodides and iodine. Iodate titrations can be performed in the presence of alcohol, saturated organic acids, and many other kinds of organic matter.

Potassium Iodide. *Determination of the percentage of* KI *by titration with potassium iodate*

This method is no longer the official method for the assay of potassium iodide.

If the concentration of hydrochloric acid does not exceed N, the reaction between potassium iodate and potassium iodide stops when the iodate has been reduced to free iodine:

$$IO_3^- + 5I^- + 6H^+ \rightarrow 3I_2 + 3H_2O \qquad (1)$$

In the presence of more concentrated hydrochloric acid (exceeding $4N$), the iodine produced in the above reaction is oxidised by iodate to the iodine cation, I^+. The high concentration of chloride ions leads to the formation of iodine monochloride, which is stabilised against hydrolysis by the hydrochloric acid present

$$2[I_2 \rightarrow 2I^+ + 2e]$$
$$IO_3^- + 6H^+ + 4e \rightarrow 3H_2O + I^+$$
$$5I^+ + 5Cl^- = 5ICl$$

This reaction may, therefore, be summarised as:

$$KIO_3 + 2I_2 + 6HCl \rightarrow KCl + 5ICl + 3H_2O \qquad (2)$$

Thus, the complete reaction between KIO_3 and KI in the presence of $4N$ HCl is expressed as follows by combining equations (1) and (2)

$$KIO_3 + 2KI + 6HCl \rightarrow 3KCl + 3ICl + 3H_2O \qquad (3)$$

In the presence of cyanide ions, iodine monochloride is converted to iodine cyanide which is less susceptible than the former to hydrolysis. Accordingly when cyanide ions are present in the solution, the acidity with respect to hydrochloric or sulphuric acid need only be just above N. Thus the use of large quantities of HCl is avoided.

The following equation applies in the presence of cyanide ions:

$$HIO_3 \text{ (from KIO}_3 \text{ and acid)} + 2HI + 3HCN \rightarrow 3ICN + 3H_2O$$
$$\therefore \quad 2KI \equiv KIO_3$$
$$\therefore \quad 332.0 \text{ g KI} \equiv 1000 \text{ ml } M$$
$$\therefore \quad 0.01660 \text{ g KI} \equiv 1 \text{ ml } 0.05M \text{ KIO}_3$$

The use of molar solutions of iodate. Iodate solutions are usually expressed in terms of molarity instead of the more usual normality. This is because potassium iodate solution is often used in the titration of solutions containing both iodide and free iodine and the equivalent of potassium iodate in its reaction with potassium iodide is different from its equivalent when considering its reaction with iodine:

$$KIO_3 \equiv 2I_2 \equiv 4e \, [\text{eq. 2}]$$
$$KIO_3 \equiv 2I^- \equiv I_2 \equiv 2e \, [\text{eq. 3}]$$

Therefore molar solutions of iodate are used to avoid confusion and ambiguity.

Method. Dissolve the sample (about 0.5 g), accurately weighed, in Water (about 50 ml). Add concentrated hydrochloric acid (60 ml) and titrate with *0.05M* potassium iodate solution. The solution will become brown as the titration proceeds due to the liberation of iodine.

$$HIO_3 + 5HI \rightarrow 3I_2 + 3H_2O$$

Then the solution becomes a lighter colour as titration is continued:

$$HIO_3 + 2I_2 + 5HCl \rightarrow 5ICl + 3H_2O$$

When the solution becomes pale yellow, add 1 ml of amaranth solution and continue the titration until the red colour changes to pale yellow.

Standardisation of Potassium Iodate Solutions

Because potassium iodate can be purchased in a high state of purity (99.9 per cent) a standard solution may be prepared by direct weighing. However, solutions may be standardised by one of the following methods:

(1) To a definite volume of potassium iodate solution add excess potassium iodide and dilute hydrochloric acid and titrate the liberated iodine with standard sodium thiosulphate solution. Conversely this method can be used to standardise sodium thiosulphate solutions.

(2) Use pure potassium iodide and perform the determination as described under the determination of potassium iodide.

Weak Iodine Solution. *Determination of the percentage w/v of I and KI*

The free iodine content is determined by titration with *0.1N* sodium thiosulphate solution. The total iodine and potassium iodide is determined on a separate portion of solution by acidification and titration with potassium iodate solution.

(a) *Titration with Sodium Thiosulphate Solution*

$$I_2 + 2Na_2S_2O_3 \rightarrow Na_2S_4O_6 + 2NaI$$
$$\therefore \quad 0.01269 \text{ g } I \equiv 1 \text{ ml } 0.1N \text{ } Na_2S_2O_3$$

(b) *Titration with Potassium Iodate Solution* (HCl and cyanide ions present)

$$2I_2 + HIO_3 + 5HCl \rightarrow 5ICl + 3H_2O$$
$$2HI \text{ (from KI)} + HIO_3 + 3HCl \equiv 3ICl + 3H_2O$$
$$\therefore \quad 0.02538 \text{ g } I \equiv 1 \text{ ml } 0.05M \text{ } KIO_3$$
$$\therefore \quad 0.01660 \text{ g } KI \equiv 1 \text{ ml } 0.05M \text{ } KIO_3$$

Calculation of I *and* KI *content in the solution.* The potassium iodate reacts with both the iodine and the potassium iodide whereas the sodium thiosulphate solution reacts only with the iodine.

From equation (a) $0.02538 \text{ g } I \equiv 2 \text{ ml } 0.1N \text{ } Na_2S_2O_3$
From equation (b) $0.02538 \text{ g } I \equiv 1 \text{ ml } 0.05M \text{ } KIO_3$

Therefore using the same weight of iodine, the burette reading of $0.1N$ $Na_2S_2O_3$ will be twice the burette reading of $0.05M$ KIO_3, the solutions being calculated as exactly $0.1N$ and exactly $0.05M$ respectively. Therefore if equal volumes of Weak Solution of Iodine are used for the sodium thiosulphate titration and for the potassium iodate titration, the volume of exactly $0.05M$ KIO_3 solution required by the free I_2 present is equal to half the volume of exactly $0.1N$ $Na_2S_2O_3$ required in the sodium thiosulphate titration.

$0.05M$ KIO_3 required for KI in sample =
$$\text{volume of } 0.05M \text{ } KIO_3 - \tfrac{1}{2} \text{ volume } 0.1N \text{ } Na_2S_2O_3$$

Method. Use 10 ml of sample for both sodium thiosulphate and the potassium iodate titrations. Drain the pipette for 30 seconds (ethanolic solution).

(a) *Sodium thiosulphate titration.* Dilute 10 ml with water (20 ml) and titrate with $0.1N$ sodium thiosulphate until colourless.

(b) *Potassium iodate titration.* Pipette the sample (10 ml), add Water (30 ml), and concentrated hydrochloric acid (50 ml), and titrate with $0.05M$ KIO_3 as described for Potassium Iodide.

Cognate Determinations

Aqueous Iodine Solution. Determination of the percentage w/v of I *and* KI. For these determinations 20 ml of the diluted sample are taken for the sodium thiosulphate titration and 10 ml of diluted sample for the potassium iodate titration. Therefore one quarter of the $0.1N$ (exact) $Na_2S_2O_3$ required must be subtracted from the $0.05M$ (exact) KIO_3 needed, to obtain the volume of $0.05M$ KIO_3 required by the potassium iodide.

Sodium Acetrizoate Injection. *Determination of the percentage of* $C_9H_5I_3NNaO_3$

The iodine is in organic combination and before it can be titrated with potassium iodate solution it must be converted to an ionised form by boiling with sodium hydroxide and zinc powder under reflux to produce zinc iodide.

Water is added and the solution filtered, and cooled. Concentrated hydro-chloric acid and solution of potassium cyanide are added and the liberated HI titrated with 0.05M KIO$_3$ in the usual manner.

Cognate Determinations

Iodised Oil Fluid Injection
Iothalamic Acid
Meglumine Diatrizoate Injection
Meglumine Iothalamate Injection
Sodium Iothalamate Injection

Phenylhydrazine Hydrochloride. *Determination of the percentage of* C$_6$H$_5$.NH.NH$_2$,HCl

This determination depends on the reactions expressed by the following equations:

$$3C_6H_5NHNH_2 + 2HIO_3 \rightarrow 3C_6H_5OH + 3N_2 + 3H_2O + 2HI$$
$$2HI + 3HCl + HIO_3 \rightarrow 3ICl + 3H_2O$$

Add

$$3C_6H_5NHNH_2 + 3HCl + 3HIO_3 \rightarrow 3C_6H_5OH + 3ICl + 3N_2 + 6H_2O$$

Divide by 3

$$C_6H_5NHNH_2 + HCl + HIO_3 \rightarrow C_6H_5OH + ICl + N_2 + 2H_2O$$
$$\therefore \quad C_6H_5NHNH_2,HCl \equiv HIO_3 \equiv 1000 \text{ ml } 0.05M \text{ KIO}_3$$
$$\therefore \quad 144.6 \text{ g } C_6H_5NHNH_2,HCl \equiv 1000 \text{ ml } M \text{ KIO}_3$$
$$\therefore \quad 0.007231 \text{ g } C_6H_5NHNH_2,HCl \equiv 1 \text{ ml } 0.05M \text{ KIO}_3$$

Method. Dissolve the sample (about 0.2 g), accurately weighed, in Water (50 ml) in a stoppered bottle. Add concentrated hydrochloric acid (50 ml) and titrate with 0.05M KIO$_3$ until the brown colour first produced becomes less intense. Add chloroform (5 ml) and continue the titration with vigorous shaking between each addition until the chloroform layer becomes colourless.

Cognate Determinations

Semicarbazide Hydrochloride. This determination is similar to that of phenylhydrazine hydrochloride.

$$3NH_2.CO.NH.NH_2 + 3HCl + 2HIO_3 \rightarrow 3NH_4Cl + 3CO_2 + 3N_2 + 3H_2O + 2HI$$
$$2HI + 3HCl + HIO_3 \rightarrow 3ICl + 3H_2O$$

Add

$$3NH_2.CO.NH.NH_2 + 6HCl + 3HIO_3 \rightarrow 3NH_4Cl + 3CO_2 + 6H_2O + 3ICl + 3N_2$$

Divide by 3

$$NH_2.CO.NH.NH_2 + 2HCl + HIO_3 \rightarrow NH_4Cl + CO_2 + 2H_2O + ICl + N_2$$
$$\therefore \quad NH_2.CO.NH.NH_2 \equiv HIO_3 \equiv 1000 \text{ ml } M \text{ KIO}_3 \equiv 20\ 000 \text{ ml } 0.05M \text{ KIO}_3$$

Stannous Chloride. *Determination of the percentage of* SnCl$_2$,2H$_2$O

This determination depends on the reactions expressed by the following equations:

$$6SnCl_2 + 12HCl + 2HIO_3 \rightarrow 6SnCl_4 + 6H_2O + 2HI$$
$$2HI + 3HCl + HIO_3 \rightarrow 3ICl + 3H_2O$$

Add

$$6SnCl_2 + 15HCl + 3HIO_3 \rightarrow 6SnCl_4 + 9H_2O + 3ICl$$

Divide by 3

$\therefore$ $2SnCl_2 + 5HCl + HIO_3 \rightarrow 2SnCl_4 + 3H_2O + ICl$

$\therefore$ $2SnCl_2,2H_2O \equiv HIO_3 \equiv 1000$ ml M KIO_3

$\therefore$ 2×225.6 g $SnCl_2,2H_2O \equiv 1000$ ml M KIO_3

$\therefore$ 0.02256 g $SnCl_2,2H_2O \equiv 1$ ml $0.05M$ KIO_3

AMMONiUM CERIC SULPHATE TITRATIONS

Ceric sulphate is a powerful oxidising agent in acid solution (1-$8N$). It is bright yellow in colour both in the solid state and in solution, and since the corresponding cerous salt formed by reduction is colourless, strong solutions are self indicating. However, usually $0.05N$ solutions are used and this concentration is too dilute for observation of the end point without the addition of a suitable indicator. The oxidation reaction is represented as follows:

$$Ce^{4+} + e \rightleftharpoons Ce^{3+}$$

As a practical oxidising agent ceric sulphate possesses a number of advantages over potassium permanganate. As the solutions are not so highly coloured, observation of the meniscus in burettes, pipettes, and other volumetric apparatus is less difficult. More important, the solutions are stable in air over long periods, they do not require protection from light and can even be heated for short periods without change in composition. Ceric sulphate, unlike potassium permanganate, can be used as an oxidising agent in the presence of high concentrations of hydrochloric acid.

Preparation of Volumetric Solutions of Ammonium Ceric Sulphate

Ammonium ceric sulphate $Ce(SO_4)_2.2(NH_4)_2SO_4,2H_2O$ contains not less than 95 per cent of the pure ceric salt, and volumetric solutions may be prepared by dissolving the appropriate weight (see below) in $2N$ sulphuric acid.

Since the oxidation reaction is given by:

$$Ce^{4+} + e \rightleftharpoons Ce^{3+}$$

$\therefore$ 632.57 g $Ce(SO_4)_22(NH_4)_2SO_4.2H_2O \equiv 1000$ ml N

$\therefore$ 31.63 g $Ce(SO_4)_22(NH_4)_2SO_4.2H_2O \equiv 1000$ ml $0.05N$ Ammonium Ceric Sulphate

Standardisation of $0.05N$ Ammonium Ceric Sulphate Solution

Prepare an $0.05N$ solution of AnalaR ferrous ammonium sulphate (19.61 g per litre). Pipette 20 ml of this solution into a 250 ml conical flask, add N sulphuric acid (20 ml) and 1 to 2 drops of o-phenanthroline-ferrous iron indicator. Titrate with approximately $0.05N$ ammonium ceric sulphate until the colour changes sharply to pale blue. [p. 102 for explanation.]

Acetomenaphthone. *Determination of the percentage of* $C_{15}H_{14}O_4$

This determination depends on the reactions expressed by the following equations:

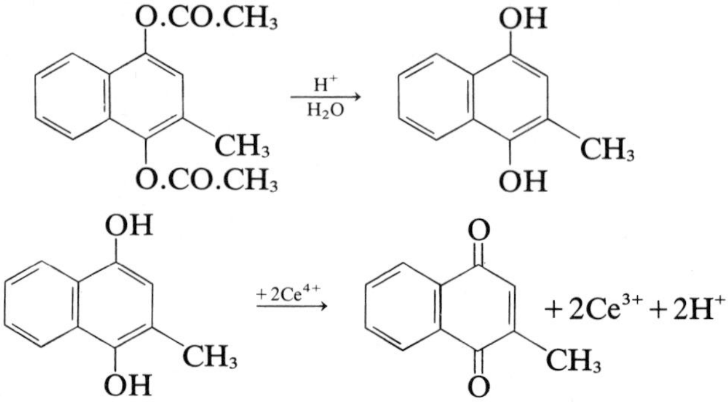

$$\therefore \quad 258.3 \text{ g } C_{15}H_{14}O_4 \equiv 2Ce^{4+} \equiv 2000 \text{ ml } N$$
$$\therefore \quad 0.006457 \text{ g } C_{15}H_{14}O_4 \equiv 1 \text{ ml } 0.05N \text{ Ammonium Ceric Sulphate}$$

Method. Accurately weigh the sample (about 0.2 g) into a 100 ml flat-bottomed Quickfit flask. Add glacial acetic acid (15 ml) and dilute hydrochloric acid (15 ml). Attach a water-cooled condenser and boil gently under reflux for 15 minutes. Cool, taking precautions to avoid oxidation of the 2-methylnapthaquinol, add 1 to 2 drops of o-phenanthroline-ferrous iron indicator and titrate with $0.05N$ ammonium ceric sulphate, taking the orange to blue colour change as the end point. Carry out a blank determination on the reagents repeating the above operations but omitting the sample.

Cognate Determinations

Acetomenaphthone Tablets

Ferrous Gluconate. *Determination of the percentage of* $C_{12}H_{22}O_{14}Fe$, *calculated with reference to the substance dried at 105° for five hours*

$$C_{12}H_{22}O_{14}Fe \equiv Fe \equiv e$$
$$\therefore \quad 446.2 \text{ g } C_{12}H_{22}O_{14}Fe \equiv 1000 \text{ ml } N$$
$$\therefore \quad 44.62 \text{ g } C_{12}H_{22}O_{14}Fe \equiv 1000 \text{ ml } 0.1N$$
$$\therefore \quad 0.04462 \text{ g } C_{12}H_{22}O_{14}Fe \equiv 1 \text{ ml } 0.1N \text{ Ammonium Ceric Sulphate}$$

Method. Accurately weigh the sample (1.5 g) into a conical flask and dissolve in a mixture of Water (75 ml) and dilute sulphuric acid (25 ml). Titrate the combined filtrate and washings immediately with $0.1N$ ammonium ceric sulphate using o-phenanthroline-ferrous iron solution as indicator.

Cognate Determinations

Ascorbic Acid Tablets
Ferrous Fumarate Tablets
Ferrous Gluconate Tablets
Ferrous Succinate, and Tablets
Ferrous Sulphate, and *Tablets*

Iron-Dextran Injection. *Determination of the percentage of* Fe^{2+}

Iron is completely reduced to the ferrous state by passing the injection diluted with acid through a Jones reductor (p. 189) prepared with electrolytically-reduced cadmium, and the eluate titrated with $0.1N$ ammonium ceric sulphate using o-phenanthroline-ferrous iron solution as indicator.

Cognate Determination

Iron-Sorbitol Injection

Ammonium Ceric Sulphate. *Determination of the percentage of* $Ce(SO_4)_2.2(NH_4)_2SO_4,2H_2O$

The determination depends upon the oxidation of sodium arsenite to arsenate by the ceric component of the salt, using ferroin as indicator. The reaction is slow in the absence of a catalyst, and osmium tetroxide is used to increase the reaction rate.

$$2Ce^{4+} + 2e \rightleftharpoons 2Ce^{3+}$$
$$NaAsO_2 + 2H_2O + 2e \rightarrow NaH_2AsO_4 + 2H^+$$
$$\therefore \quad 2 \times 632.57 \text{ g } Ce(SO_4)_2.2(NH_4)_2SO_4,2H_2O \equiv 2000 \text{ ml } N \text{ } NaAsO_2$$
$$\therefore \quad 0.06326 \text{ g } Ce(SO_4)_2.2(NH_4)_2SO_4,2H_2O \equiv 1 \text{ ml } 0.1N \text{ } NaAsO_2$$

Method. Dissolve the sample (about 1 g accurately weighed) in dilute sulphuric acid (50 ml). Add 1 per cent osmium tetroxide solution (0.1 ml) and titrate with $0.1N$ sodium arsenite using ferroin as indicator.

TITANOUS CHLORIDE TITRATIONS

Titanous chloride is a reducing agent, and can be used to determine ferric salts, azo dyes, nitro and nitroso compounds, and quinones. It is rapidly oxidised in air, especially in the presence of sunlight, and consequently the volumetric solution must be stored under hydrogen in dark-coloured bottles. A suitable apparatus for storage and titration is shown in Fig. 21. An atmosphere of carbon dioxide or hydrogen must be maintained above the liquid during the course of any titration to prevent oxidation of the reagent. Titanous sulphate solutions are more stable than titanous chloride solutions and it is recommended that the sulphate be used in preference to the chloride.

Preparation of 0.1N Titanous Chloride

Boil commercial titanous chloride solution (15 to 20 per cent; 400 ml) for 1 minute with concentrated hydrochloric

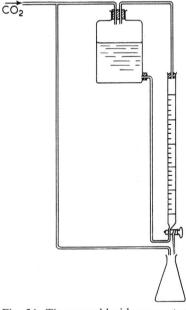

Fig. 21. Titanous chloride apparatus

acid (100 ml) to remove traces of sulphides (decomposed to H_2S). Displace the air over the solution with carbon dioxide, and cool. Transfer to an aspirator, and dilute to 5 litres with freshly boiled and cooled water. Fill the burette and displace air above the solution with carbon dioxide.

Standardisation of 0.1N Titanous Chloride Solution

Titanous chloride and sulphate solutions must be standardised immediately before use, against AnalaR ferric ammonium sulphate (AnalaR iron is better).

Method. Prepare an 0.1N solution of ferric ammonium sulphate in 10 per cent sulphuric acid. Pipette this solution (25 ml) into a conical flask, pass a current of carbon dioxide through the liquid to replace the air in the flask, and maintain a stream of carbon dioxide through the liquid throughout the titration. Add 10 per cent ammonium thiocyanate solution (10 ml). The solution will become an intense red colour. Run in the titanous chloride solution from the burette until the red colour becomes less intense. Then titrate slowly allowing about 10 seconds between each drop near the end point, because the reaction is slow when only a little ferric salt remains. Titrate until the red colour disappears. The stream of carbon dioxide must be maintained throughout the titration. The standardisation depends upon the reaction expressed by the following equation:

$$Fe^{3+} + Ti^{3+} \rightarrow Fe^{2+} + Ti^{4+}$$

Methylene Blue. *Determination of the percentage of* $C_{16}H_{18}ClN_3S,2H_2O$

This determination depends upon the reactions expressed by the following equations:

$$C_{16}H_{18}ClN_3S \equiv 2e$$
$$\therefore \quad 355.9 \text{ g } C_{16}H_{18}ClN_3S,2H_2O \equiv 2000 \text{ ml } N \text{ TiCl}_3$$
$$\therefore \quad 0.01599 \text{ g } C_{16}H_{18}ClN_3S,2H_2O \equiv 1 \text{ ml } N \text{ TiCl}_3$$

Method. Dissolve the sample about 0.5 g, accurately weighed in water (100 ml). Add hydrochloric acid (10 ml), heat to boiling, and pass a current of CO_2 through the solution. Titrate with 0.1N TiCl$_3$ until the clear blue colour disappears, leaving a reddish-grey solution.

Cognate Determinations

Indigocarmine
Nitranilic Acid

Determinations Involving the Use of Excess Titanous Chloride and Back Titration with Ferric Ammonium Sulphate

This method is employed when the substance to be determined is not so easily reduced as the substances previously mentioned. It is used for the determination of nitro and nitroso compounds and certain organic dyestuffs. A solution of the substance is made, CO_2 passed in to remove air, and a current maintained throughout the determination. Excess titanous chloride is added, the solution boiled for a certain time to effect reduction of the substance, cooled, and the excess titanous chloride titrated with $0.1N$ ferric ammonium sulphate using ammonium thiocyanate as indicator. A blank determination is then performed.

The dyestuffs *Brilliant Green* and *Crystal Violet* are determined by this method.

9 Argentimetric Titrations

Quantitative precipitation can be used for volumetric determinations, provided that the point at which precipitation is complete can be determined. Thus when silver nitrate solution is run into a solution of sodium chloride a precipitate of silver chloride is formed:

$$NaCl + AgNO_3 \rightarrow AgCl \downarrow + NaNO_3$$

The end point is the point at which all the chloride has been precipitated as silver chloride. However, detection of complete precipitation by observing the point at which addition of reagent causes no further precipitation is too tedious in practice, and use is usually made of a chemical reaction to give a coloured precipitate or coloured solution at the end point. For instance in the above example potassium chromate is added to the solution. As the silver nitrate solution is added the chloride is precipitated as silver chloride, but when all the chloride has been precipitated, the next drop of silver nitrate causes precipitation of red chromate indicating that the end point has been reached (p. 91).

Preparation and Standardisation of *0.1N* Silver Nitrate Solution

The standardisation depends upon the reactions expressed by the following equations:

$$AgNO_3 + NaCl \rightarrow AgCl \downarrow + NaNO_3$$
$$\therefore \quad AgNO_3 \equiv NaCl \equiv H$$
$$\therefore \quad 169.89 \text{ g } AgNO_3 \equiv 58.45 \text{ g } NaCl \equiv 1000 \text{ ml } N$$
$$\therefore \quad 0.01699 \text{ g } AgNO_3 \equiv 0.005845 \text{ g } NaCl \equiv 1 \text{ ml } \textit{0.1N } AgNO_3$$

Method. AnalaR silver nitrate contains 99.9 to 100 per cent $AgNO_3$ and can therefore be used as a primary standard. An *0.1N* solution can be prepared by dissolving 16.989 g of silver nitrate in Water and adjusting to 1000 ml. A solution may be standardised against a standard solution of sodium chloride using potassium chromate as indicator (Mohr's Method).

Dry some AnalaR sodium chloride at about 300° for 2 hours, cool in a desiccator and use to prepare an accurate *0.1N* solution of sodium chloride. Pipette 25 ml of the *0.1N* sodium chloride solution into a conical flask, add potassium chromate solution (5 per cent w/v; 1 ml), and titrate with the silver nitrate solution. *Do not dilute* with Water during the titration, but shake continuously. If these instructions are carried out, coagulation of the precipitated silver chloride usually occurs just before the end point. Continue to add the silver nitrate dropwise shaking the flask well, until a permanent *faint reddish-brown* colour is obtained.

Mohr's method is only applicable if the solution to be titrated is neutral but acid solutions may be determined if they are first neutralised with chloride-free calcium carbonate or sodium bicarbonate.

DIRECT TITRATION WITH SILVER NITRATE

Sodium Chloride

Sodium chloride may be determined using the method described for the standardisation of silver nitrate solutions (above) but approximately 0.2 g accurately weighed portions should be titrated. The sample to be titrated is not dried, but the percentage of NaCl is calculated with reference to the substance dried at 130°. A separate portion of the sample is dried in a widemouthed weighing bottle in order to get the necessary data for the calculation (see Chalk Determination for method of calculation).

Cognate Determinations

Potassium Chloride Injection
Potassium Chloride Tablets
Sodium Chloride and Dextrose Injection
Sodium Chloride Injection

Acetyl chloride. This is determined in two ways: (*a*) by hydrolysis with excess sodium hydroxide and back titration with standard H_2SO_4:

$$CH_3COCl + 2NaOH \rightarrow CH_3COONa + NaCl + H_2O$$

and (*b*) by titration of the resulting neutral solution with silver nitrate solution, as for Sodium Chloride.

Potassium Cyanide. *Determination of the percentage of KCN*

This determination depends upon the reaction expressed by the following equations:

$$AgNO_3 + KCN \rightarrow AgCN + KNO_3 \qquad (1)$$

The silver cyanide first formed dissolves in the potassium cyanide present in the solution:

$$AgCN + KCN \rightarrow KAg(CN)_2 \qquad (2)$$

When all the potassium cyanide has been converted to $KAg(CN)_2$ the next drop of silver nitrate solution would react thus:

$$KAg(CN)_2 + AgNO_3 \rightarrow 2AgCN \downarrow + KNO_3$$

However, because dilute solution of ammonia is added to the solution the precipitate of silver cyanide is not formed since silver cyanide is soluble in ammonia solution:

$$AgCN + 2NH_3 \rightarrow Ag(NH_3)_2CN$$

Therefore one drop of solution of potassium iodide is added as indicator to detect the completion of the reaction (2) above—a precipitate of silver iodide being formed with the next drop of silver nitrate added. Silver iodide is soluble in solutions of simple cyanides but is insoluble in solutions of complex cyanides.

∴ from eqs. (1) and (2):

$$2KCN + AgNO_3 \rightarrow KAg(CN)_2 + KNO_3$$
$$\therefore \quad 2KCN \equiv AgNO_3$$
$$\therefore \quad 2 \times 65.12 \text{ g } KCN \equiv 1000 \text{ ml } N \text{ } AgNO_3$$
$$\therefore \quad 0.01302 \text{ g } KCN \equiv 1 \text{ ml } 0.1N \text{ } AgNO_3$$

Method. Dissolve the sample (about 0.5 g) accurately weighed in Water (50 ml). Add dilute solution of ammonia (5 ml) and potassium iodide solution (1 drop). Titrate with *0.1N* AgNO₃ solution until a faint *permanent* precipitate appears which may be readily observed by viewing the solution against a black paper background.

Chloral Hydrate. *Determination of the percentage of* $CCl_3.CH(OH)_2$
This determination depends upon the reactions expressed by the following equation:

$$CCl_3CH(OH)_2 + NaOH \rightarrow CHCl_3 + HCOONa + H_2O \qquad (1)$$
$$\therefore \quad CCl_3.CH(OH)_2 \equiv NaOH \equiv H$$
$$\therefore \quad 165.4 \text{ g } C_2H_3O_2Cl_3 \equiv 1000 \text{ ml } N$$
$$\therefore \quad 0.1654 \text{ g } C_2H_3O_2Cl_3 \equiv 1 \text{ ml } N \text{ } NaOH$$

Since the chloroform produced also reacts with the alkali to some extent, addition of alkali followed by back titration does not give the correct result.

$$CHCl_3 \xrightarrow{\text{alkali}} H.COOH + 3HCl$$
$$\downarrow 4NaOH \qquad (2)$$
$$H.COONa + 3NaCl + 4H_2O$$

Therefore in reaction (2) $CHCl_3 \equiv 4NaOH$
The ionised chloride produced in this side reaction can be determined by titration with *0.1N* silver nitrate solution, and a correction applied to the alkali titration reading to allow for this side reaction. From eq. (2)

$$CHCl_3 \equiv 4NaOH \equiv 4000 \text{ ml } N \text{ solution.}$$

Also:
$$3NaCl \equiv 30\ 000 \text{ ml } 0.1N$$

Thus $\frac{2}{15}$ of the volume of *0.1N* AgNO₃ required ($\frac{2}{15}$ of 30 000 = 4000) will give the volume of *N* NaOH which reacted with chloroform according to eq. (2).
The sample is dissolved in excess *N* NaOH. After standing for two minutes (reaction 1), the excess alkali is back titrated with *N* H₂SO₄ using phenolphthalein as indicator. Because reaction (2) has occurred to some extent, the neutral solution is titrated with *0.1N* AgNO₃ using potassium chromate as indicator. A volume equivalent to $\frac{2}{15}$ of the amount of *0.1N* (exact) AgNO₃ required gives the volume of *N* NaOH taken up by reaction (2). This value plus the amount of *N* H₂SO₄ required by the previous back titration when subtracted from the volume of *N* NaOH originally added, gives the amount of *N* NaOH required by the sample in the reaction expressed by eq. (1).

Cognate Determination

Dichloralphenazone

Potassium Iodide

Bromides and iodides cannot be titrated directly with silver nitrate by Mohr's method, since the colour of the precipitate interferes with the end point. Potassium iodide can, however, be determined by direct titration with silver nitrate using iodine and starch as indicator.

$$KI + AgNO_3 \rightarrow KNO_3 + AgI \downarrow$$
$$KI \equiv AgNO_3 \equiv H$$
$$\therefore \quad 166 \text{ g KI} \equiv 1000 \text{ ml } N \text{ AgNO}_3$$
$$\therefore \quad 0.0166 \text{ g KI} \equiv 1 \text{ ml } 0.1N \text{ AgNO}_3$$

The titration is simply a double decomposition. At the end-point, excess silver nitrate reduces the indicator, iodine, to iodide ion, destroying the purple starch-iodide complex and leaving a pale yellow solution.

Method. Dissolve the sample (about 0.3 g), accurately weighed in water (10 ml). Add iodide-free starch solution (10 ml, containing 1 drop of ethanolic iodine solution in 50 ml), and titrate with $0.1N$ AgNO$_3$ until the blue colour is dispersed and a pale yellow solution remains.

AMMONIUM THIOCYANATE TITRATION OF SILVER SALTS

Ammonium thiocyanate reacts with silver nitrate in nitric acid solution:

$$NH_4SCN + AgNO_3 \rightarrow AgSCN \downarrow + NH_4NO_3$$

The thiocyanate solution is always used in the burette and is run into the silver nitrate solution which has been acidified with nitric acid (nitrous acid-free, because nitrous acid gives a red colour with thiocyanic acid). The end point is detected by use of ferric ammonium sulphate since ferric ions give a deep red colour (ferric thiocyanate) with a trace of thiocyanate ion. The temperature of the solution must be kept below 25° because at higher temperatures the colour of the ferric thiocyanate complex fades.

$$\therefore \quad NH_4SCN \equiv AgNO_3 \equiv H$$

Standardisation of Approximately *0.1N* NH₄SCN Solution

Ammonium thiocyanate solution is standardised against $0.1N$ silver nitrate solution.

Method. Pipette a standard $0.1N$ AgNO$_3$ solution (25 ml) into a conical flask. Add concentrated nitric acid (5 ml), and ferric ammonium sulphate solution (10 per cent w/v; 3 ml). Dilute with Water (100–150 ml) since this sharpens the end point. Titrate with ammonium thiocyanate solution. A white precipitate of silver thiocyanate is formed as the thiocyanate solution is added and then a reddish-brown colour is produced which disappears upon shaking leaving a white precipitate of silver thiocyanate. The end point is the permanent faint reddish brown colour which does not disappear upon shaking.

AMMONIUM THIOCYANATE TITRATION OF MERCURY COMPOUNDS

The determination of mercury salts with ammonium thiocyanate depends upon the reactions expressed by the following equation:

$$Hg(NO_3)_2 + 2NH_4SCN \rightarrow Hg(SCN)_2 + 2NH_4NO_3$$

This reaction is quantitative in presence of nitric acid and the end point can be detected by using ferric ammonium sulphate as indicator.

Mercuric Acetate. *Determination of the percentage of* $C_4H_6O_4Hg$

$$(CH_3.CO_2)_2.Hg + 2NH_4SCN \rightarrow Hg(SCN)_2 + 2CH_3.CO_2NH_4$$
$$\therefore \quad 318.8 \text{ g } C_4H_6O_4Hg \equiv 2NH_4SCN \equiv 2000 \text{ ml } N$$
$$\therefore \quad 0.01594 \text{ g } C_4H_6O_4Hg \equiv 1 \text{ ml } 0.1N \text{ NH}_4SCN$$

Method. Dissolve the sample (0.7 g) in Water (40 ml) and dilute nitric acid (5 ml). Add ferric alum indicator (5 ml) and titrate with *0.1N* NH₄SCN, taking the first permanent brownish-red colour as the end point.

Cognate Determinations

Mercuric Nitrate
Mercuric Sulphate
Mercury. This determination depends on the reaction expressed by the following equations:

$$Hg \rightarrow Hg(NO_3)_2$$
$$Hg(NO_3)_2 + 2NH_4SCN \rightarrow Hg(SCN)_2 + 2NH_4NO_3$$

The mercury is dissolved in a solution of nitric acid (1:1), and heated gently until colourless. The solution is diluted with water and titrated with *0.1N* NH₄SCN using ferric ammonium sulphate as indicator.

Sulthiame. The sample is precipitated with standard mercuric perchlorate solution, and the excess mercuric perchlorate titrated with *0.1N* NH₄SCN using ferric ammonium sulphate as indicator.

Yellow Mercuric Oxide. *Determination of the percentage of* HgO

The determination depends upon the reactions expressed by the following equations:

$$HgO + 2HNO_3 \rightarrow Hg(NO_3)_2 + H_2O$$
$$Hg(NO_3)_2 + 2NH_4SCN \rightarrow Hg(SCN)_2 + 2NH_4NO_3$$
$$\therefore \quad HgO \equiv Hg(NO_3)_2 \equiv 2NH_4SCN \equiv 2000 \text{ ml } N$$
$$\therefore \quad 216.6 \text{ gHgO} \equiv 2000 \text{ ml } N$$
$$\therefore \quad 0.01083 \text{ g HgO} \equiv 1 \text{ ml } 0.1N \text{ NH}_4SCN$$

Method. Dissolve the sample (about 0.4 g), accurately weighed, in concentrated nitric acid (5 ml) and Water (10 ml) and dilute to about 150 ml with Water. Titrate with *0.1N* NH₄SCN using ferric ammonium sulphate as indicator. The end point is a faint brownish-red colour which does not disappear on shaking.

Mersalyl Acid. *Determination of the percentage of* Hg

The mercury must be freed from organic combination before it can be titrated with ammonium thiocyanate. The sample is refluxed with formic acid and zinc dust. A zinc amalgam is obtained. The excess of zinc powder and the amalgam are filtered off, washed with water and dissolved in nitric acid. Nitrous acid and reducing substances are removed by the addition of a little urea followed by sufficient $0.1N$ potassium permanganate to produce a faint pink colour. The slight excess of potassium permanganate is removed with a few drops of hydrogen peroxide solution, and the mercuric nitrate solution titrated with $0.1N$ NH_4SCN using ferric ammonium sulphate as indicator.

$$\longrightarrow Hg/Zn \longrightarrow Hg(NO_3)_2$$

O.CH₂COOH ... CO.NH.CH₂.CH(OCH₃)CH₂.HgOH

$$Hg(NO_3)_2 + 2NH_4SCN \rightarrow Hg(SCN)_2 + 2NH_4NO_3$$

Cognate Determinations

Mersalyl Injection. Determine as for Mersalyl Acid.
Phenylmercuric Nitrate. Determine as for Mersalyl Acid.
Thiomersal. Organic matter is destroyed by heating with concentrated sulphuric acid.

AMMONIUM THIOCYANATE—SILVER NITRATE TITRATIONS (VOLHARD'S METHOD)

Ammonium thiocyanate solution is used in conjunction with $0.1N$ silver nitrate solutions in the assay of substances which react with silver nitrate but which cannot be determined by direct titration with silver nitrate solutions. Excess standard silver nitrate solution is added together with concentrated nitric acid, and the excess silver nitrate titrated with $0.1N$ ammonium thiocyanate solution. This is called Volhard's method.

Bromides and iodides can be determined by Volhard's method without filtering off the silver halide formed, but in the case of chlorides it is usual either to filter off the silver chloride or to coagulate the precipitate by means of nitrobenzene or preferably dibutyl phthalate, which is non-toxic, because silver chloride reacts slowly with ammonium thiocyanate:

$$AgCl + NH_4SCN \rightarrow AgSCN \downarrow + NH_4Cl$$

and makes the end point rather flat since the end point involves the production of red ferric thiocyanate complex with the thiocyanate ions.

Ammonium Chloride. *Determination of the percentage of* NH_4Cl

An aqueous solution of this substance is slightly acid and therefore the end point is indistinct using direct titration with silver nitrate solution. Therefore Volhard's method is used. The determination (no longer official; see p. 138) depends upon the reactions expressed by the following equations:

$$NH_4Cl + AgNO_3 \rightarrow AgCl \downarrow + NH_4NO_3$$
$$AgNO_3 + NH_4SCN \rightarrow AgSCN \downarrow + NH_4NO_3$$
$$\therefore \quad NH_4Cl \equiv AgNO_3 \equiv H$$
$$\therefore \quad 53.5 \text{ g } NH_4Cl \equiv 1000 \text{ ml } N$$
$$\therefore \quad 0.00535 \text{ g } NH_4Cl \equiv 1 \text{ ml } 0.1N \text{ AgNO}_3$$

Method. Dissolve the sample (about 0.2 g), accurately weighed, in Water (35 ml), add dilute nitric acid (15 ml), dibutyl phthalate (5 ml) and *0.1N* silver nitrate (50 ml by pipette) and shake vigorously for half to one minute. Add ferric ammonium sulphate solution (5 ml) and titrate with *0.1N* NH₄SCN shaking well until a reddish-brown colour, which does not fade in five minutes, is obtained.

Cognate Determinations

Anhydrous Aluminium Chloride
Carbromal
Compound Sodium Lactate Injection. Determination of total Cl.
Cyclophosphamide Injection. Determination of NaCl.
Potassium Chloride
Propantheline Tablets
Sodium Chloride
Sodium Chloride Tablets
Thiamine Hydrochloride. The total Cl content is determined by Volhard's method (after filtration because the halide is a chloride). Chlorine present as HCl is determined by titration with standard sodium hydroxide solution using bromothymol blue as indicator.
2,3,5-Triphenyltetrazolium Chloride

Barium Chloride. *Determination of percentage of* $BaCl_2,2H_2O$

This determination depends upon the reactions expressed by the following equation:

$$BaCl_2 + 2AgNO_3 \rightarrow Ba(NO_3)_2 + 2AgCl \downarrow$$
$$\therefore \quad BaCl_2 \equiv 2AgNO_3 \equiv 2000 \text{ ml } N$$
$$\therefore \quad 244.4 \text{ g } BaCl_2,2H_2O \equiv 20\ 000 \text{ ml } 0.1N$$
$$\therefore \quad 0.01222 \text{ g } BaCl_2,2H_2O \equiv 1 \text{ ml } 0.1N \text{ AgNO}_3$$

Method. Accurately weigh the sample (about 0.5 g) into a 250 ml stoppered titration flask. Dissolve in Water (50 ml), add concentrated nitric acid (10 ml), *0.1N* silver nitrate solution (50 ml), and dibutyl phthalate (3 ml). Stopper and shake vigorously for one minute, to coat the particles of silver chloride with dibutyl phthalate. Titrate with *0.1N* NH₄SCN using ferric alum as indicator.

Chlorbutol. *Determination of the percentage of* $CCl_3.C(CH_3)_2.OH$

Organically combined chlorine is converted by hydrolysis with sodium hydroxide to ionic chloride which can be determined by Volhard's method, in the presence of nitrobenzene.

$$CCl_3.C(CH_3)_2.OH + 3NaOH \rightarrow C(OH)_3.C(CH_3)_2.OH + 3NaCl$$

$$\downarrow NaOH$$

$$(CH_3)_2.C(OH).COONa$$
$$3NaCl + 3AgNO_3 \rightarrow 3NaNO_3 + 3AgCl$$
$$\therefore \quad C_4H_7OCl_3 \equiv 3NaCl \equiv 3AgNO_3 \equiv 3H \equiv 3000 \text{ ml } N$$
$$\therefore \quad 186.5 \text{ g } C_4H_7OCl_3 \equiv 3000 \text{ ml } N$$
$$\therefore \quad 0.006216 \text{ g } C_4H_7OCl_3 \equiv 1 \text{ ml } 0.1N \text{ AgNO}_3$$

3,5-Dinitrobenzoyl Chloride

This is required to contain not less than 98 per cent $C_7H_3O_5N_2Cl$ as deter-mined by the following methods:

(*a*) by titration in pyridine with N NaOH to phenolphthalein, when the following reaction occurs:

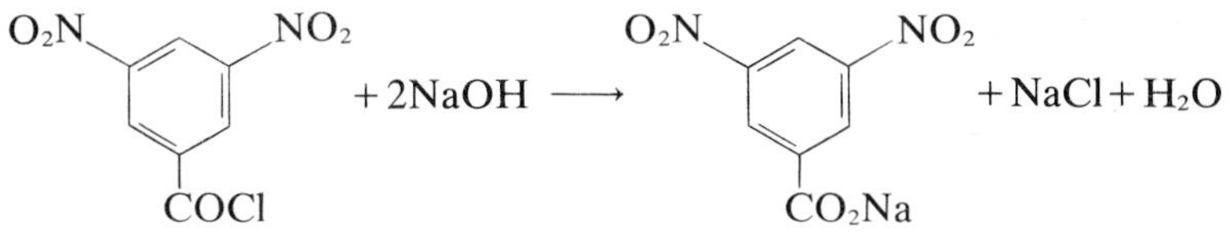

The pyridine does not interfere with the titration, because it does not give an alkaline reaction with phenolphthalein.

(*b*) Titration of the neutralised liquid from (*a*), which contains ionised chloride, by Volhard's method.

Cognate Determination

Nitrobenzoyl Chloride

Dibromopropamidine Isethionate

Organically combined bromine is converted to bromide ion by refluxing with sodium and amyl alcohol, and then titrated by Volhard's method.

Ethionamide

Treatment with acid decomposes the thioamide link. Neutralisation with ammonia gives ammonium sulphide. Addition of silver nitrate gives a precipi-tate of silver sulphide. Excess silver nitrate in the filtrate is titrated with $0.1N$ NH$_4$SCN.

Gamma Benzene Hexachloride

A preliminary hydrolysis is carried out with ethanolic potassium hydroxide, to convert organically combined chlorine to potassium chloride. The latter is determined by Volhard's method after acidification of the hydrolysate with nitric acid.

Cognate Determinations

Chlorambucil
Chlorambucil Tablets
Dicophane
Mustine Hydrochloride
Triphenylmethyl Chloride

Potassium Chlorate. *Determination of the percentage of* $KClO_3$

This determination depends upon the reduction of the chlorate to chloride by means of nitrous acid and then the determination of the chloride by Volhard's method with filtration.

$$KClO_3 \equiv KCl \equiv AgNO_3 \equiv H$$
$$\therefore \quad 122.6 \text{ g } KClO_3 \equiv 1000 \text{ ml } N$$
$$\therefore \quad 0.01226 \text{ g } KClO_3 \equiv 1 \text{ ml } 0.1N \text{ } AgNO_3$$

Method. Dissolve the sample (about 0.3 g), accurately weighed, in Water (10 ml) in a stoppered bottle. Add sodium nitrite (1 g) dissolved in Water (10 ml) followed by nitric acid (20 ml). Stopper the bottle and set aside for 10 minutes. The chlorate is reduced to chloride. Add Water (100 ml) and a little potassium permanganate solution until the solution is pale pink, to oxidise the nitrous acid. Decolourise with a trace of ferrous sulphate and add urea (0.1 g) to make sure that nitrous acid is completely absent. Add $0.1N$ $AgNO_3$ (30 ml), filter off the silver chloride, wash and titrate the filtrate and washings with $0.1N$ NH_4SCN using ferric ammonium sulphate as indicator.

Triclophos Sodium. *Determination of the percentage of* Cl

The organically combined chlorine is converted to ionised chloride by ignition with anhydrous sodium carbonate before being titrated by Volhard's method.

$$C_2H_3Cl_3NaO_4P \rightarrow 3NaCl(\equiv 3AgNO_3)$$

Cognate Determination

Cloxacillin Sodium. Determination of Cl

Sodium Bromide

This is determined by Volhard's method, without filtration, but allowance must be made for any chloride present as impurity. Because the limit test involves using Volhard's method the actual determination also consists of a Volhard's titration, so that by subtraction of the former from the latter the volume of $0.1N$ $AgNO_3$ required by the bromide can be found.

Limit test for chloride. Dissolve the sample (about 1 g), accurately weighed, in water (75 ml) and concentrated nitric acid (25 ml). Boil for one minute the bromine. Cool the drawing a rapid current of air through the solution to remove the bromine. Cool the solution for 20 minutes. Passing air through the solution during this period converts the bromide to elemental bromine which is swept away by the current of air.

$$2HBr + \bar{O} \rightarrow H_2O + Br_2 \downarrow$$

The chloride present is not oxidised under these conditions which must be rigidly adhered to. The HCl in the resultant solution is determined by Volhard's method. Add $0.1N$ AgNO$_3$ (5 ml) and dibutyl phthalate (5 drops) and back titrate the excess with $0.1N$ NH$_4$SCN using ferric ammonium sulphate as indicator. Do not shake vigorously here at the end point because the presence of chloride causes the end point to fade. (See p. 92 for further details of method and apparatus).

Determination of the percentage of NaBr. The determination depends upon the reaction expressed by the following equation:

$$NaBr + AgNO_3 \rightarrow AgBr \downarrow + NaNO_3$$
$$\therefore \quad NaBr \equiv AgNO_3 \equiv H$$
$$\therefore \quad 102.9 \text{ g NaBr} \equiv 1000 \text{ ml } N$$
$$\therefore \quad 0.01029 \text{ g NaBr} \equiv 1 \text{ ml } 0.1N \text{ AgNO}_3$$

Method. Dissolve the sample (about 0.4 g), accurately weighed, in Water (100 ml) and concentrated nitric acid (5 ml). Add $0.1N$ AgNO$_3$ (50 ml) and back titrate with $0.1N$ NH$_4$SCN using ferric ammonium sulphate as indicator. Correct for the chloride present as found by the chloride limit test.

Suppose

    1 g sample contains total halide $\equiv$ x ml $0.1N$ AgNO$_3$
and
    1 g sample contains chloride $\equiv$ y ml $0.1N$ AgNO$_3$
    $\therefore$ 1 g sample contains NaBr $\equiv (x - y)$ ml $0.1N$ AgNO$_3$

Ammonium Reineckate. *Determination of the percentage of* NH$_4$-[Cr(NH$_3$)$_2$(SCN)$_4$],H$_2$O

The sample is decomposed by boiling with sodium hydroxide solution until a green precipitate of chromic hydroxide is obtained:

$$NH_4[Cr(NH_3)_2(SCN)_4] + 4NaOH \rightarrow 4NaSCN + 3NH_3 + Cr(OH)_3 + H_2O$$

After acidification with dilute nitric acid excess of silver nitrate is added and the excess back titrated with standard ammonium thiocyanate.

Potassium Thiocyanate. *Determination of the percentage of* KSCN

This is determined by adding excess standard silver nitrate solution and back titrating with $0.1N$ NH$_4$SCN.

$$KSCN + AgNO_3 \rightarrow KNO_3 + AgSCN$$

Thiourea. *Determination of the percentage of* CH$_4$N$_2$S

This substance is determined by a modified Volhard method, the sulphur being precipitated as silver sulphide.

$$NH_2.CS.NH_2 \equiv Ag_2S \equiv 2H$$

A solution of the substance is heated with excess $0.1N$ AgNO$_3$ and solution of ammonia, cooled, and excess dilute nitric acid added. This causes all the sulphur present to be precipitated as silver sulphide. The precipitate is filtered off and washed and the excess silver nitrate in the filtrate and washings titrated with $0.1N$ NH$_4$SCN using ferric ammonium sulphate as indicator.

10 Complexometric Analysis

INTRODUCTION

A complexing agent is any electron-donating ion or molecule, usually called a *ligand*, which, by its ability to form one or more covalent or dative bonds with the metal ion, produces a complex which has different properties from those of the free metal ion. Thus, the metal may not be precipitated from the complex by the usual metal ion precipitants. The decomposition potential of the complex at the dropping mercury cathode, also, is usually more negative than that of the free metal ion and, where the ion can exist in different valency states, the redox potential is reduced if the complexing agent combines more avidly with the cation of higher valency. Moreover, if the higher valency state of the ion is stabilised by complex formation, it will be more difficult to reduce. For example, if potassium iodide solution is acidified with acetic acid and treated with cupric sulphate solution, iodine is liberated and cuprous iodide precipitated. If, however, sodium tartrate is added to the cupric sulphate solution first, there is no reduction to the cuprous state and no iodine liberated. This is because the cupric ion is stabilised as the cupritartrate complex, which is not reduced by iodide. Similarly, an acidic solution of ferric chloride is not reduced by potassium iodide if excess sodium edetate (see below) is first added. Moreover, ferrous iron in the presence of excess edetate is actually oxidised by iodine to the ferric complex.

The stability of complexes varies greatly, and the greater the stability, the more marked will be the difference in properties from those of the original cations. When potassium iodide is added to a suspension of red mercuric iodide, the latter will pass into solution as the colourless complex $2K^+(HgI_4)^{2-}$, the dissociation of which is so slight that there are insufficient mercuric ions to precipitate as mercuric oxide in the presence of alkali. Potassium ferricyanide $K_3[Fe(CN)_6]$, in which the iron is in the ferric state, does not yield ferric hydroxide in the presence of alkali in spite of the sparing solubility of the latter (solubility product $= 1.1 \times 10^{-36}$) showing that the dissociation of the complex into free ferric ions is negligible. Ferric Ammonium Citrate is a complex since it does not yield ferric hydroxide with ammonia, although it does so with sodium hydroxide. The removal by ignition of organic radicals such as citrate, tartrate and lactate in group analysis before passing to group III is essential because of their tendency to prevent precipitation of certain metal ions under the usual conditions owing to complex formation. Great difference in stability of complexes is illustrated in the group II separation of copper and cadmium. in which hydrogen sulphide is passed into a mixture of their complex cyanides, $K_3[Cu(CN)_4]$ (copper stabilised in the cuprous state) and $K_2[Cd(CN)_4]$. Sufficient free cadmium ions are present for the solubility product of cadmium sulphide to be reached, and it is precipitated; the cuprous complex on the other hand is extremely stable so that cuprous sulphide is not precipitated under these conditions, although its solubility product is extremely low (2×10^{-47}).

Neutral molecules with lone pairs of electrons, e.g. $\ddot{N}H_3$, will also form complexes as illustrated by the familiar cupric ammonium ion, $(Cu(NH_3)_4)^{2+}$, and cobaltammine, $Co[(NH_3)_6]^{3+}$. Generally speaking the electron-donating atoms are N, S, O and the halogens, and possibly carbon in such groups as CO and $(CN)^-$.

Bonding in Complexes

The bonds are either ordinary covalent bonds in which both the metal and the ligand contribute one electron each, or co-ordinate bonds in which both electrons are contributed by the ligand. Thus, the ferricyanide ion may be considered to consist of three ordinary covalent bonds and three co-ordinate bonds ((Fig. 22A), although in the complex the bonds are identical hybrid bonds (Fig. 22B) which have been shown to be directed towards the apices of a regular octahedron (Fig. 22C).

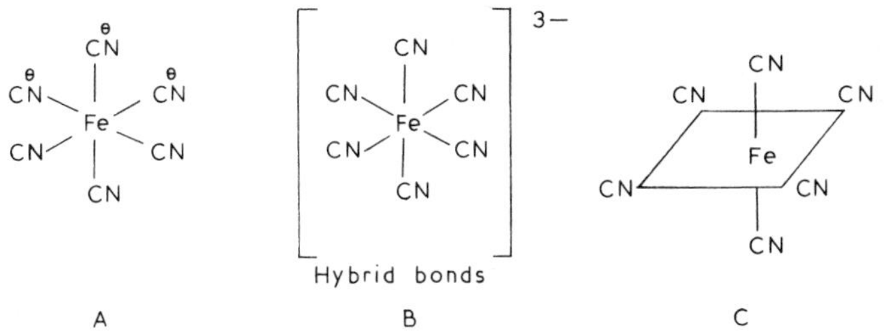

Fig. 22

The negative charge on the complex ion is equal to the total number of the negative groups *minus* the valency of the metal ion. When neutral groups only are involved, the charge on the complex is positive and is equal to the valency of the metal ion, e.g. $[Cu(NH_3)_4]^{2+}$.

Werner's Co-ordination Number and Electronic Structure of Complex Ions

Werner (1891) first noticed that for each atom there is an observed maximum number of small groups which can be accommodated around it. This number, which is called Werner's co-ordination number, depends purely upon steric factors and is in no way related to the valency of the ion. Thus, although the valency shell of the elements of the third period is theoretically capable of expanding up to 18 electrons, and that of the fourth to 32 electrons, there is in practice a limit to the number of small groups which can be accommodated owing to limitations of space around the ion. In the second period, the maximum number is 4 and in the third period it is 6, and with certain elements such as tungsten and molybdenum it is 8. For example, in the $[BF_4]^-$ ion, the octet is completed and the maximum co-ordination number is reached, but in the $[AlF_6]^-$ ion the outer shell contains 12 electrons and cannot expand to the

Table 24. Electronic configurations of metal ions and complex ions

Metal ion	Electronic configuration	Complex ion	Electronic configuration of complex ion
Co^{3+}	2, 8, 14	$[Co(NH_3)_6]^{3+}$	2, 8, 18, 8
Cu^{2+}	2, 8, 17	$[Cu(NH_3)_4]^{2+}$	2, 8, 17, 8
Cu^+	2, 8, 18	$[Cu(CN)_4]^{3-}$	2, 8, 18, 8
Hg^{2+}	2, 8, 18, 32, 18	$[Hg(CNS)_4]^{2-}$	2, 8, 18, 32, 18, 8
Fe^{2+}	2, 8, 14	$[Fe(CN)_6]^{4-}$	2, 8, 18, 8
Fe^{3+}	2, 8, 13	$[Fe(CN)_6]^{3-}$	2, 8, 17, 8

maximum number of 18 electrons since the maximum co-ordination number has been reached.

Within the limits imposed by Werner's co-ordination number, there is a tendency for the metal to attain or approach inert gas structure, and this is probably the driving force for complex formation. The examples in Table 24 illustrate this point.

Chelating Agents

Complexes involving simple ligands, that is, those forming only one bond are described as *co-ordination compounds*. Ligands having more than one electron-donating group are called *chelating agents* (from a Greek word meaning *claw*), since they combine with metal ions in a manner reminiscent of a crab's claw to form cyclic structures. Thus, 1,2-diaminoethane behaves like two ammonia molecules and forms *chelates* with copper and cobalt ions:

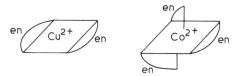

en = ethylenediamine ($H_2NCH_2CH_2NH_2$)

The copper ion is common to two five-membered rings and the cobalt ion to three five-membered rings. The latter has no plane of symmetry and thus exists in two optically-active forms. There is no fundamental difference between co-ordination and chelate compounds except that in the latter, ring size influences stability. Five- or six-membered rings are the most stable and are readily formed, although both larger and smaller rings may also be formed.

Many organic compounds will chelate metals if they contain groups with an easily replaceable proton (—COOH, phenolic and enolic OH), or neutral groups offering a lone pair of electrons (NH_2, CO and alcoholic OH), and the structures of the molecules are such as to permit the formation of stable rings. The greater the number of rings which can be formed, the more stable the chelate is likely to be. Most rings formed in chelates involve the highest valency state of a metal since these are more stable than those involving lower valency states.

The solubility of metal chelates in water depends upon the presence of hydrophilic groups such as COOH, SO_3H, NH_2 and OH. When both acidic and basic groups are present, the complex will be soluble over a wide range of pH. When hydrophilic groups are absent, the solubilities of both the chelating agent and the metal chelate will be low, but they will be soluble in organic solvents. The term *sequestering agent* is generally applied to chelating agents which form water-soluble complexes with bi- or poly- valent metal ions. Thus, although the metals remain in solution, they fail to give normal ionic reactions. Ethylenediamine tetra-acetic acid is a typical sequestering agent, whereas dimethylglyoxime and salicylaldoxime are chelating agents, forming insoluble complexes.

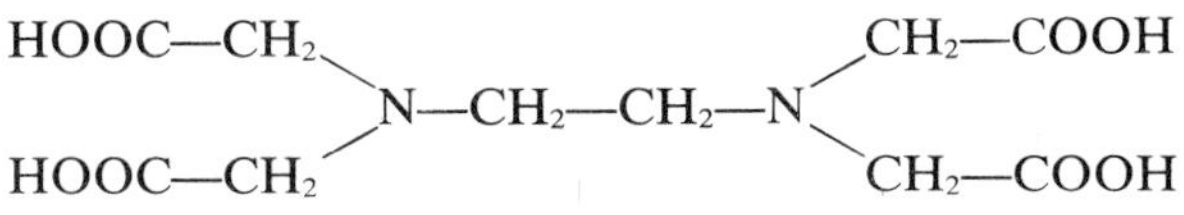

HOOC—CH$_2$
HOOC—CH$_2$ N—CH$_2$—CH$_2$—N CH$_2$—COOH
CH$_2$—COOH

Ethylenediamine tetra-acetic acid

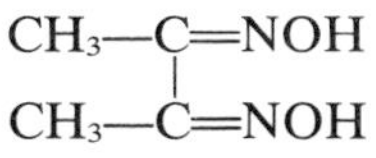

CH$_3$—C=NOH
CH$_3$—C=NOH

Dimethylglyoxime

CH=N—OH
OH

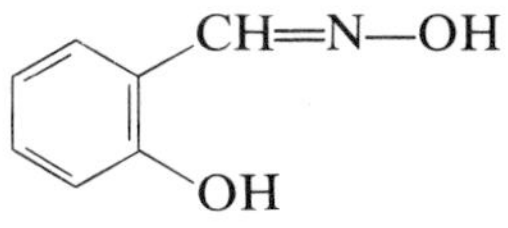

Salicylaldoxime

As a sequestering agent, ethylenediamine tetra-acetic acid reacts with most polyvalent metal ions to form water-soluble complexes which cannot be extracted from aqueous solutions with organic solvents. Dimethylglyoxime and salicylaldoxime form complexes, which are insoluble in water, but soluble in organic solvents; for example, nickel dimethylglyoxime has a sufficiently low solubility in water to be used as a basis for a gravimetric assay.

Nature and Stability of Metal Complexes of Ethylenediamine Tetra-acetic Acid

Ethylenediamine tetra-acetic acid forms complexes with most cations in a 1 : 1 ratio, irrespective of the valency of the ion:

$$M^{2+} + [H_2X]^{2-} \rightarrow [MX]^{2-} + 2H^+$$
$$M^{3+} + [H_2X]^{2-} \rightarrow [MX]^{-} + 2H^+$$
$$M^{4+} + [H_2X]^{2-} \rightarrow [MX] + 2H^+$$

where M is a metal and $[H_2X]^{2}$ is the anion of the disodium salt (*disodium edetate*) which is most frequently used. The structures of these complexes with di-, tri- and tetra-valent metals contain three, four and five rings respectively:

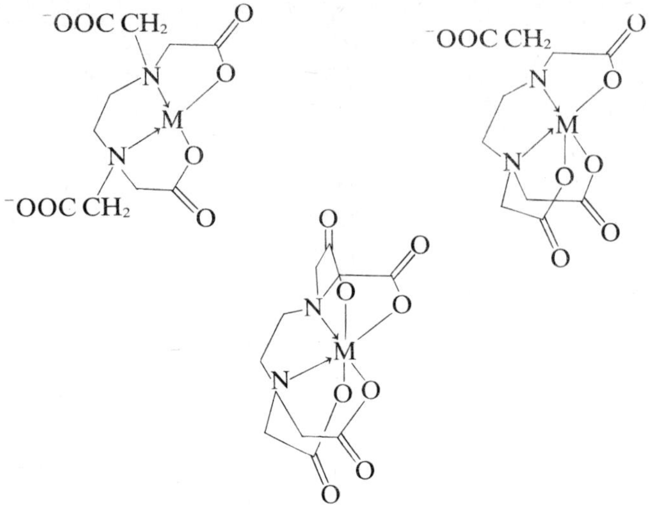

Effect of pH on Complex Formation

Edetic acid ionises in four stages ($pK_1 = 2.0$, $pK_2 = 2.67$, $pK_3 = 6.16$ and $pK_4 = 10.26$) and since the actual complexing species is Y^{4-}, complexes will form more efficiently and be more stable in alkaline solution. If, however, the solubility product of the metal hydroxide is low, it may be precipitated if the hydroxyl ion concentration is increased too much. On the other hand, at lower pH values when the concentration of Y^{4-} is lower, the stability constant of the complexes will not be so high. Complexes of most divalent metals are stable in ammoniacal solution. Those of the alkaline earth metals decompose below pH 7, whilst those of the more firmly bound divalent metals such as copper, lead and nickel are stable down to pH 3 and hence can be titrated selectively in the presence of alkaline earth metals. Trivalent metal complexes are usually still more firmly bound and stable in strongly acid solutions, e.g. the cobalt (Co^{3+}) edetate complex is stable in concentrated hydrochloric acid. Although most complexes are stable over a fair range of pH, solutions are usually buffered at a pH at which the complex is stable and at which the colour change of the indicator is most distinct.

Colour of Complexes

There is always a change in the absorption spectrum when complexes are formed and this forms the basis of many colorimetric assays.

Stability of Complexes

The general equation for the formation of a 1:1 chelate complex, MX, is,

$$M + X \rightleftharpoons MX$$

Table 25. Stability constants of edetate complexes

Cation	Log K	Cation	Log K
Ba^{2+}	7.8	Co^{2+}	15.5
Mg^{2+}	8.7	Co^{3+}	26.0
Ca^{2+}	10.6	Al^{3+}	15.5
Cr^{2+}	13.0	Zn^{2+}	16.5
Cr^{3+}	24.0	Pb^{2+}	17.0
Mn^{2+}	13.5	Ni^{2+}	18.0
Fe^{2+}	14.3	Cu^{2+}	18.0
Fe^{3+}	25.1	Hg^{2+}	21.0

where M is the metal ion, and X the chelating ion

$$\therefore \quad \text{Stability constant}\,(K) = \frac{[MX]}{[M][X]}$$

where [ ] represents activities. Increase in temperature causes a slight increase in the ionisation of the complex and a slight lowering of K. The presence of electrolytes having no ion in common with the complex decreases K, whilst the presence of ethanol increases K, probably due to the suppression of ionisation. Table 25 gives values of the logarithms of stability constants of edetate complexes of metals of pharmaceutical interest.

Titration of Metal Ions Using Disodium Edetate

Since edetic acid is only sparingly soluble in water (about 0.2 per cent) the disodium salt is usually used (solubility about 10 per cent). When sodium edetate solution is run into a solution of a metal ion buffered to promote efficient complex formation, the *rate of change* of concentration of metal ion is slow at first, but increases very rapidly as the amount of sodium edetate added approaches one equivalent, in the same way as hydrogen ion concentration changes during the titration of a strong acid with a strong base. Fig. 23 shows a plot of log $[M^{n+}]$ and pM {log $\frac{1}{[M^{n+}]}$ or, $-\log[M^{n+}]$} against equivalents of sodium edetate added.

End Point Detection in Complexometric Titration using pM Indicators

The end point of complexometric titrations is shown by means of pM indicators. The concept of pM arises as follows:
  If K is the stability constant,

$$K = \frac{[MX]}{[M][X]}$$

then

$$[M] = \frac{[MX]}{[X]\,K}$$

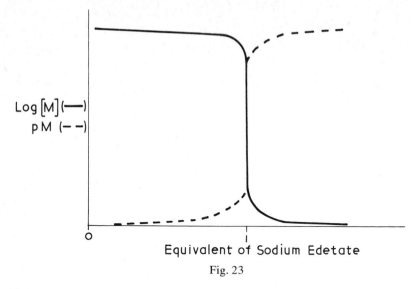

Fig. 23

or

$$\log [M] = \log \frac{[MX]}{[X]} - \log K$$

and

$$pM = \log \frac{[X]}{[MX]} - pK$$

Therefore if a solution is made such that $[X] = [MX]$, $pM = -pK$ (or $pM = pK'$ where $K' =$ dissociation constant). This means that in a solution containing equal activities of metal-complex and free chelating agent, the concentration of metal ions will remain roughly constant and will be buffered in the same way as are hydrogen ions in a pH buffer. Since, however, chelating agents are also bases, equilibrium in a metal-buffer solution is often greatly affected by a change in pH. In general, for chelating agents of the amino acid type (e.g. edetic acid, ammoniatriacetic acid etc.), it may be said that when $[X] = [MX]$, pM increases with pH until about pH 10, when it attains a constant value. This pH is therefore usually chosen for carrying out titrations of metals with chelating agents in buffered solution.

The pM indicator is a dye which is capable of acting as a chelating agent to give a dye-metal complex. The latter is different in colour from the dye itself, and also has a lower stability constant than the chelate-metal complex. The colour of the solution, therefore, remains that of the dye-complex until the end point, when an equivalent amount of sodium edetate has been added. As soon as there is the slightest excess of edetate, the metal-dye complex decomposes to produce free dye; this is accomplished by a change in colour.

The colours of dyes and of the metal complexes vary with pH. This fact, together with complex stability, must be considered when deciding at which pH to carry out a titration. It is also essential to use a buffer solution to maintain the required pH during the titration.

pM Indicators

Murexide (ammonium purpurate) is mainly used for the titration of calcium at pH 12. Calcium is bound in an eight-membered ring in the murexide chelate. The larger ring confers a lower stability than the usual five- and six-membered ring chelates. Since magnesium-murexide is less stable than the calcium complex, calcium can be titrated in the presence of magnesium. This is applied in water analysis (p. 246).

Murexide (violet at pH 12) Calcium-murexide

Copper, cobalt, nickel and cerium form yellow complexes with murexide in alkaline solution and give clear colour changes when titrated with sodium edetate. Owing to the instability of the solution, murexide is stored as a 0.2 per cent dispersion in powdered sodium chloride.

Mordant Black 2 (Eriochrome Black T, Solochrome Black T). This indicator has wide application. It is blue at about pH 10 and most of its complexes are reddish. Below pH 6.3 and above pH 11.5, the dye itself is reddish like the complexes. It is, therefore, necessary to carry out the titration in the presence of a buffer at pH 10. The reaction with a divalent metal cation may be expressed as follows:

Blue (pH 10) Pink

Magnesium, calcium, cadmium, zinc, manganese, lead, lanthanum and mercury (Hg^+) may be titrated directly using this indicator. For barium and strontium, it is better to add excess sodium edetate and back-titrate with a standard magnesium chloride solution.

Cobalt, nickel, copper, aluminium, silver, titanium and platinum form more stable complexes with the dye than with the edetate and hence the former cannot be used as an indicator. Iron (Fe^{3+}), cerium (Ce^{4+}) and vanadate ions oxidise the dye, whilst stannous and titanous ions reduce it. All these ions interfere and must be removed or masked (see below). For the same reason as murexide, Mordant Black 2 is used as a freshly-prepared 0.2 per cent dispersion in sodium chloride.

Catechol Violet. The catechol groups of the molecule form weak, highly-coloured complexes with a wider range of metals than Mordant Black 2 or murexide in neutral, acid and alkaline solution. These complexes may contain one or two metal atoms; all are blue in alkaline solution, and the more stable ones are also blue in acid solution except at low pH values when they change to purple or red. For example, the thorium (Th^{4+}) complex is blue down to pH 3 and the bismuth (Bi^{3+}) complex is blue down to pH 1.5. The indicator itself is red in acid solution below pH 1.5 when it changes colour to yellow owing to loss of a proton from the sulphonic acid group and thence to violet at pH 6 owing to loss of a further proton and finally to reddish-purple at pH 9 owing to the loss of a further proton. In alkaline solution, clear end points are shown with magnesium, manganese, iron (Fe^{2+}), cobalt, nickel, zinc, cadmium and calcium. Between pH 4 and 7, copper, lead and lanthanum give clear end points, whilst bismuth, thorium and antimony can be titrated in the pH range 1.5 to 4 and hence selective titrations may be carried out. Catechol violet also has the advantage that its aqueous solutions are highly stable.

Xylenol Orange. This is an acid-base indicator in which two imino-acetic acid groups have been introduced into each cresol ring *ortho* to the phenolic and quinonoid oxygen respectively to promote the formation of the complex. As an acid-base indicator, xylenol orange has the same characteristics as cresol red, i.e. lemon-yellow in acid solution and red in alkaline solution. The metal complexes are red. Hence, its use is restricted to titrating metals whose edetate complexes are stable in acid solution, e.g. bismuth, thorium, lead, lanthanum, cadmium and mercury. The stability of the complexes, however, varies with pH. Thus, from pH 1 to 3 bismuth and thorium can be titrated, from pH 4 to 5 lead and zinc, and from pH 5 to 6 cadmium and mercury can be titrated. Hexamine alone or with varying amounts of nitric, hydrochloric, or acetic acid is a suitable buffering agent.

Methyl Thymol Blue is derived from thymol blue. It can be used for bismuth in strongly acid solution (N nitric acid) and for lead, zinc, cadmium, mercury (Hg^{2+}) and cobalt (Co^{3+}) in weakly acid solution (hexamine buffer). In alkaline solution, lead, zinc, cadmium, cobalt (Co^{2+}), magnesium and manganese (Mn^{2+}) can be determined using an ammonium chloride-ammonia buffer at pH 10, and in still stronger alkali ($0.05N$ sodium hydroxide) calcium, strontium and barium can be titrated. A 1 per cent aqueous solution of the dye is stable and the colour changes from blue to yellow in acid solution and blue to grey in alkaline solution. It is used in the official limit test for calcium in Sulpho-bromophthalein Sodium.

Alizarin Fluorine Blue (alizarin complexone) may be used in acid solution at pH 4.3 for the titration of lead, zinc, cobalt (Co^{2+}), mercury (Hg^{2+}) and copper (Cu^{2+}) when the colour changes from red to yellow.

It is used in the determination of *fluorine* in Triamcinolone and Fluocinolone Acetonides, and in Fluocortolone Hexanoate and Pivalate. The determination depends upon the fact that the red cerous complex changes to blue in the presence of fluoride ions. The organically combined fluorine is converted to inorganic fluoride by burning the compound in oxygen and absorbing the gases in water. The complex is formed by the addition of alizarin fluorine blue and cerous nitrate in the presence of a sodium acetate-acetic acid buffer and its intensity measured spectrophotometrically at 610 nm.

Sodium Alizarinesulphonate gives a bluish-red lake with aluminium and thorium ions at about pH 4 and is yellow at this pH in the absence of these ions.

It has been used as an indicator for the determination of fluorine in Triam-cinolone by titration with standard thorium nitrate solution.

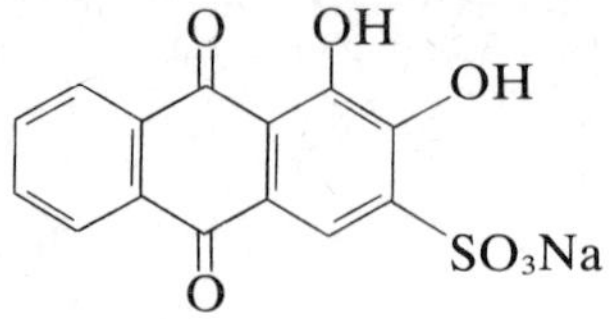

Diphenylcarbazone is used as an indicator for mercury in the determination of methylthiouracil (p. 254). It forms a violet-coloured complex with mercuric ions, but the reaction is only specific for mercury in *0.2N* nitric acid, provided that chromates and molybdates are absent. In neutral or slightly acid solution, copper, iron, cobalt and other ions give coloured complexes. Chlorides reduce sensitivity to mercury due to the formation of $(HgCl_4)^{2-}$ ions.

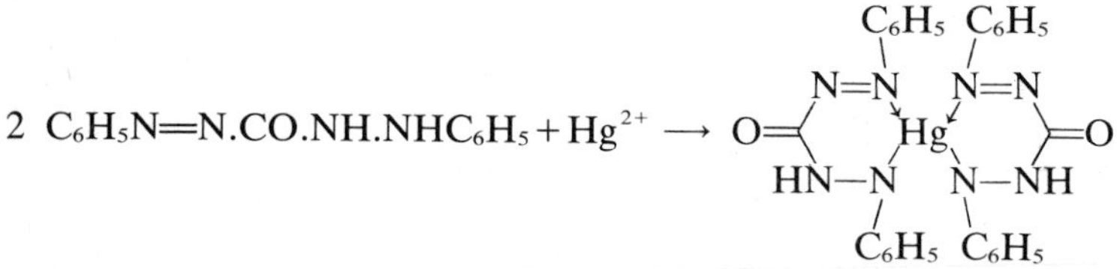

$$2 \ C_6H_5N{=}N.CO.NH.NHC_6H_5 + Hg^{2+} \longrightarrow$$

Tiron (Disodium, 1,2-dihydroxyphenyl-3,5-disulphonate) forms a blue colour with ferric iron between pH 2 and 5, a violet colour between pH 5.7 and 7 and a red complex in alkaline solution. The blue complex in acid solution is a suitable indicator for titration with disodium edetate, the colour changes being greenish-blue through green to yellow. Titrations are best carried out at 40° to 50°C. The red ferric complex in alkaline solution is stable and specific for ferric iron; it is sensitive to 1 in 200 million. It is not used as an indicator in edetate-iron titrations, but it can be used as the basis of a colorimetric assay for ferric iron.

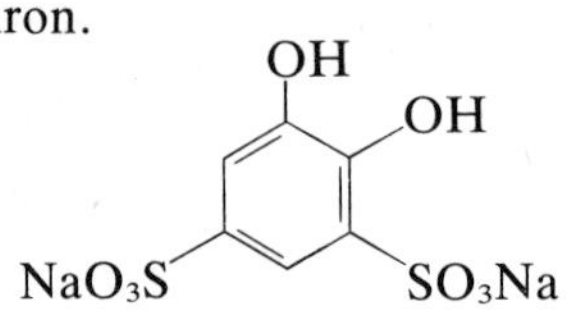

Tiron

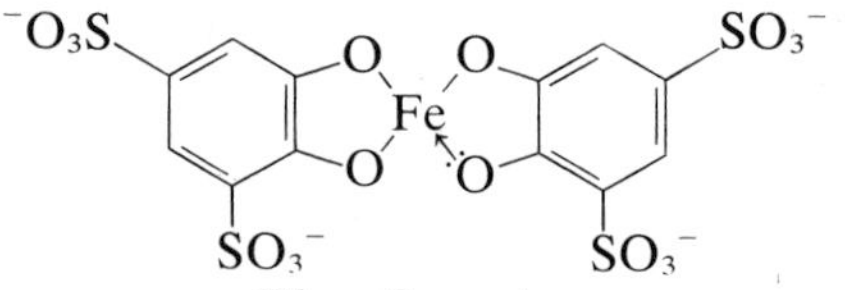

Blue Complex

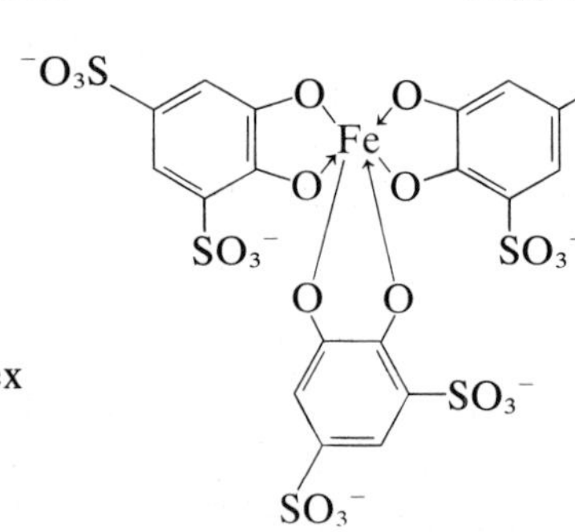

Red Complex

Physical Methods of End Point Detection in Complexometric Titrations

Spectrophotometric detection. The change in absorption spectrum when a metal ion or a complexing agent is converted to the metal complex, or when one complex is converted to another can usually be detected more accurately and in more dilute solution by spectrophotometric than by visual methods. Thus, in disodium edetate titrations an accurate end point can be obtained using $0.001M$ solutions. An indicator giving a colour change in the visible region is generally employed, although coloured ions may be titrated without an indicator.

It is sometimes possible to use an end point in the ultraviolet region for ions and complexes which are colourless in the visible region.

Amperometric titration. The effect of complex formation on the half-wave potential of an ion is to render it more negative. If the electrode potential is adjusted to a value between that of the half-wave potential of the free cation and that of the complex, and disodium edetate solution is added slowly, the diffusion current will fall steadily until it equals the residual current, that is, until the last trace of free cation has been complexed. This is the end point, and the amount of standard disodium edetate solution added is equivalent to the amount of metal present.

Potentiometric titration. Since disodium edetate reacts preferentially with the higher valency state of an ion, it will reduce the redox potential according to the equation

$$E = E_0 + \log_e \frac{[Ox]}{[Red]}$$

where E = the potential of the electrode,

E_0 = the standard electrode potential,

$[Ox]$ = activity of ions in the oxidised state and

$[Red]$ = activity of ions in the reduced state.

This method is of limited application owing to the lack of suitable indicator electrodes. Iron (Fe^{3+}) and copper (Cu^{2+}) however can be titrated in this way. Back-titration of excess disodium edetate with ferric chloride in acid solution is possible for some ions.

High frequency titration. This method is particularly suitable for dilute solutions, in some cases with concentrations as low as $0.0002M$. The ions may be titrated directly in buffered solution or excess reagent can be added to the unbuffered solution and the liberated protons titrated with standard alkali. Since buffer solutions and other extraneous electrolytes reduce the sensitivity of the titration, their concentration must be kept to a minimum.

General Principles Involved in Disodium Edetate Titrations

Direct titration. A suitable buffer solution and indicator are added to the metal ion solution and the solution titrated with standard disodium edetate until the indicator just changes colour. A *blank* titration may be performed omitting the sample as a check on the presence of traces of metallic impurities in the reagents. Greater accuracy may be obtained if the amount of indicator used and volumes of solution are controlled and the colour of the solution at the end point is matched with that from a blank titration or a slightly over-titrated sample.

Back-titration. This procedure is necessary for metals which precipitate as hydroxides from solution at the pH required for titration, for insoluble substances (e.g. lead as sulphate, calcium as oxalate), for substances which do not react instantaneously with disodium edetate, and for those metal ions which form more stable complexes with disodium edetate than with the desired indicator. Excess of standard sodium edetate and a suitable buffer solution is added to the metal solution or suspension. The solution is heated to effect complex formation, cooled, and the disodium edetate not required by the sample back-titrated with magnesium or zinc chloride (or sulphate) using a suitable indicator.

Replacement of one complex by another. When direct titration or back-titration do not give sharp end points, the metal may be determined by the displacement of an equivalent amount of magnesium or zinc from a less stable edetate complex according to the equation:

$$M^{2+} + MgX^{2-} \rightarrow MX^{2-} + Mg^{2+}$$

Calcium, lead and mercury can be determined satisfactorily using Mordant Black 2 by this method.

Alkalimetric titration of metals. In this method, protons from disodium edetate are displaced by a heavy metal, and titrated with standard alkali, according to the equation:

$$M^{n+} + H_2X^{2-} \rightarrow MX^{(n-4)} + 2H^+$$

The titration is carried out in unbuffered solution. A visual pH indicator may be used, but a potentiometric method of detecting the end point is also suitable, especially when the colour of the complex would mask that of a pH indicator.

Masking and Demasking Agents

Owing to the wide range of cations complexed by disodium edetate, the selectivity of the method is poor and metal impurities may be titrated with the ion it is desired to determine. When it is required to assay selectively one or more ions in a mixture of cations and to eliminate the effects of possible impurities which would add to the titre, *masking agents* are used. These act either by precipitation or by formation of complexes more stable than the interfering ion-edetate complex. It is important that any colour due to auxiliary complexes or precipitates should not obscure the end point.

Masking by precipitation. Many heavy metals (e.g. cobalt, copper and lead) can be separated either in the form of insoluble sulphides using sodium sulphide or as insoluble complexes using thioacetamide. These are filtered, decomposed and titrated with disodium edetate. Other common precipitating agents are sulphate for lead and barium, oxalate for calcium and lead, fluoride for calcium, magnesium and lead, ferrocyanide for zinc and copper, and cupferron (ammonium N-nitrosophenylhydroxylamine) and 8-hydroxy-quinoline for many heavy metals. Thioglycerol ($CH_2SH.CHOH.CH_2OH$) is used to mask copper by precipitation in the assay of lotions containing copper and zinc. If it is desired to mask a *trace* of ions, which interfere, filtration is usually unnecessary since the rate of reaction of sodium edetate with the insoluble complexes is very slow.

Ammonium Fluoride. Fluoride will mask aluminium, iron and titanium by complex formation.

Ascorbic acid is a convenient reducing agent for ferric iron which is then masked by complexing as the very stable ferrocyanide complex. This latter is more stable and less intensely coloured than the ferricyanide complex.

Dimercaprol (2,3-Dimercaptopropanol) ($CH_2SH.CHSH.CH_2OH$). Cations of mercury, cadmium, zinc, arsenic, antimony, tin, lead and bismuth react with dimercaprol in weakly acid solution to form precipitates which are soluble in alkaline solution. All these complexes are stronger than the corresponding edetate complexes, and are almost colourless. Cobalt, copper and nickel form intense yellowish-green complexes with the reagent under the above conditions. Cobalt and copper, but not nickel are displaced from their edetate complexes by dimercaprol.

Potassium cyanide. This reacts with silver, copper, mercury, iron, zinc, cadmium, cobalt and nickel ions to form complexes in alkaline solution which are more stable than the corresponding edetate complexes, so that other ions such as lead, magnesium, manganese and the alkaline earth metals can be determined in their presence. Of the metals in the first group mentioned, zinc and cadmium can be *demasked* from their cyanide complexes by aldehydes such as formaldehyde or chloral hydrate (due to the preferential formation of a cyanohydrin), and selectively titrated.

Potassium iodide is used to mask the mercuric ion as $(HgI_4)^{2-}$, and is specific for mercury. It can be used in the assay of mercuric chloride.

Tiron (disodium catechol-3,5-disulphonate) will mask aluminium and titanium as colourless complexes. Iron forms highly-coloured complexes and is best masked as its ferrocyanide complex.

Triethanolamine [$N(CH_2.CH_2.OH)_3$]. This forms a colourless complex with aluminium, a yellow complex with iron (Fe^{3+}), the colour of which is almost discharged by adding sodium hydroxide solution, and a green manganese (Mn^{3+}) complex which oxidises Mordant Black 2. For these reasons, if murexide is used in the presence of iron and manganese, it is best to mask them with triethanolamine; similarly, Mordant Black 2 can be used in the presence of triethanolamine-aluminium complex.

pH control. A very simple method of masking is based upon the fact that the alkaline earth metals do not form edetate complexes below pH 7, whilst most transition elements form edetate complexes stable down to pH 3. Certain metal ions such as tin (Sn^{4+}), iron (Fe^{3+}), cobalt (Co^{3+}) and thorium (Th^{4+}) can be selectively titrated at still lower pH values.

PRACTICAL EXERCISES

Purification of Disodium Edetate

Commercial samples of disodium edetate may be purified for use as a *primary standard* by adding ethanol to a saturated aqueous solution until the first permanent precipitate appears; filter and add an equal volume of ethanol; filter the precipitated disodium edetate, wash with acetone and ether, and dry to constant weight at 80°C. Drying may require four days. The official material contains not less than 98 per cent of the dihydrate and is assayed as described under Disodium Edetate.

Standardisation of Approximately *0.01M* Disodium Edetate

Buffer solution. Dissolve 67.5 g of ammonium chloride in 570 ml of strong ammonia solution, dilute to 900 ml with Water and add a solution of 0.616 g of magnesium sulphate and 0.93 g of disodium edetate in 50 ml of Water; add sufficient water to produce 1000 ml. *Note.* The small amount of magnesium edetate complex is included in the buffer to render the end point sharper in calcium titrations. Since the calcium edetate complex is more stable than the magnesium edetate complex, magnesium ions will be liberated in the presence of excess calcium ions which will be completely complexed before the magnesium ions as the titration proceeds. At the end point, the magnesium, present as the dye-chelate, is being titrated, and since this has a more distinct colour than the calcium dye-chelate, the end point will be sharper.

Standard calcium chloride solution, 0.01M. Dissolve 1.001 g of AnalaR calcium carbonate in water (25 ml) containing the minimum quantity of dilute hydrochloric acid. Boil off carbon dioxide, cool, and transfer quantitatively to a 1000 ml flask. Dilute with Water to 1000 ml.

Indicator. Dissolve 0.5 g Solochrome Black WDFA and 4.5 g of hydroxylamine hydrochloride in 1000 ml of ethanol.

Method. Pipette 20 ml of standard calcium chloride solution into a conical flask, add 1 ml of buffer solution and 3 drops (from a dropping pipette) of indicator and titrate with the disodium edetate solution until the colour changes from wine-red to clear blue, i.e. until there is no further change in colour, adding 2 or 3 drops of chelating agent in excess to ensure slight over-titration. Keep the solution for matching purposes and repeat the process carefully until the colours exactly match. This technique ensures reproducibility to within 0.2 per cent even with these dilute solutions. A *blank* titration should be performed on the reagents, i.e. omitting the metal, and if more than one drop is required, this should be subtracted from the result.

Standardisation of Approximately *0.05M* Disodium Edetate

This may be carried out as above using 5 ml of buffer solution and titrating with an *0.1N* calcium chloride solution prepared as described above, but using 1.251 g of calcium carbonate and diluting to 250 ml. The official procedure differs in that a solution is used for titration containing a standard amount of zinc (as sulphate) and 0.1 g of the official indicator, Solochrome Black Mixture (1 part in 500 of Solochrome Black WDFA in sodium chloride).

DIRECT TITRATIONS WITH DISODIUM EDETATE

Bismuth Nitrate. *Determination of the percentage of* $Bi(NO_3)_3,5H_2O$

The determination depends upon the reactions expressed by the following equation:

$$Bi^{3+} + [H_2X]^{2-} \rightarrow [BiX]^- + 2H^+$$
$$\therefore \quad Bi(NO_3)_3,5H_2O \equiv Bi^{3+} \equiv Na_2H_2X,2H_2O$$
$$\therefore \quad 485.0 \text{ g } Bi(NO_3)_3,5H_2O \equiv 20\ 000 \text{ ml } 0.05M$$
$$\therefore \quad 0.02425 \text{ g } Bi(NO_3)_3,5H_2O \equiv 1 \text{ ml } 0.05M \text{ disodium edetate}$$

Method. Dissolve about 1 g of sample, accurately weighed, in a mixture of glycerol (20 ml) and Water (20 ml). Add sulphamic acid (0.1 g) and titrate with *0.05M* disodium edetate using catechol violet solution as indicator. The colour changes from that of the deep blue bismuth-indicator complex to the yellow colour of the indicator at the pH of the titration (pH 2 to 3).

Note. The glycerol assists solution of the bismuth nitrate by reducing hydrolysis and preventing the formation of basic nitrates. The sulphamic acid (NH_2SO_3H) ensures that the pH is about 2 to 3 at which the bismuth-catechol violet complex is pure blue (it is red below pH 1.5) and the indicator itself is yellow, so that the colour change is clear. It also removes traces of nitrite or oxides of nitrogen which may oxidise the indicator. Buffer solutions containing halides cannot be used owing to the formation of bismuth tetrahalogeno-complexes which may interfere with the assay.

Cognate Determinations

Bismuth Carbonate; Bismuth Oxynitrate; Bismuth Subnitrate; Ointments containing Bismuth Subnitrate

The sample is dissolved in nitric acid, and the pH adjusted, if necessary, with ammonia until the bismuth-catechol violet complex is blue. The amount of sulphamic acid is increased to 0.2 g as oxides of nitrogen are formed when the sample is dissolved in the nitric acid, and therefore more likely to be present.

Bismuth Oxychloride; Bismuth Sodium Tartrate; Bismuth Subgallate

These compounds are ignited with sulphuric acid at a temperature of not more than 500° to remove chloride; this interferes in the titration due to the formation of tetrachloro-complexes. After ignition, they are titrated as for Bismuth Carbonate.

Anhydrous Calcium Chloride. *Determination of the percentage of* $CaCl_2$

$$Ca^{2+} + [H_2X]^{2-} \rightarrow [CaX]^{2-} + 2H^+$$
$$\therefore \quad CaCl_2 \equiv Ca^{2+} \equiv Na_2H_2X,2H_2O$$
$$\therefore \quad 111.0 \text{ g } CaCl_2 \equiv 20\,000 \text{ ml } 0.05M$$
$$\therefore \quad 0.00555 \text{ g } CaCl_2 \equiv 1 \text{ ml } 0.05M \text{ disodium edetate}$$

Method. Dissolve about 0.25 g of sample, accurately weighed, in Water (50 ml), add *0.05M* magnesium sulphate (5 ml) and strong ammonia-ammonium chloride solution (10 ml) and titrate with *0.05M* disodium edetate using Mordant Black mixture as indicator until the last trace of pink just disappears and the solution is pure blue. From the volume required subtract the volume of *0.05M* magnesium sulphate added. The difference is the amount of disodium edetate required by the sample.

Strong Ammonia-Ammonium Chloride Solution contains ammonium chloride (67.5 g) and strong ammonia solution (650 ml) per litre.

Note. Since the colour change at the end point with calcium alone and Mordant Black 2 is not very distinct, a known volume of standard magnesium sulphate solution is added to the sample before titration. The magnesium ions are not complexed by edetate in the presence of calcium ions since the calcium edetate complex is more stable than that of the magnesium. When all the calcium ions have been complexed, then the magnesium ions are titrated, so that at the end point the colour change will be from that of the magnesium-indicator complex to that of the indicator at the pH of the titration. The ammonia-ammonium chloride solution buffers the solution at about pH 10 at which the colour change is best observed.

The end point is best observed if the Mordant Black mixture is weighed (say 0.1 g) and the colour matched with that obtained in a rough determination in which the sample is slightly over-titrated.

Cognate Determinations

Calcium Chloride
Calcium Gluconate
Calcium Gluconate Injection
Calcium Lactate
Calcium Lactate Tablets

Nickel Chloride. *Determination of the percentage of* $NiCl_2, 6H_2O$

This is titrated with $0.1M$ disodium edetate, in the presence of sodium acetate, using Mordant Red 7 solution as indicator until the solution changes colour from amethyst to green.

$$Ni^{2+} + [H_2X]^{2-} \rightarrow [NiX]^{2-} + 2H^+$$
$$\therefore \quad NiCl_2 \equiv Ni^{2+} \equiv Na_2H_2X, 2H_2O$$
$$\therefore \quad 237.7 \text{ g } NiCl_2, 6H_2O \equiv 10\,000 \text{ ml } M \text{ disodium edetate}$$
$$\therefore \quad 0.02377 \text{ g } NiCl_2, 6H_2O \equiv 1 \text{ ml } M \text{ disodium edetate}$$

Lanthanum Nitrate. *Determination of the percentage of* $La(NO_3)_3, 6H_2O$

This is titrated with $0.05M$ disodium edetate in a hexamine-nitric acid buffer solution using xylenol orange as indicator until the colour changes from red to lemon-yellow.

$$La^{3+} + [H_2X]^{2-} \rightarrow [LaX]^- + 2H^+$$
$$\therefore \quad La(NO_3)_3, 6H_2O \equiv La^{3+} \equiv Na_2H_2X, 2H_2O$$
$$\therefore \quad 430.0 \text{ g } La(NO_3)_3, 6H_2O \equiv 20\,000 \ 0.05M$$
$$\therefore \quad 0.02615 \text{ g } La(NO_3)_3, 6H_2O \equiv 1 \text{ ml } 0.05M \text{ disodium edetate}$$

Lead Acetate. *Determination of the percentage of* $C_4H_6O_4Pb, 3H_2O$

This is titrated with $0.05M$ disodium edetate in a hexamine-acetic acid buffer at pH 5–6 using xylenol orange as indicator until the colour changes from red to lemon-yellow.

$$Pb^{2+} + [H_2X]^{2-} \rightarrow [PbX]^{2-} + 2H^+$$
$$\therefore \quad (CH_3COO)_2Pb, 3H_2O \equiv Pb^{2+} \equiv Na_2H_2X, 2H_2O$$
$$\therefore \quad 379.3 \text{ g } (CH_3COO)_2Pb, 3H_2O \equiv 20\,000 \text{ ml } 0.05M$$
$$\therefore \quad 0.01897 \text{ g } (CH_3COO)_2Pb, 3H_2O \equiv 1 \text{ ml } 0.05M \text{ disodium edetate}$$

Cognate Determinations

Lead Monoxide
Strong Lead Subacetate Solution

Magnesium Sulphate. *Determination of the percentage of* $MgSO_4$

This determination depends upon the reactions expressed by the following equation:

$$Mg^{2+} + [H_2X]^{2-} \rightarrow [MgX]^{2-} + 2H^+$$
$$\therefore \quad MgSO_4 \equiv Mg^{2+} \equiv Na_2H_2X,2H_2O$$
$$\therefore \quad 120.38 \text{ g } MgSO_4 \equiv 20\ 000 \text{ ml } 0.05M$$
$$\therefore \quad 0.006019 \text{ g } MgSO_4 \equiv 1 \text{ ml } 0.05M \text{ disodium edetate}$$

Method. Dissolve about 0.4 g of sample, accurately weighed in Water (50 ml), add strong ammonia-ammonium chloride solution (10 ml) and titrate with *0.05M* disodium edetate using Mordant Black mixture (0.1 g) as indicator until the pink tint is discharged from the blue. It is best to use a comparison solution containing an over-titrated sample as recommended under Anhydrous Calcium Chloride. The result is calculated with reference to the sample dried to constant weight at 300°.

Note. The colour change of the indicator is clearest at the pH of the buffer solution employed, which is about pH 10.

Cognate Determinations

Magnesium Carbonate in Compound Powders.

Magnesium Carbonate in Compound Powders and Suspensions containing Kaolin. The filtrate from the determination of *acid-insoluble matter* is treated with ascorbic acid and potassium cyanide to mask iron impurity as ferrocyanide, and with triethanolamine to mask traces of aluminium derived from the kaolin. The solution is titrated with *0.05M* disodium edetate in an ammonia buffer with Mordant Black 2 as indicator.

Magnesium Oxide in Sterilizable Maize Starch (Absorbable Dusting Powder). The combined filtrate and washings from the determination of *acid-insoluble ash* are treated with potassium cyanide to mask heavy metal impurities such as copper and iron, and then titrated as for Magnesium Sulphate.

Magnesium Powder. The sample is dissolved in dilute hydrochloric acid and then titrated as for Magnesium Sulphate.

Magnesium Stearate

Magnesium Sulphate Paste

Magnesium Trisilicate. After preliminary treatment with perchloric acid to free the magnesium from the silica, an aliquot of the combined filtrate and washings is treated for Magnesium Sulphate.

Disodium Edetate. *Determination of the percentage of* $C_{10}H_{14}N_2Na_2O_8,2H_2O$

This determination depends upon the reactions expressed by the following equation:

$$Zn^{2+} + [H_2X]^{2-} \rightarrow [ZnX]^{2-} + 2H^+$$
$$\therefore \quad Na_2H_2X \equiv Zn$$
$$\therefore \quad 372.2 \text{ g } C_{10}H_{14}N_2Na_2O_8,2H_2O \equiv 65.37 \text{ g Zn}$$
$$\therefore \quad 1.423 \text{ g } C_{10}H_{14}N_2Na_2O_8,2H_2O \equiv 0.25 \text{ g Zn}$$

Method. To 100 ml of a solution of zinc sulphate in Water containing the equivalent of 0.25 g of zinc, add strong ammonia-ammonium chloride solution (10 ml), *0.05M* disodium edetate (5 ml) and *0.05M* magnesium sulphate (5 ml). Dissolve about 4 g of the sample, accurately weighed, in sufficient Water to produce 100 ml and use this solution to titrate the zinc sulphate solution using Mordant Black mixture (0.1 g) as indicator.

Note. Since the stability constant of the zinc edetate complex (log $K = 16.5$) is higher than that of the magnesium complex (log $K = 8.7$), the 5 ml of $0.05M$ disodium edetate will form the zinc complex leaving the magnesium ions free, but the result of the titration will be unaffected, since the magnesium and edetate ions are added in equivalent amounts. During titration, the zinc ions will be complexed before the magnesium ions which are included because the colour change from that of the magnesium-indicator complex to that of the indicator is clearer than the corresponding change involving the zinc complex. The change is from pink to clear blue, that is, just after the last trace of pink has disappeared.

Zinc Sulphate. *Determination of the percentage of* $ZnSO_4,7H_2O$

This determination depends upon the reactions expressed by the following equation:

$$Zn^{2+} + [H_2X]^{2-} \rightarrow [ZnX]^{2-} + 2H^+$$
$$\therefore \quad ZnSO_4,7H_2O \equiv Zn^{2+} \equiv Na_2H_2X,2H_2O$$
$$\therefore \quad 287.5 \text{ g } ZnSO_4,7H_2O \equiv 20\,000 \text{ ml } 0.05M$$
$$\therefore \quad 0.01438 \text{ g } ZnSO_4,7H_2O \equiv 1 \text{ ml } 0.05M \text{ disodium edetate}$$

The sample is titrated with $0.05M$ disodium edetate in a hexamine-hydrochloric acid buffer at about pH 5, using xylenol orange solution as indicator, until the last trace of red disappears and the solution is pure yellow.

Cognate Determinations

Zinc Oxide in Dusting Powders with Talc and Starch

The sample is treated with hot dilute nitric acid, cooled, filtered and an aliquot of the filtrate titrated as for Zinc Sulphate. Traces of aluminium from the talc do not interfere, since aluminium complexes form only slowly with disodium edetate and xylenol orange.

Zinc Undecenoate; Zinc Undecenoate Ointment. The zinc is separated from organic combination by boiling with dilute hydrochloric acid. After filtering and washing the residue, the combined filtrate and washings are neutralised with dilute ammonia, and titrated as for Zinc Sulphate.

Bacitracin Zinc. *Determination of the percentage of* Zn

This determination depends upon the reactions expressed by the following equation:

$$Zn^{2+} + [H_2X]^{2-} \rightarrow [ZnX]^{2-} + 2H^+$$
$$\therefore \quad Zn^{2+} \equiv Na_2H_2X,2H_2O$$
$$\therefore \quad 65.37 \text{ g } Zn \equiv 1000 \text{ ml } M \text{ disodium edetate}$$
$$\therefore \quad 0.0006537 \text{ g } Zn \equiv 1 \text{ ml } 0.01M \text{ disodium edetate}$$

Method. Dissolve the sample (about 0.2 g), accurately weighed, in Water (20 ml). Add strong ammonia-ammonium chloride solution (3 ml) and titrate with $0.01M$ disodium edetate, using Mordant Black mixture as indicator.

Allied Determinations

Zinc Oxide in Zinc Paste Bandages.

The sample is ignited to destroy organic matter, the residue dissolved in dilute nitric acid, neutralised and an aliquot titrated as for Zinc Chloride.

Zinc Oxide in Zinc Paste Bandages containing calcium chloride.
After boiling the sample with water, adding nitric acid, filtering and washing, the calcium is precipitated as oxalate from an aliquot by the addition of
ammonium chloride (to keep the zinc in solution), ammonium oxalate and
excess dilute ammonia solution. The combined filtrate and washings are
determined as for Zinc Chloride.

Zinc Oxide in Dusting Powders with Alum and Talc.
The sample is shaken with dilute sulphuric acid to ensure solution of
the zinc oxide and alum. Aluminium is precipitated as hydroxide by the
addition of ammonium chloride and excess ammonia solution. Sufficient
triethanolamine is then added to re-dissolve the precipitated aluminium hydroxide, and to mask it from reaction with disodium edetate and the indicator,
Mordant Black 2. The determination is completed by titration as for Zinc
Chloride.

DIRECT TITRATION OF POLY-METALLIC SYSTEMS WITH DISODIUM EDETATE

Bismuth, Calcium and Magnesium Carbonates and Sodium Bicarbonate in Compound Powders and Compound Liquid Mixtures

Bismuth is determined directly as for Bismuth Carbonate, since calcium and
magnesium do not interfere under the conditions of the assay.

Magnesium. Bismuth, if present, is removed by precipitation as the oxychloride. Calcium is removed by precipitation as oxalate from an aliquot of the
filtrate, and the combined filtrate and washings titrated as for Magnesium
Sulphate.

Calcium. A further aliquot of the filtrate obtained in the magnesium
titration is titrated, without prior removal of calcium, as for Magnesium
Sulphate. The titre is equivalent to both calcium and magnesium, and the
calcium titre is obtained by difference.

Sodium bicarbonate. The determination of sodium bicarbonate necessitates
a correction for the appreciable amounts of calcium and magnesium bicarbo-
nates which survive the attempt to decompose them by boiling, filtering twice,
and cooling. The filtrate is titrated with $0.5N$ hydrochloric acid using screened
methyl orange (methyl orange-xylene cyanol FF) as indicator, strong
ammonia-ammonium chloride added and the solution titrated with $0.05M$
disodium edetate using Mordant Black 2 solution as indicator. One-fifth of
this titration is deducted from the volume of $0.5N$ hydrochloric acid required.
The difference is equivalent to the sodium bicarbonate in the sample. The
correction factor of one-fifth is derived as follows:

$$Ca/Mg(HCO_3)_2 + 2HCl \rightarrow Ca/MgCl_2 + H_2O + CO_2$$
$$\therefore \quad Ca/Mg(HCO_3)_2 \equiv 2HCl \equiv 4000 \text{ ml } 0.5N \text{ HCl}$$

but

$$Ca/Mg(HCO_3)_2 \equiv 1 \text{ mole disodium edetate}$$
$$\equiv 20\ 000 \text{ ml } 0.05M \text{ disodium edetate}$$

hence the equivalent of calcium and magnesium present, when expressed in terms of $0.5N$ HCl is equal to 4000/20 000 i.e. one-fifth of the titre obtained in the disodium edetate titration.

Determination of the Hardness of Water

This involves the determination of the calcium and magnesium ions present, the *total hardness* being expressed as parts per million or grains per gallon of calcium and magnesium salts together calculated as calcium carbonate. This is determined by titration with $0.01M$ disodium edetate in buffered alkaline solution using solochrome black solution as indicator since it forms chelates with both Ca^{2+} and Mg^{2+}. To determine calcium hardness only, murexide is used as indicator in strongly alkaline solution since, under these conditions, it chelates with Ca^{2+} only. If the water contains the alkaline-earth metals only (as is usually the case), the buffer used in the above procedure is satisfactory for total hardness; if, however, heavy metals, e.g. Cu, Fe, Sn, Zn, Pb and Al are present in traces, they will lead to errors, and an alternative buffer containing sulphide is recommended.

Total Hardness

Buffer solution. Dissolve sodium edetate (0.93 g) and magnesium sulphate (0.616 g) in strong ammonia-ammonium chloride solution (1000 ml).

Method. Add buffer solution (2 ml) to the sample (100 ml). Add 6 drops (from a dropper) of Mordant Black 2 indicator solution, and titrate with $0.01M$ disodium edetate until there is no further colour change. The indicator changes colour from pink through mauve to blue, and titrant should be added dropwise towards the end point. Add a few drops of titrant in excess, and repeat the titration, taking the colour of the over-titrated sample as the end point.

$$\text{Total Hardness} = \frac{\text{Volume of } 0.01M \text{ disodium edetate}}{\text{Volume of Sample}} \times 1000 \text{ p.p.m. } CaCO_3$$

Calcium Hardness

Buffer solution. Potassium cyanide solution (10 per cent; 1 ml); sodium sulphide solution (10 per cent; 1 ml); dilute with N NaOH to 100 ml.

Indicator. Murexide (0.2 per cent) in finely-powdered AnalaR sodium chloride.

Method. Add N NaOH (5 ml) and Murexide indicator (0.2 g) to the sample (100 ml). Titrate with $0.01M$ disodium edetate until the pink colour changes to purple, and addition of excess titrant causes no further change. Repeat and obtain an accurate end point by matching.

BACK TITRATIONS WITH DISODIUM EDETATE

Alum. *Determination of the percentage of* $KAl(SO_4)_2, 12H_2O$

The determination depends upon the reaction expressed by the following equations:

$$Al^{3+}[H_2X]^{2-} \rightarrow [AlX]^- + 2H^+$$
$$\therefore \quad KAl(SO_4)_2,12H_2O \equiv Al^{3+} \equiv Na_2H_2X,2H_2O$$
$$\therefore \quad 474.4 \text{ g } KAl(SO_4)_2,12H_2O \equiv 20\ 000 \text{ ml } 0.05M$$
$$\therefore \quad 0.02372 \text{ g } KAl(SO_4)_2,12H_2O \equiv 1 \text{ ml } 0.05M \text{ disodium edetate}$$

Similarly for Ammonia Alum 0.02267 g $NH_4Al(SO_4)_2,12H_2O \equiv 1$ ml $0.05M$

Method. Dissolve about 1.7 g of sample, accurately weighed, in sufficient Water to produce 100 ml. To 20 ml of this solution, add *0.05M* disodium edetate (30 ml) and Water (100 ml), heat on a water-bath for 10 minutes and cool. Add hexamine (1 g) and titrate with *0.05M* lead nitrate using xylenol orange solution (0.4 ml) as indicator. The colour will change from that of the indicator (yellow at the pH of the titration) to reddish-purple, the colour of the lead complex of the indicator.

Note. The alums are highly hydrated and sampling errors are reduced by making a solution and taking a one-fifth aliquot part for the assay. The aluminium edetate complex is stable (log $K = 15.5$) but it is only formed slowly. Hence, the sample is heated with excess disodium edetate to ensure complete complex formation. Hexamine acts as a buffer, stabilising the pH between 5 and 6, the optimum pH for the titration of the disodium edetate not required by the aluminium with *0.05M* lead nitrate using xylenol orange as indicator.

Cognate Determinations

Aluminium Hydroxide Gel; Dried Aluminium Hydroxide Gel

These preparations require preliminary dissolution in hydrochloric acid. The solution is then adjusted to about pH 5 (methyl red), heated with excess disodium edetate solution, hexamine added, and back-titrated as under Alum.

Aluminium Hydroxide and *Magnesium Trisilicate* in Tablets.

Aluminium is determined as for Alum, after extraction with either hydrochloric acid or perchloric acid, filtering and neutralising.

Aluminium Powder and *Zinc Oxide* in Paraffin Pastes.

The liquid paraffin is dissolved in chloroform. The residue is dissolved in hydrochloric acid, an aliquot neutralised with sodium hydroxide and the aluminium determined as for Alum. Zinc is also determined on a second aliquot after masking the aluminium with triethanolamine by titration as for Zinc Sulphate.

Calcium Phosphate. *Determination of the percentage of* $Ca_3(PO_4)_2$

Since calcium phosphate is insoluble, the sample is dissolved with the aid of heat in excess hydrochloric acid; an aliquot of the solution is treated with excess *0.05M* disodium edetate, strong ammonia-ammonium chloride solution added to bring the pH to 10, and the disodium edetate not required by the sample is back-titrated with *0.05M* zinc chloride using Mordant Black 2 as indicator. This procedure illustrates an alternative way of avoiding the indistinct colour change with calcium alone and Mordant Black 2.

Cognate Determination
Calcium and *Vitamin D Tablets*

Sodium Fluoride. *Determination of the percentage of* NaF

The determination depends upon the reaction expressed by the following equations:

$$2F^- + Pb^{2+} \rightarrow PbF_2 \downarrow$$
$$Pb^{2+} + [H_2X]^{2-} \rightarrow [PbX]^{2-} + 2H^+$$

Method. To the sample (80 mg), accurately weighed, in Water (45 ml) add sodium chloride (0.2 g) and alcohol (95 per cent; 20 ml). Heat to boiling and add *0.05M* lead nitrate initially dropwise and then more rapidly with constant stirring. Heat to coagulate the precipitate, cool to room temperature, filter and wash the residue with small volumes of alcohol (20 per cent). To the filtrate and washings, add hexamine (1 g) and titrate the excess lead nitrate with *0.05M* disodium edetate using xylenol orange solution as indicator. The end point is yellow.

TITRATIONS INVOLVING DISPLACEMENT OF ONE COMPLEX BY ANOTHER

Mercuric Chloride. *Determination of the percentage of* $HgCl_2$

The determination depends upon the reactions expressed by the following equation:

$$HgCl_2 + [H_2X]^{2-} \rightarrow [HgX]^{2-} + 2HCl$$
$$\therefore \quad HgCl_2 \equiv Na_2H_2X, 2H_2O$$
$$\therefore \quad 271.5 \text{ g } HgCl_2 \equiv 20\,000 \text{ ml } 0.05M$$
$$\therefore \quad 0.01358 \text{ g } HgCl_2 \equiv 1 \text{ ml } 0.05M \text{ disodium edetate}$$

Method. Dissolve the sample (about 0.3 g), accurately weighed, in Water (100 ml), and *0.05M* disodium edetate (about 40 ml), strong ammonia-ammonium chloride solution (5 ml) and Mordant Black 2 as indicator and titrate with *0.05M* zinc chloride until the colour changes to purple. Add potassium iodide (3 g) and allow to stand 2 minutes. Titrate again with *0.05M* zinc chloride until a purple colour is obtained a second time.

Note. The disodium edetate solution need not be measured accurately, but there must be an excess present as is shown by the fact that the indicator is blue. The excess edetate is then titrated with *0.05M* zinc chloride until the indicator just changes to purple, the colour of the zinc-indicator complex. Advantage is then taken of the fact that the mercuri-iodide complex is more stable than the mercuri-edetate complex. The former is obtained in the presence of excess potassium iodide:

$$[HgX]^{2-} + 4I^- \rightarrow [HgI_4]^{2-} + X^{4-}$$

The two minutes standing time ensures that the reaction is complete, as shown by the re-appearance of the blue colour of the free indicator. The edetate thus liberated is equivalent to the mercuric chloride in the sample. The method has the advantage that it is specific for mercury.

Sodium Calcium Edetate. *Determination of the percentage of* $C_{10}H_{12}N_2O_8CaNa_2$

The determination depends upon the reactions expressed by the following equations:

$$[CaX]^{2-} + 2H^+ \rightarrow Ca^{2+} + [H_2X]^{2-}$$
$$[H_2X]^{2-} + Pb^{2+} \rightarrow [PbX]^{2-} + 2H^+$$
$$\therefore \quad Na_2[CaX] \equiv [H_2X]^{2-} \equiv Pb^{2+}$$
$$\therefore \quad 374.3 \text{ g } C_{10}H_{12}N_2O_8CaNa_2 \equiv 20\ 000 \text{ ml } 0.05M$$
$$\therefore \quad 0.01871 \text{ g } C_{10}H_{12}N_2O_8CaNa_2 \equiv 1 \text{ ml } 0.05M \text{ lead nitrate}$$

Method. Dissolve the sample (about 0.5 g) accurately weighed in Water (90 ml). Add hexamine (7 g) and dilute hydrochloric acid (5 ml), and titrate with *0.05M* lead nitrate using xylenol orange as indicator.

Note. The hexamine-hydrochloric acid buffer has a pH of about 4, and at this pH the calcium edetate complex breaks down. The liberated edetate can then be titrated with lead ions, which form a stable complex at this pH.

MISCELLANEOUS COMPLEXOMETRIC METHODS

Volumetric Determination of Sulphate

Many methods for the determination of sulphate (and barium) have been described. Direct titration of sulphate at pH 10 with standard barium chloride solution gives a poor end point with Mordant Black 2. This may be improved by adding excess standard barium chloride solution, filtering off the barium sulphate, and determining the barium not precipitated as sulphate by adding excess disodium edetate, and back-titrating the disodium edetate not required by the barium with standard magnesium chloride solution. Alternatively, sulphate may be determined by precipitation as barium sulphate. The precipitate is washed and dissolved by heating in about twice the theoretical amount of standard disodium edetate solution at pH 10, cooled, and the disodium edetate not required by the barium determined by back-titrating with magnesium chloride solution using Mordant Black 2 as indicator.

Yet another method depends upon the fact that tetrahydroxyquinone forms a red complex with barium ions.

Method. To the sample containing the equivalent of 1 to 10 mg of Na_2SO_4 in 25 ml of Water add very dilute sodium hydroxide solution until the solution is just alkaline to phenolphthalein, and then add very dilute hydrochloric acid until it is just neutral to phenolphthalein. Add 25 ml of ethanol (95 per cent) previously neutralised to phenolphthalein and 0.1 g of tetrahydroxyquinone indicator and titrate with standard solution of barium chloride until the first sign of red coloration appears. Repeat the operation omitting the sample. The difference between the two titrations represents the amount of solution of barium chloride required by the sample; each ml of standard solution of barium chloride is equivalent to 0.0025 g of sulphate calculated as Na_2SO_4.

Note. The reaction at the end point is very slow. Allow about 30 seconds between each drop when approaching the end point which is the first trace of red or pink coloration in the yellow solution. A little practice is required before reproducible results are obtained. It is desirable to keep the volume of standard barium chloride solution to between 2 and 3 ml, otherwise the precipitate of barium sulphate tends to obscure the end point.

Standard solution of barium chloride. Dissolve 4.220 g of $BaCl_2,2H_2O$ in 1000 ml of Water.

Tetrahydroxyquinone indicator: 1 per cent of tetrahydroxyquinone in sucrose.

Cognate Determination

Sulphate in Eosin. Eosin is removed by precipitation with phosphoric acid, an aliquot of the filtrate exactly neutralised, an equal volume of neutralised alcohol (95 per cent) added and the solution titrated with standard barium chloride solution.

Determination of Iron using Tiron

Method. Prepare a solution of the sample containing not more than 1 mg of ferric iron in 100 ml. Transfer a 5 ml aliquot to a 50 ml graduated flask, add 1 ml of Tiron reagent and make up to volume with either the buffer at pH 4 or that at pH 9.5. The intensity of the colour produced may be compared visually or spectrophotometrically with those produced from known amounts of iron under the same conditions. The absorption maxima of the red and blue complexes are at 480 and 620 nm respectively.

Tiron reagent solution (approx. $0.0075M$). Dissolve 0.25 g of Tiron in 100 ml of Water.

Buffer solution, pH 4.0. Dissolve 68 g of sodium acetate, $NaC_2H_3O_2,3H_2O$, and 33.3 ml of $12N$ HCl in sufficient Water to produce 1000 ml. Alternatively, phthalate buffer may be used, but this reduces the sensitivity to 1 in 20 000 000.

Buffer solution, pH 9.5. Dissolve 71.6 g disodium phosphate, $Na_2HPO_4,12H_2O$, and 4 ml of N NaOH in sufficient Water to produce 1000 ml.

Determination of Lead Using Dithizone

Method. Place the sample containing 1 to 100 micrograms of lead in a separating funnel, add 20 ml of dilute ammonia solution, 1 ml of 10 per cent KCN solution and 5 ml of 0.005 per cent dithizone in chloroform and shake vigorously. Allow to separate, shake the lower layer with a mixture of 5 ml of dilute ammonia and 1 ml of the KCN solution to remove excess of dithizone. The intensity of the red colour in the chloroform layer may be compared visually or spectrophotometrically with that produced by known amounts of lead under the same conditions. The absorption maximum of lead dithizonate is at 525 nm which is close to the position of minimum absorption of dithizone itself.

11 Miscellaneous Methods

SODIUM NITRITE TITRATIONS

Aromatic primary amines react with sodium nitrite in acid solution (i.e. nitrous acid) to form diazonium salts:

$$C_6H_5.NH_2 + NaNO_2 + HCl \rightarrow C_6H_5.N_2Cl + NaCl + 2H_2O$$

Under controlled conditions the reaction is quantitative, and can be used for the determination of most substances containing a free primary amino group, as in sulphanilamide and other sulpha drugs.

Observation of the end point depends upon the detection of the small excess of nitrous acid which is then present. This can be demonstrated visually using starch-iodide paper or paste as external indicator:

$$KI + HCl \rightarrow HI + KCl$$
$$2HI + 2HNO_2 \rightarrow I_2 + 2NO + 2H_2O$$

The iodine liberated reacts with starch to form a blue colour.

Alternatively the end point may be detected electrometrically, using a pair of bright platinum electrodes immersed in the titration liquid. Electrode polarisation occurs when a small voltage (30–50 mV) is applied across the electrodes, and no current flows through the sensitive galvanometer included in the circuit, during the course of the titration. Liberation of excess nitrous acid at the end point depolarises the electrodes, current flows in the galvanometer, and a permanent deflection of the galvanometer needle is observed. This is known as the dead-stop end point. The electrodes must be clean, otherwise the end point is sluggish. Cleaning can be accomplished by immersing the electrodes in boiling nitric acid containing a little ferric chloride for about 30 seconds, and then washing with water. A blank determination is necessary; the difference between the two titrations represents the volume of sodium nitrite solution equivalent to the aromatic amine.

Sodium Aminosalicylate. *Determination of the percentage of $C_7H_6NNaO_3$ calculated with reference to the substance dried to constant weight at 105°*

This determination depends upon the reaction expressed by the following equation:

$$\therefore \quad 175.2 \text{ g } C_7H_6NNaO_3 \equiv NaNO_2 \equiv 1000 \text{ ml } M$$
$$\therefore \quad 0.01752 \text{ g } C_7H_6NNaO_3 \equiv 1 \text{ ml } 0.1M \text{ NaNO}_2$$

Method 1. Visual end point

Weigh the sample (about 2.5 g), accurately, into a funnel in the mouth of a 250 ml graduated flask. Wash through with concentrated hydrochloric acid (50 ml) and Water. Dissolve, add potassium bromide (5 g) and adjust to volume. Pipette 50 ml into a conical flask, cool to below 15° (cold tap) and titrate slowly with $0.1M$ NaNO$_2$ solution shaking continuously until a distinct blue colour is produced when a drop of titrated solution is placed on a thin film of starch-iodide paste spread on a porcelain tile five minutes after the last addition of the $0.1M$ NaNO$_2$ solution. The latter should be added in quantities of 0.1 ml towards the end of the titration. Note that sometimes a blue colour results even after only a few ml of sodium nitrite have been added. The reaction is slow in starting but the solution behaves normally as the titration proceeds. A blue colour is produced on the starch iodide paste before the end point is reached, but at this stage it takes time to develop. Only at the end point does the blue colour appear on the paste instantaneously.

Method 2. The electrometric end point (see Part 2)

Accurately weigh the sample (about 0.5 g) into a 250 ml beaker and dissolve in hydrochloric acid (10 ml) and Water (75 ml). Insert a pair of bright platinum electrodes into the solution, connected through a sensitive galvanometer and a suitable potentiometer to a 2 V battery in such a way as to produce a potential drop of between 30 and 50 mV across the electrodes. Titrate slowly with $0.1M$ NaNO$_2$, stirring continuously (mechanical stirrer), until a permanent deflection of the galvanometer is observed at the end point.

Cognate Determinations

Benzocaine

Calcium Aminosalicylate

Chloramphenical. The nitro group is just reduced with zinc and hydrochloric acid, and the reduction product titrated with $0.1M$ NaNO$_2$ using starch-iodide paper.

Dapsone

Primaquine Phosphate

Primaquine Phosphate Tablets

Procainamide Hydrochloride

Procainamide Injection

Procaine Hydrochloride

Sodium Aminosalicylate Tablets

Sulphacetamide Eye Ointment

Sulphacetamide Sodium

Sulphadiazine

Sulphadiazine Injection

Sulphadiazine Tablets

Sulphadimethoxine

Sulphadimethoxine Tablets

Sulphadimidine

Sulphadimidine Injection

Sulphadimidine Sodium

Sulphadimidine Tablets

Sulphamethizole

Sulphamethizole Tablets

Sulphamethoxydiazine

Sulphamethoxydiazine Tablets
Sulphamethoxypyridazine
Sulphamethoxypyridazine Tablets
Sulphapyridine
Sulphapyridine Tablets

Succinylsulphathiazole. *Determination of the percentage of* $C_{13}H_{13}O_5N_3S_2$ *calculated with reference to the substance dried to constant weight at* 105°

This substance is determined similarly to sulphaguanidine after a preliminary hydrolysis of the protecting acyl group. This determination depends upon the reactions expressed by the following equations:

$$HOOC.CH_2CH_2.CO.NH-\langle\bigcirc\rangle-SO_2.NH.[\text{thiazole}]$$

$$\longrightarrow \quad \begin{matrix} CH_2.COOH \\ | \\ CH_2.COOH \end{matrix} + H_2N-\langle\bigcirc\rangle-SO_2NH.[\text{thiazole}]$$

$$H_2N-\langle\bigcirc\rangle-SO_2NH-[\text{thiazole}] + NaNO_2 + 2HCl$$

$$\longrightarrow \quad ClN_2-\langle\bigcirc\rangle-SO_2NH.[\text{thiazole}] + NaCl + 2H_2O$$

Method. Dissolve the sample (about 0.5 g), accurately weighed, in a mixture of concentrated hydrochloric acid (33 ml) and Water (66 ml). Reflux for one hour to hydrolyse the substance as shown by the equation above, and then complete the assay as described under Sodium Aminosalicylate.

Cognate Determinations

Phthalylsulphathiazole
Phthalylsulphathiazole Tablets
Succinylsulphathiazole Tablets

Isocarboxazid. *Determination of the percentage of* $C_{12}H_{13}N_3O_2$

This determination depends upon the rapid breakdown of Isocarboxazid in acid solution to benzylhydrazine, which reacts quantitatively with nitrous acid to yield the benzylazide.

$$\xrightarrow{H_2O}$$

$$+\; C_6H_5.CH_2NH.NH_2$$

$$C_6H_5.CH_2.NH.NH_2 + HNO_2 \rightarrow C_6H_5.CH_2.N_3 + 2H_2O$$
$$231.3 \text{ g } C_{12}H_{13}N_3O_2 \equiv NaNO_2 \equiv 1000 \text{ ml } M$$
$$\therefore \quad 0.02313 \text{ g } C_{12}H_{13}N_3O_2 \equiv 1 \text{ ml } 0.1M \text{ NaNO}_2$$

MERCURIC ACETATE TITRATIONS

Methylthiouracil. *Determination of the percentage of $C_5H_6ON_2S$ calculated with reference to the substance dried to constant weight at 105°*

This determination depends upon the conversion of methylthiouracil to a water-soluble disodium derivative with sodium hydroxide, and titration of the latter in an acetate buffer system with mercuric acetate:

$$\therefore \quad 2 \times 142.2 \text{ g } C_5H_6ON_2S \equiv Hg(O.COCH_3)_2 \equiv 20\ 000 \text{ ml } 0.05M$$
$$\therefore \quad 0.01422 \text{ g } C_5H_6ON_2S \equiv 1 \text{ ml } 0.05M \text{ mercuric acetate}$$

Method. Accurately weigh the sample (about 0.35 g) and warm with *0.1N* NaOH (50 ml) and Water (200 ml) to dissolve. Cool, add sodium acetate (10 g), and acetic acid until the solution is just acid to litmus. Titrate with *0.05M* mercuric acetate solution using a freshly prepared solution (1 ml) of diphenylcarbazone (0.5 per cent w/v) in ethanol (95 per cent) as indicator. The end point is indicated by the formation of the mercury-diphenylcarbazone complex which is rose-violet in colour. This colour should persist for 2 to 3 min.

A separate portion of the sample must be dried, and the moisture loss at 105° determined.

Cognate Determinations

Methylthiouracil Tablets
Penicillamine
Penicillamine Capsules
Penicillamine Hydrochloride
Penicillamine Tablets
Propylthiouracil
Propylthiouracil Tablets

DETERMINATION OF METHOXYL

Chloromethoxyacridone. *Determination of the percentage of* $C_{14}H_{10}O_2NCl$

The proportion of methoxyl radical in the sample is determined by a modification of the Zeisel method. The reaction is expressed by the following equations:

$$R.OCH_3 + HI \rightarrow ROH + CH_3I$$
$$CH_3I + CH_3COOH \rightarrow CH_3COO.CH_3 + HI$$
$$HI + 3Br_2 + 3H_2O \rightarrow HIO_3 + 6HBr$$
$$(Br_2 + H.COOH \rightarrow 2HBr + CO_2) \text{ Removal of excess bromine}$$

Add KI,
$$HIO_3 + 5HI \rightarrow 3I_2 + 3H_2O$$
$$\therefore \quad C_{13}H_7ONCl(OCH_3) \equiv CH_3I \equiv HIO_3 \equiv 6I \equiv 60\,000 \text{ ml } 0.1N$$
$$\therefore \quad 259.7 \text{ g } C_{13}H_7ONCl(OCH_3) \equiv 60\,000 \text{ ml } 0.1N$$
$$\therefore \quad 0.004328 \text{ g } C_{14}H_{10}O_2NCl \equiv 1 \text{ ml } 0.1N \text{ Na}_2S_2O_3$$

Method. Transfer an accurately weighed quantity of sample, approximately equivalent to 50 mg of methyl iodide to the flask. Add a short piece of glass rod to ensure even boiling, melted phenol (2.5 ml) and hydriodic acid (5 ml). Assemble the apparatus as in Fig. 24 after half-filling the scrubber S with a 25 per cent solution of sodium acetate. Place in receiver A about 6 ml and in receiver B about 4 ml of 10 per cent solution of potassium acetate in glacial acetic acid to which 4 drops of bromine have been added. Pass a slow stream of carbon dioxide through the capillary side arm of the flask. This sweeps out methyl iodide from the flask as it is formed. Carefully heat the flask using a mantled microbunsen in such a way that the vapours of the boiling liquid rise half way up the condenser. Continue heating thus for 30 minutes. Wash the contents of the two receivers A and B into a conical flask containing 25 per cent w/v aqueous sodium acetate and make up the total volume to 125 ml with Water. Add 6 drops of formic acid and rotate the flask until the bromine colour is discharged. Add a further 12 drops of formic acid and allow to stand 1 to

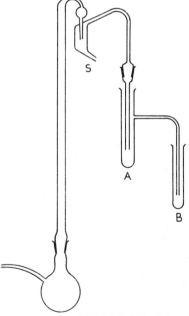

Fig. 24. Methoxyl
determination apparatus

2 minutes. Add potassium iodide (1 g), acidify with dilute sulphuric acid and titrate the liberated iodine with $0.1N$ $Na_2S_2O_3$. Carry out a blank determination.

Cognate Determinations

Methicillin Injection
Methicillin Sodium
Methylcellulose 450

DETERMINATION OF ALCOHOL IN LIQUID GALENICALS

Distillation Procedure (BPC)

The traditional process for the determination of alcohol in liquid galenicals by distillation and physical measurement is no longer official in the British Pharmacopoeia, but is described and recommended, with one or two exceptions, for galenical preparations in the British Pharmaceutical Codex. This process consists of the following steps:

(1) Removal of volatile impurities, if any.
(2) Distillation of all the ethyl alcohol and some of the water.
(3) Dilution of the distillate to four times the original volume and determination of the specific gravity of the resulting solution.
(4) A test on the distillate to determine whether the refractive index is in agreement with the recorded specific gravity. The distillate must comply with the limit test for absence of methyl alcohol.

The alcohol content is calculated from the specific gravity using the table (quadruple bulk) in the British Pharmaceutical Codex.

Three general methods are described in the B.P.C. for the determination of alcohol in galenicals, the choice of method depending upon the nature of the other substances also present.

Method 1. *No other volatile substances present*
Measure 25 ml of the galenical in a graduated flask at a temperature of 20°. Transfer to a distillation flask (500 ml) with about 100 to 150 ml of Water. Add pumice powder to prevent bumping and distil the liquid through a short fractionating column, collecting about 90 ml of distillate in a 100 ml graduated flask. Adjust the temperature of the distillate to 20° and dilute to 100 ml with water at the same temperature. Determine the specific gravity at 20° and the refractive index of the solution at 20°. Refer to the alcohol (quadruple bulk) table. The refractive index calculated on the basis of the observed specific gravity should not differ from the observed refractive index by more than 0.00007. If the distillate complies with this requirement, read off the alcohol content from the table. If it does not comply some impurity is present, and 75 ml of the distillate should be treated with sodium chloride and light petroleum as in Method 2. Distil about 70 ml and dilute to 75 ml, and then re-check the specific gravity and refractive index for mutual agreement.

Acid solutions, such as Morphine Hydrochloride Solution should be neutralised with sodium hydroxide before distillation.

Method 2. *Simple solutions containing other volatile substances*
Measure 25 ml of the galenical at 20° in a graduated flask. Transfer to a separator with about 100 ml of Water. Saturate the solution with sodium chloride and shake

vigorously with an equal volume of light petroleum (b.p. 40 to 60°) for two to three minutes. The sodium chloride helps to salt out unwanted substances which dissolve in the light petroleum. This solvent is used because it is completely insoluble in water and does not extract ethyl alcohol from aqueous solutions. Allow to separate for 15–30 minutes and run the lower aqueous layer into a distillation flask. Wash the organic solution in a separator by shaking vigorously with saturated sodium chloride solution (25 ml) and run the aqueous phase into the distillation flask. In some cases a double separation is used, in which the aqueous liquors are extracted with a second volume of light petroleum. The same volume of wash liquor is used for both petroleum extracts. Make the mixed aqueous solutions just alkaline with $0.1N$ NaOH (use solid phenol-phthalein as indicator). Distil as in Method 1.

This method may be applied to Camphorated Opium Tincture, Chloroform Spirit, and Concentrated Peppermint Water.

Method 3. *Complex solutions containing other volatile substances*
Measure 25 ml of the galenical as before, and transfer to a 500 ml distillation flask with water (100 to 150 ml). Add pumice powder and distil about 100 ml. Treat the distillate by Method 2.

This method is used for galenicals such as Coal Tar Solution, which contain other volatile substances or material which precipitates on dilution with water; these combinations would complicate the petroleum ether extractions.

Limit Tests to be Applied to the Distillate

Test for methyl alcohol. Adjust to approximately 10 per cent ethyl alcohol by the addition of either water or of Alcohol (95 per cent). To 5 ml add solution of potassium permanganate (3 per cent; 2 ml) in dilute phosphoric acid (13.5 per cent) and set aside for 10 minutes. Add a solution of oxalic acid (5 per cent in 50 per cent sulphuric acid) (2 ml) and Schiff's reagent (5 ml). Set aside for thirty minutes at room temperature. The solution should remain colourless.

Note. Some preparations give a positive reaction due to the presence of methyl compounds, and in this case the test is not significant.

Test for the absence of isopropyl alcohol. Add 1 ml of the distillate to 2 ml of a solution of mercuric sulphate in dilute sulphuric acid. No precipitate should be formed when the mixture is heated to boiling.

Gas-liquid Chromatographic Procedure

The g.l.c. method is carried out on a 1.5 m column of porous polymer beads (Porapak Q or Chromosorb 101) at 125°, with nitrogen as carrier gas, and flame ionisation detector. *n*-Propyl alcohol is used as an internal standard, and the column calibrated with a solution containing *n*-propyl alcohol (5 per cent v/v) and dehydrated ethanol (5 per cent v/v) in Water. This is compared with a second chromatogram prepared from the sample suitably diluted with Water to contain between 4 and 6 per cent of ethanol, to ensure that no impurity is present with the same retention time as the internal standard. Alcohol content is then determined from a third chromatogram prepared from the sample, diluted as before, containing the internal standard (5 per cent v/v).

Determination of Industrial Methylated Spirit

The g.l.c. method is applied to the determination of both ethanol and methanol in Industrial Methylated Spirit, and preparations prepared with it. The method is essentially that used for determination of ethyl alcohol, except that a fourth solution containing the same internal standard, n-propyl alcohol (5 per cent v/v), together with methanol (0.25 per cent v/v), and instrument amplifier gain is increased when the methanol peak is recorded. The total alcohol content must be within the range specified in the monograph for the preparation under examination, and the methanol content must reasonably conform with that calculated on the basis that Industrial Methylated Spirit has been used.

DETERMINATION OF CINEOLE

Cineole, like most ethers, is relatively unreactive, and its determination in volatile oils depends upon the formation of an addition compound with an equimolecular proportion of o-cresol. When pure, this addition compound

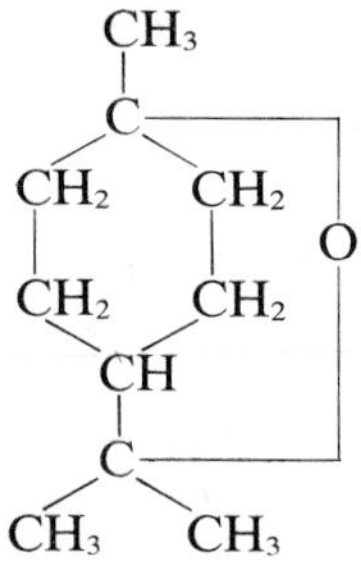

melts at 55.2°, and its melting point is depressed by the presence of impurities. Table 26 (reproduced here from the British Pharmacopoeia) correlates the freezing point of the cineole-o-cresol mixture with various percentages of cineole in Eucalyptus Oil.

Table 26

Freezing point (°)	% w/v of cineole	Freezing point (°)	% w/v of cineole
41	68.6	49	84.2
42	70.5	50	86.3
43	72.3	51	88.8
44	74.2	52	91.3
45	76.1	53	93.8
46	78.0	54	96.3
47	80.0	55	99.3
48	82.1	55.2	100.0

Intermediate values can be obtained by interpolation.

Method. Dry the oil by shaking with anhydrous calcium chloride. Weigh precisely 3 g of the oil into a hard glass test tube (80 mm × 15 mm) supported vertically on the balance pan by inserting it in a cork with hole as in the diagram (Fig. 25A). It is convenient to add the oil from a dry teat pipette. Add precisely 2.1 g of melted *ortho*-cresol. Warm the tube gently until the contents are completely molten and insert

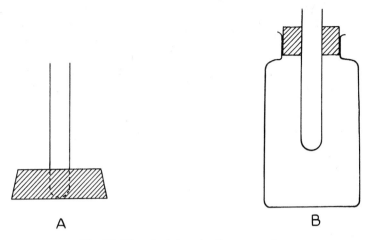

Fig. 25. Cineole determination apparatus

the tube in a bored cork in a wide-mouthed bottle (Fig. 25B). Allow to cool slowly, stirring continuously by rubbing the inside of the tube with the bulb of a thermometer graduated in fifths of a degree. Note the highest temperature recorded on the thermometer whilst crystallisation is occurring. Remelt and repeat the process until two consecutive identical results are obtained.

DETERMINATION OF PHENOLS AS ALKALI-SOLUBLE MATTER

Clove Oil

Eugenol, which is present in the oil together with acetyl eugenol and other phenolic substances to the extent of 85 to 90 per cent w/v is determined by treating the oil with a solution of potassium hydroxide and measuring the material not dissolved. A Cassia flask is used. It has a capacity of about 150 ml (Fig. 26). The neck of the flask is long, having a capacity of 10 ml and being of such diameter that it is not less than 15 cm in length. It is graduated in tenths of a ml. Before use the flask should be cleaned with sulphuric acid and well rinsed with water to free it from grease.

Method. Plase 80 ml of 5 per cent aqueous potassium hydroxide in the flask and pipette in 10 ml of the sample. The phenolic eugenol dissolves in the potassium hydroxide and non-phenolic material (terpenes) forms a separate layer on the surface of the liquid. Shake thoroughly at five minute intervals during half an hour. Gradually add more aqueous potassium hydroxide, so that the undissolved oil is slowly raised into the graduated neck of the flask. Allow to stand for not less than 24 hours and record the volume of undissolved oil. The volume of undissolved oil should be between 1 and 1.5 ml.

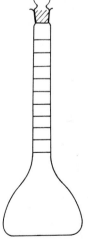

Fig. 26. Cassia flask

AQUAMETRY

Water, in small quantities, is generally controlled by a loss on drying under specified conditions. When present in appreciable amount, the water content may often be readily determined by titration with *Karl Fischer Reagent* (K.F.R.) which contains iodine, sulphur dioxide, anhydrous methanol and anhydrous pyridine. The reaction of K.F.R. is stated to proceed in two stages:

$$3\ C_5H_5N + SO_2 + I_2 + H_2O \longrightarrow 2\ C_5H_5N.HI + C_5H_5N\begin{smallmatrix}SO_2\\|\\O\end{smallmatrix}$$

$$C_5H_5N\begin{smallmatrix}SO_2\\|\\O\end{smallmatrix} + CH_3OH \longrightarrow C_5H_5N\begin{smallmatrix}SO_4.CH_3\\ \\H\end{smallmatrix}$$

Thus each molecule of iodine is equivalent to one molecule of water.

Apparatus

Many forms of apparatus for the determination of water by the Karl Fischer method are available commercially. Only one design will be considered here and the apparatus described is suitable for materials having water contents of about 0.1 per cent.

Figure 27 illustrates the main features of the apparatus. The reaction vessel should have a working capacity of at least 60 ml, the side tube being for the

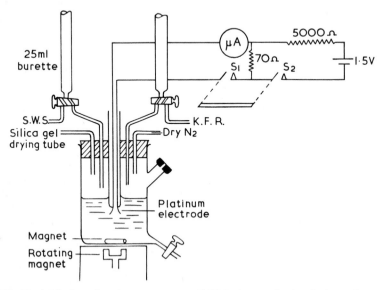

Fig. 27. Karl Fischer titration apparatus. S.W.S. is standard solution of water in Methanol. The dry N_2 inlet and outlet may be omitted.

insertion of liquid samples. The galvanometer ($100\ \mu A$ full scale deflection) should have a resistance of less than $100\ \Omega$. Other apparatus required includes 10 and 20 ml pipettes, a weighing pipette, all of which should be fitted with hypodermic needles (1 mm bore); volumetric flasks (100 ml) closed with latex vaccine caps, two or three drying bottles (containing freshly activated silica gel) and a final wash-bottle containing K.F.R. for drying the nitrogen supply.

Note 1. Precautions must be taken to prevent atmospheric moisture from entering the apparatus, by using suitable guard tubes filled with freshly activated silica gel.

Note 2. Lubricate the stop-cocks with silicone grease, and not petroleum jelly which reacts with K.F.R.

Note 3. (*a*) When titrating with K.F.R. only the cathode is polarised, the anode is depolarised by iodide ions. At the end point free iodine depolarises the cathode, and therefore, the current rises rapidly.

(*b*) When titrating with the water-methanol reagent the electrodes are both depolarised, hence the current is large. At the end point (when all the iodine is reduced) the iodide ions polarise the cathode, and therefore, the current decreases rapidly almost to zero. The zero of the meter scale is reached only when an excess of the water-methanol reagent is added.

Preparation of K.F.R.

Method. Into a glass-stoppered flask (about 750 ml capacity) place anhydrous methanol (400 ml, containing $\not> 0.03$ per cent water) and pure dry pyridine (80 g). Immerse the flask in a freezing mixture and slowly pass sulphur dioxide into the cold solution, with continuous agitation, until the increase in weight is 20 g. Care must be taken at all times to prevent access of moisture. Finally add iodine (45 g), shake to dissolve, and allow the mixture to stand for 24 hours before use. The reagent prepared in this manner will have a water equivalent of 3.5 mg/ml approximately.

Water-Methanol Reagent

Prepare a solution containing 0.15 per cent w/v water in anhydrous methanol.

Standardisation of the Water-Methanol Reagent

Run into the reaction vessel sufficient K.F.R. to cover the electrodes. Pass dry nitrogen through the apparatus, switch on the magnetic stirrer and leave for about 15 minutes to remove traces of moisture. Close switch S_1/S_2 and run in sufficient water-methanol reagent to reduce the meter reading to about $10\ \mu A$ and leave for a further 15 minutes. The galvanometer reading should remain constant throughout this period, thus indicating that the apparatus is dry and free from air leaks. Add more water-methanol reagent to give a meter reading of about $2\mu A$. Partially drain the reaction vessel, if necessary, and run in exactly 10 ml of K.F.R. and titrate with the water-methanol reagent until the meter again reads $2\ \mu A$. Repeat until successive titrations agree to within 0.10 ml and open switch S_1/S_2. Calculate the strength of the water-methanol reagent in terms of the K.F.R., expressing the result as follows:

$$1 \text{ ml water-methanol reagent} = M \text{ ml K.F.R.}$$

Standardisation of K.F.R.

Karl Fischer reagent is not completely stable and requires standardisation with a *standard solution of water in methanol* (S.W.S.).

Method. Two thoroughly dry volumetric flasks (100 ml), fitted with vaccine caps and filled with dry nitrogen, are required. Fill one flask (A) with anhydrous methanol, using a hypodermic needle, and allow the displaced nitrogen to escape through another needle protected by a guard tube. Similarly, half fill the other flask (B) with methanol, add about 0.5 g of distilled water, accurately weighed (w g) from the weighing pipette and adjust to volume with methanol. Remove the needles and shake thoroughly.

Allow most of the liquid in the reaction vessel to drain and transfer to it, by means of a dry pipette, 10 ml of methanol from flask A. Pass dry nitrogen through the apparatus, switch on the magnetic stirrer and close switch S_1/S_2. Add the K.F.R. until a definite brown colour persists for at least 20 seconds, when the meter should read about 80 μA (more or less depending on the size and distance apart of the electrodes). Record accurately the volume of K.F.R. added (v_1 ml). Titrate with the water-methanol reagent until the meter gives a small positive reading of about 2 μA. Record the volume (v_2). Repeat until successive titrations do not differ by more than 0.10 ml.

Repeat the titration using 10 ml of the *standard solution of water in methanol* from flask B. Record accurately the volume of K.F.R. (v_3) and water-methanol reagent (v_4) added. Open switch S_1/S_2.

Calculate the water equivalent (F), in mg water per ml, of the K.F.R. from the formula:

$$F = \frac{100w}{(v_3 - v_1) + M(v_2 - v_4)}$$

The water content of the water-methanol reagent is MF mg/ml and this value may be used for subsequent re-standardisation of the K.F.R.

Determination of Water Content of a Liquid

Method. Into the dry reaction vessel, through which nitrogen is passing, place about 20 ml of methanol and run in sufficient K.F.R. to give a reading of over 20 μA when switch S_1/S_2 is closed. Slowly add water-methanol reagent until the meter reads about 2 μA. Pipette into the reaction vessel 10 or 20 ml of the sample (v_1 ml) and run in a measured excess of K.F.R. (v_2 ml). Titrate with the water-methanol reagent until the meter reads about 2 μA (v_3 ml). Open switch S_1/S_2.

Calculate the percentage w/w of H_2O in the sample from the formula:

$$\text{per cent w/w } H_2O = \frac{F(v_2 - Mv_3)}{10\, v_1 S}$$

where F = the water equivalent of the K.F.R. (mg/ml).
      M = the volume of K.F.R. equivalent to 1 ml water-methanol reagent.
      S = the specific gravity of the sample.

Note: for samples insoluble in methanol, solvents such as dioxan, pyridine and benzene-methanol mixtures may be used.

Determination of Water in a Solid Sample

Method. Prepare a 10 per cent w/w solution of the sample in anhydrous methanol, or other suitable solvent and note the volume of solvent required. Transfer 10 or 20 ml of the solution to the reaction vessel, add excess K.F.R. and titrate with the water-methanol reagent, as above. Similarly, titrate 10 or 20 ml of solvent alone.

$$\text{per cent w/w } H_2O = \frac{F}{10\ w} \left[(v_1 - v_2 M) - \frac{V(v_3 - v_4 M)}{100} \right]$$

where w = weight of sample (g) placed in the reaction vessel.

v_1 = volume of K.F.R. added to the sample (ml).

v_2 = volume of water-methanol reagent used in the back-titration of the sample.

v_3 = volume of K.F.R. added to the solvent blank.

v_4 = volume of water-methanol reagent used in the back titration of the solvent blank.

V = volume of solvent in 100 ml solution of sample.

F = water equivalent of the K.F.R. (mg/ml).

M = volume of K.F.R. equivalent to 1 ml water-methanol reagent.

The method is used to control the water content of
Betamethasone Sodium Phosphate
Cloxacillin Sodium
Cyclophosphamide
Cyclophosphamide Injection
Prednisolone Sodium Phosphate
Procaine Penicillin
Fortified Procaine Penicillin Injection
Sodium Fusidate

OXYGEN FLASK COMBUSTION

Flask combustion analysis as developed by Schöniger (1955) provides a rapid method for the determination of organically-combined halogens. Complete combustion of the organic compound in oxygen destroys organic matter, releasing the halogen, which is absorbed in the combustion flask into sodium hydroxide solution to yield, in the case of iodo-compounds, a mixture of iodide and iodate. Subsequent oxidation (with bromine) and then acidification in the presence of iodide gives a six-fold yield of free iodine which is determined by titration with sodium thiosulphate.

The method is used widely to determine iodine in organo-iodine compounds, e.g. Acetrizoic Acid, chlorine in Dichlorophen, and flourine in Fluocinoline Acetonide. It is also used as an identity test to confirm the presence of chlorine in Diloxamide Furoate and Hydrochlorthiazide. In other instances, flask combustion halogen determination forms the basis of limit tests designed to limit contamination of Cortisone Acetate and similar steroids by fluorinated intermediates.

Flask combustion analysis has also been adapted to the determination of elemental sulphur, which is oxidised to sulphur trioxide absorbed in water (and peroxide to ensure complete oxidation) and titrated. The same principle is used to limit sulphurous contaminants in spironolactone. Other applications of flask combustion include the determination of organo-mercury compounds, and the combustion of water in animal tissue and blood samples prior to measurement of their carbon-14 and tritium content.

Apparatus

The apparatus (Fig. 28) consists of an iodine-flask (500 ml) conforming to British Standard Specification (BS 2735 : 1956), specially modified by fusion into the stopper of a platinum wire, to which is attached a platinum gauze basket of specified dimensions. The gauze complies with the dimensions of a No. 36 sieve.

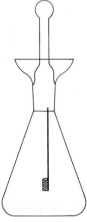

Fig. 28. Oxygen flask combustion apparatus

Di-iodohydroxyquinoline. *Determination of the percentage* $C_9H_5I_2NO$, *calculated with reference to the dried substance*

The determination depends upon the reactions expressed by the following equations:

$$3 \quad \underset{\substack{I \\ \\ OH}}{\overset{I}{\bigcirc\!\!\!\bigcirc_N}} \quad \xrightarrow{O_2/Pt} \quad 3I_2$$

$$3I_2 + 6NaOH \rightarrow 5NaI + NaIO_3 + 3H_2O$$

$$5[NaI + 3Br_2 + 3H_2O \rightarrow NaIO_3 + 6HBr]$$

$$6[NaIO_3 + 5KI + 6HBr \rightarrow 3I_2 + NaBr + 5KBr]$$

$$I_2 + 2Na_2S_2O_3 \rightarrow 2NaI + Na_2S_4O_6$$

$\therefore \quad 3 \times 397$ g $C_9H_5I_2NO \equiv 6I \equiv 6NaIO_3 \equiv 36I \equiv 36\,000$ ml N

$\therefore \quad 0.6616$ mg $C_9H_5I_2NO \equiv 1$ ml $0.02N$ $Na_2S_2O_3$

Method. Accurately weigh the sample (20 mg) onto a strip of filter paper (Whatman No. 1) about 5 cm by 3 cm which has been folded into three along its length. Wrap the sample in the filter paper by folding in the outer thirds and rolling up the folded strip. Place the roll in the platinum gauze cup, and insert a narrow strip of filter paper into the roll to act as a fuse. Displace air from the flask with oxygen, moisten the neck of the flask with Water, place Water (10 ml) and N sodium hydroxide (2 ml) in the flask and fill it with oxygen. Light the end of the fuse, insert the stopper immediately and *hold it firmly in place.* When the sample is burning vigorously, tilt the flask to prevent incompletely burned material from falling into the liquid. As soon as combustion is complete, shake the flask vigorously for 5 minutes. Place a few ml of Water in the cap, carefully withdraw the stopper, and rinse the stopper and wire assembly with Water. Add an excess of acetic bromine solution (5–10 ml) and allow to stand for two minutes; the bromine in potassium acetate-acetic acid buffer solution completes the oxidation of iodide to iodate. Add formic acid (0.5–1 ml) to destroy excess bromine (reduction to hydrobromic acid) and sweep out any bromine vapour from the flask with a current of air. Add potassium iodide (1 g) and titrate with $0.02N$ $Na_2S_2O_3$ using starch mucilage as indicator. Carry out a loss on drying on a separate sample at 105°.

Cognate Determinations

Diodone Injection. Weigh the sample, which is liquid, onto about 15 mg of ashless filter paper flock contained in one half of a methylcellulose capsule, seal the capsule inserting the filter paper fuse between the two parts. Place the capsule in the platinum gauze basket and ignite. Determine the weight per ml of the sample and calculate the percentage w/v of iodine in the sample.

Iophendylate Injection. Treat as described for Diodone Injection.
Liothyronine Sodium
Propyliodone
Propyliodone Oily Injection
Thyroid
Thyroid Tablets
Thyroxine Sodium

Dichlorophen. *Determination of the percentage* $C_{13}H_{10}Cl_2O_2$

Organically-combined chlorine is converted to chlorine by oxygen flask combustion and the halogen absorbed in sodium hydroxide to yield a mixture of chloride and hypochlorite.

$$Cl_2 + 2NaOH = NaCl + NaOCl$$

Treatment with hydrogen peroxide converts hypochlorite to chloride with release of oxygen. The resulting chloride solution is precipitated quantitatively with silver nitrate, and excess precipitant back-titrated with ammonium thiocyanate solution.

$$NaOCl + H_2O_2 \rightarrow NaCl + H_2O + Cl_2$$
$$2NaCl + 2AgNO_3 \rightarrow 2AgCl \downarrow + 2NaNO_3$$
$$\therefore \quad C_{13}H_{10}Cl_2O_2 \equiv Cl_2 \equiv 2AgNO_3$$
$$\therefore \quad 269.1 \text{ g } C_{13}H_{10}Cl_2O_2 \equiv 2000 \text{ ml } N$$
$$\therefore \quad 2.691 \text{ mg } C_{13}H_{10}Cl_2O_2 \equiv 1 \text{ ml } 0.02N \text{ AgNO}_3$$

Cognate Determination

Dichlorophen Tablets

Fluocinolone Acetonide. *Determination of fluorine*

Organically-combined fluorine is converted to fluorine by combustion in oxygen in a silica or soda glass flask. Borosilicate glass must be avoided as fluorine reacts with boron present in hard glass. Fluorine is absorbed into water and determined colorimetrically at 610 nm using alizarin fluorine blue and cerous nitrate in an acetate buffer, in comparison with a sodium fluoride standard.

Sulphur Ointment. *Determination of the percentage of* S

The determination depends upon combustion of the sulphur to sulphur dioxide and oxidation with hydrogen peroxide of the sulphurous acid (formed in presence of water) to sulphuric acid.

$$S \equiv SO_2 \equiv H_2SO_4 \equiv 2H \equiv 2000 \text{ ml } N$$
$$\therefore \quad 0.3206 \text{ mg S} \equiv 1 \text{ ml } 0.02N \text{ NaOH}$$

The determination of sulphur in sulphur ointment by the flask combustion method may give low results, and the alternative method given on p. 195 is recommended.

Method. Accurately weigh the sample (50 mg) onto a small piece of greaseproof paper. Wrap and fold in a filter paper strip. Insert the roll together with a filter paper fuse into the platinum gauze basket as described for Di-iodohydroxyquinoline. Displace air from the flask with oxygen, moisten the neck of the flask with water, place Water (25 ml) and strong hydrogen peroxide solution (2 ml), previously neutralised with *0.02N* NaOH in the flask, and fill it with oxygen. Light the end of the fuse, insert the stopper immediately and *hold it firmly in place.* When the sample is burning vigorously, tilt the flask to prevent incompletely burned material from falling into the liquid. As soon as combustion is complete, shake the flask vigorously for 5 minutes. Place a few ml of Water in the cap, carefully withdraw the stopper, and rinse the stopper and wire assembly with water, using in all about 100 ml of water. Shake for one minute, and concentrate to 50 ml by boiling. Cool and titrate with *0.02N* NaOH using methyl red solution as indicator.

INDICATOR EXTRACTION TITRATIONS

The formation of highly-coloured complexes between cations and inorganic or organic compounds is well known. It forms the basis of sensitive tests for many metallic ions as the complexes may often be extracted by organic solvents e.g. zinc or lead with dithizone, copper with diethyldithiocarbamate and ferric iron with thiocyanate. Similarly the formation of a blue chloroform-soluble base cobaltithiocyanate forms the basis of a test for the identification of Poldine Methylsulphate. Amine salts and quaternary ammonium compounds may also be identified and determined by formation of a salt or complex with complex metal ions or indicators and extraction of the base-complex into an appropriate solvent. Thus, Prudhomme (1938) found that quinine formed a chloroform-soluble complex with acid dyes such as eosin, and Auerbach (1943) described the use of bromophenol blue for the determination of some quaternary ammonium compounds. Under carefully controlled conditions the extinction (p. 43) of the organic extract may be used as a measure of the amine salt or quaternary ammonium compound. Similar principles apply to the determination of anionic compounds such as alkylaryl sulphonates, but methylene blue must be used instead of the acid indicators.

Although absorptiometry is widely used and is particularly advantageous for small concentrations of active material, titrimetric procedures can also be applied, as for example in the titration of surface-active materials. Anionic and cationic surface-active agents are mutually incompatible and it is possible to titrate one against the other in the presence of a suitable indicator. The indicator forms a chloroform-soluble complex with one of the substances, and the end-point is indicated either by a transfer of colour from chloroform to aqueous phase, or by a change of colour in the chloroform phase.

In these titrations, many amine hydrochlorides act in the same way as cationic surface-active agents and this is evident in the following technique which is closely akin to indicator extraction methods. When an excess of potassium iodide solution is shaken with a quaternary ammonium compound and chloroform, the quaternary ammonium iodide is removed by the chloroform. The excess iodide may be titrated with potassium iodate. If the extraction is made from slightly alkaline solution, only quaternary ammonium compounds are measured, whereas if the aqueous layer is slightly acid, non-quaternary cationic amine impurities are also included.

It is not possible to apply one particular method to all compounds and several factors contribute to this failure. These include:

(a) *The stability of the complex*, which may be so great that the titrant fails to liberate the indicator at the end-point, e.g. the hexamethonium-dimethyl yellow complex is very stable.

(b) *The relative solubilities of the compound in water and chloroform.* It is important for a sharp end-point that the difference in solubility in the two phases is significant, e.g. Dicyclomine Hydrochloride is very soluble in chloroform, whereas Benactyzine Hydrochloride is less so. The former compound is titratable with sodium lauryl sulphate and the end-point is sharp, but that for Benactyzine Hydrochloride is vague.

(c) *Molecular weight.* Compounds of low molecular weight often differ from those of high molecular weight in the readiness with which complexes are formed.

(d) *The pH of the aqueous phase.* Different complexes may require different pH values for extraction into organic phases.

The official methods described below are based on those of Carkhuff and Boyd (1954, Dicyclomine Hydrochloride) and Brown (1953, Benzalkonium Chloride).

Atropine Sulphate Injection. *Determination of the percentage of* $(C_{17}H_{23}NO_3)_2,H_2SO_4,H_2O$

The injection solution diluted with Water and acidified with dilute sulphuric acid is titrated with sodium sulphosuccinate in the presence of chloroform, using dimethyl yellow—solvent blue as indicator. At the outset, the alkaloid sulphate is present in the strongly acid aqueous layer, and the chloroform takes the indicator colour. During the titration, neutral atropine sulphosuccinate, which is both water-soluble and chloroform-soluble, passes into the chloroform layer, but is strongly ion-paired, and hence without effect on the indicator. At the end-point, however, excess sulphosuccinate anion complexes with the indicator to form a chloroform-soluble ion pair which gives the sharp colour change from green to pinkish-grey seen in the chloroform layer.

Method. Dilute the injection solution ($\equiv 10$ mg Atropine Sulphate) with Water (25 ml) in a conical flask. Add dilute sulphuric acid (5 ml), chloroform (75 ml), and dimethyl yellow—solvent blue 19 solution (5 ml). Titrate with standard dioctyl sodium sulphosuccinate solution, shaking vigorously throughout until the chloroform layer changes colour from green to pinkish-grey.

Cognate Determinations

Atropine Sulphate Tablets
Codeine Phosphate Syrup

Dicyclomine Tablets. *Determination of the weight of* $C_{19}H_{35}NO_2,HCl$ *per tablet of average weight*

The determination depends upon the reaction of Dyclomine Hydrochloride with a standard sodium lauryl sulphate solution, in the presence of chloroform, using dimethyl yellow as the extraction indicator.

Method. Weigh and powder 20 tablets. To an accurately weighed quantity of the powder equivalent to about 30 mg of Dicyclomine Hydrochloride in a 100 ml glass-stoppered cylinder, add Water (20 ml) and shake the mixture. Add dilute sulphuric acid (10 ml), dimethyl yellow solution (1 ml) and chloroform (40 ml), and shake well to obtain a bright yellow chloroform layer on allowing to stand. Titrate the mixture with *0.004M* sodium lauryl sulphate, shaking vigorously and allowing the layers to separate after each addition, until a permanent orange-pink colour appears in the chloroform layer.

$$0.001384 \text{ g}/C_{19}H_{35}NO_2, HCl \equiv 1 \text{ ml } 0.004M \text{ sodium lauryl sulphate}$$

Sodium Lauryl Sulphate. *Determination of the percentage of sodium alkyl sulphates calculated as* $C_{12}H_{25}OSO_3Na$

Method. Dissolve the sample (about 0.25 g accurately weighed) in Water and dilute to 100 ml in a volumetric flask. Transfer the solution (10 ml) to a 100 ml glass-stoppered cylinder, add chloroform (40 ml), dilute sulphuric acid (10 ml) and dimethyl yellow solution (1 ml) and titrate with *0.004M* benzethonium chloride, shaking vigorously and allowing the layers to separate after each addition, until a full yellow colour is obtained in the chloroform layer (Note).

$$0.001154 \text{ g } C_{12}H_{25}OSO_3Na \equiv 1 \text{ ml } 0.004M \text{ benzethonium chloride}$$

Note. The determination of the end-point is often assisted by observing the appearance of the drops that form on the surface of the aqueous layer during separation.

Cognate Determinations

Cetrimide Emulsifying Ointment
Emulsifying wax

Method. Dissolve about 0.25 g accurately weighed, as completely as possible in chloroform (40 ml) in a 100 ml glass-stoppered cylinder, add Water (20 ml), and proceed as described for Sodium Lauryl Sulphate from the words 'dilute sulphuric acid (10 ml). . .'

Procyclidine Tablets. *Determination of the percentage of* $C_{19}H_{29}NO,HCl$

Procyclidine Hydrochloride is extracted from the powdered tablets by heating with Water. The base is extracted from an aliquot of a cooled solution (appropriately buffered) as the chloroform-soluble bromocresol purple complex. The indicator base-complex is re-extracted back into aqueous sodium hydroxide, and determined by absorptiometry at 592 nm.

Cognate Determination

Benzhexol Tablets

Benzalkonium Chloride Solution. *Determination of the percentage w/v benzalkonium chloride calculated as* $C_{22}H_{40}ClN$

In this determination, the quaternary salt is extracted as the quaternary ammonium iodide into chloroform. Excess potassium iodide in the aqueous phase is then determined by titration with potassium iodate.

Method. Accurately weigh a sample (Note 1) containing about 0.5 g of anhydrous benzalkonium chloride and transfer with the aid of Water (35 ml) to a 500 ml glass-stoppered conical separator containing chloroform (25 ml). Add N sodium hydroxide (0.5 ml) (Note 2) followed by solution of potassium iodide (5 per cent w/v, exactly 10 ml) (Note 3). Stopper the separator and shake well. Allow the two layers to separate and run off the lower chloroform layer through a loosely-packed plug of absorbent cotton wool (about 0.5 g) placed in a small glass funnel to absorb traces of entrained aqueous liquid. Discard the chloroform layer and repeat the extraction with chloroform (3 × 10 ml) running the chloroform layer each time through the same cotton wool filter. Finally wash the filter with a further 5 ml of chloroform and allow to drain.

Add to the separator, hydrochloric acid (40 ml), previously chilled (Note 4) and titrate the mixture in the separator with *0.05M* potassium iodate until the solution becomes light brown in colour; add chloroform (5 ml), stopper and shake. Continue the titration, with shaking, until the chloroform becomes colourless and the supernatant liquid is clear yellow. Remove the cotton wool from the funnel and add the cotton wool directly to the contents of the separator. Wash the glass funnel with Water (2 or 3 ml) receiving the washings in the separator. Stopper, shake and complete titration if necessary.

Perform a blank by titrating a mixture of Water (20 ml), solution of potassium iodide (exactly 10 ml) and hydrochloric acid (40 ml), adding chloroform (5 ml) when the mixture becomes light brown in colour. The differences between the two titrations represents the *0.05M* potassium iodate equivalent to the anhydrous benzalkonium chloride in the amount of sample taken for assay

$$0.0354 \text{ g } C_{22}H_{40}ClN \equiv 1 \text{ ml } 0.05M \text{ KIO}_3$$

Note 1. The quantity stated is for solutions containing 30–100 per cent of benzalkonium chloride. For solutions containing less than 30 per cent, the volume of liquid used should not exceed 35 ml. If the volume is less, add Water to 35 ml.

Note 2. The amount of alkali used is normally sufficient to liberate all the amines that may be present as impurities, but more alkali should be added if it is suspected that contamination is excessive. If the procedure is carried out in the presence of *0.1N* hydrochloric acid (0.5 ml, or sufficient to ensure amine impurities are present as hydrochlorides) the method will give a result which includes amine impurities.

Note 3. The solution of potassium iodide should be measured by pipette.

Note 4. The addition of chilled hydrochloric acid prevents undue warming-up of the solution on mixing.

Cognate Determinations

Cetrimide
Cetylpyridinium Bromide
Domiphen Bromide

Sodium Tetraphenylborate Titrations

A *0.02M* solution of Sodium Tetraphenylborate may be used to titrate a slightly alkaline solution of some quaternary ammonium compounds, such as cetrimide, in the presence of chloroform and bromophenol blue. At the end-point, the blue chloroform-soluble cetrimide/indicator complex is decomposed and the chloroform layer becomes colourless.

The reagent is also very useful for precipitating, at pH 3.7, the tetraphenylborates of alkaloids and synthetic organic bases. The analytical application of the reagent has been studied by Johnson and King (1962) and the assay described below is as reported by them.

Atropine Sulphate. *Determination of the percentage of* $(C_{17}H_{23}NO_3)_2,H_2SO_4$ *calculated with reference to the substance dried at 110° for four hours*

Note that this method is not official, but it may be applied to a large number of other basic nitrogen compounds either alone or in pharmaceutical preparations such as eye-drops, injections, tablets and suppositories. The following reagents are required.

Bromophenol blue solution of the British Pharmacopoeia.

Buffer solution pH 3.7. Dissolve anhydrous sodium acetate (10 g) in Water (about 300 ml), add bromophenol blue solution (1 ml) and sufficient glacial acetic acid (35 to 40 ml) until the indicator changes from blue to a pure green. Dilute to 500 ml with Water. Dilute with an equal volume of Water before use.

0.005M Cetylpyridinium chloride. Dissolve cetylpyridinium chloride (1.80 g) in ethanol (95 per cent, 10 ml) and dilute to one litre with Water. Store in an amber bottle.

0.01M Sodium tetraphenylborate. Dissolve sodium tetraphenylborate (3.42 g) in Water (50 ml), add moist aluminium hydroxide gel (0.5 g) (Note 1) and shake for 20 minutes. Dilute to 300 ml with Water, dissolve sodium chloride (Note 2) (16.6 g) in the solution and stand for 30 minutes. Filter through two thicknesses of No. 42 Whatman filter paper under suction to yield a clear filtrate. Wash the filter with Water, add the washings to the first filtrate, dilute to 1 litre with Water, and adjust to pH 8.0 to 9.0 with *0.1N* sodium hydroxide. Store the solution in an amber bottle.

0.01M Potassium chloride. Dissolve potassium chloride (analytical reagent grade, 0.1491 g), previously dried at 150° for 1 hour, in buffer solution (200 ml).

Method. Transfer the sample (about 0.3 g, accurately weighed) to a 100 ml volumetric flask and dissolve in buffer solution. Dilute to volume with more buffer solution. Transfer the solution (10 ml) to a dry beaker, add *0.01M* sodium tetraphenylborate (15 ml) with swirling of contents of beaker and allow to stand for 5 minutes. Filter through a dry sintered-glass funnel (porosity 4) under gentle suction into a dry flask. Transfer the filtrate (20 ml) to a 150 ml flask, add bromophenol blue solution (0.5 ml), accurately measured (Note 3) and titrate with *0.005M* cetylpyridinium bromide to a blue end-point (*a* ml). To a further 15 ml of the sodium tetraphenylborate solution, add undiluted buffer solution (4 ml) and bromophenol blue solution (0.5 ml) accurately measured and titrate as described above to the same end-point (*b* ml). The difference in titres, $(b - 5a/4)$ ml, is equivalent to the volume of $0.005M$ sodium tetraphenylborate solution precipitated by the atropine. At the same time, determine the molarity of the cetylpyridinium chloride solution by pipetting $0.01M$ potassium chloride (10 ml) into a dry beaker and continuing as above from the words 'add $0.01M$ sodium tetraphenylborate solution (15 ml). . .'. Let the titration be c ml.

The molarity of the cetylpyridinium chloride solution is given by the relationship

$$M = \frac{10M'}{\left(b - \frac{5c}{4}\right)}$$

where $M' =$ molarity of potassium chloride solution

0.001692 g $(C_{17}H_{23}NO_3)_2,H_2SO_4 \equiv 1$ ml $0.005M$ cetylpyridinium chloride

Note 1. The solution tends to be cloudy and addition of the gel assists in obtaining a clear filtrate.

Note 2. The precipitated tetraphenylborates are semi-colloidal, and the presence of sodium chloride allows ready filtration of the mixture.

Note 3. The volume of indicator must be accurately measured as it introduces a blank of about 0.1 ml.

Cognate Determinations

Note that the method is not official for these substances and a selection only is given

Cocaine Hydrochloride
Hyoscine Hydrobromide
Pethidine Hydrochloride
Pethidine Tablets
Procaine Hydrochloride

References

Auerbach, M. E., *Anal. Chem.* (1943) **15**, 492.
Brown, E. R., *J. Pharm. Pharmac.* (1963) **15**, 379.
Carkhuff, E. D. and Boyd, W. F., *J. Amer. Pharm. Ass. Sci. Edn.* (1954) **43**, 240.
Johnson, C. A. and King, R. E., *J. Pharm. Pharmac.* (1962) **14**, 77T.
Prudhomme, R. O., *Bull. soc. path. exot.* (1938) **31**, 929.

12 Practical Gravimetric Analysis

INTRODUCTION

Gravimetric analysis is a procedure for isolating and weighing an element or compound in as pure a form as possible; the element or compound is separated from a definite portion (weight or volume) of the substance being examined, and the weight of the constituent in the sample calculated from the weight of the product.

In pharmaceutical analysis, the product to be weighed is obtained by one of the following procedures:

  (i) volatilisation or ignition,
  (ii) solvent extraction,
  (iii) precipitation from solution.

Procedures involving volatilisation, ignition and solvent extraction are described in detail in Chapters 1, 4 and 13. Gravimetric procedures involving precipitation from solution are based on the quantitative precipitation of the anion or cation to be determined either in the form of an insoluble compound of definite composition, or as an insoluble compound which leaves a residue of definite composition upon ignition. The techniques involved, i.e. precipitation, filtration, washing of the precipitate, and drying or ignition of the residues to constant weight, are described in Chapter 4.

Selected examples of typical gravimetric determinations involving precipitation are described in the following pages.

DETERMINATION OF ALUMINIUM

Aluminium is determined volumetrically by complexometric titration. The gravimetric method, however, still provides a useful alternative procedure. Aluminium can be precipitated from a solution of the aluminium salt by the addition of ammonium hydroxide in the presence of ammonium chloride.

$$Al^{3+} + 3OH^- \rightarrow Al(OH)_3 \downarrow$$

The gelatinous precipitate is filtered, washed with dilute ammonium nitrate solution, converted to the oxide by ignition and weighed as Al_2O_3.

Sources of error in the determination are the solubility of the aluminium hydroxide if too large an excess of ammonium hydroxide is used, the co-precipitation of metal hydroxides which are normally soluble in ammonium hydroxide, the hygroscopic character of the ignited oxide, and incomplete thermal decomposition of the hydroxide. If a gravimetric method for the determination of aluminium is required, precipitation of its complex with 8-hydroxyquinoline (oxine) from ammoniacal solution or acetic acid—acetate buffer provides a reliable method.

$$Al^{3+} + 3 \quad\longrightarrow\quad Al + 3H^+$$

The precipitate, which is crystalline, is easy to filter. It is washed with water, dried at 130 to 150° and weighed as the complex. The disadvantages of this method are the lack of selectivity of oxine so that all metals except the alkalis, alkaline earths and magnesium must be absent. Another disadvantage of this method is the tendency of aluminium oxinate to adsorb oxine.

Alum. *Determination of the percentage of* Al

$$KAl(SO_4)_2, 12H_2O \equiv Al(C_9H_6NO)_3$$
$$\therefore \quad 26.98 \text{ g Al} \equiv 459.4 \text{ g Al}(C_9H_6NO)_3$$
$$\therefore \quad 0.05871 \text{ g Al} \equiv 1 \text{ g Al}(C_4H_6NO)_3$$

Method. Dissolve the sample (about 0.3 g) accurately weighed, in Water (about 150 ml) containing *0.1N* hydrochloric acid (1 ml) in a 400 ml beaker and heat the solution to about 60°. Add the oxine reagent (25 ml of a 2 per cent solution of 8-hydroxyquinoline in *2N* acetic acid) and then add a *2N* solution of ammonium acetate slowly from a pipette until a precipitate forms if this has not already occurred. Then add a further portion (50 ml) of ammonium acetate solution with stirring. Allow the liquid to stand for 1 hour with frequent stirring and filter through a No. 3 or 4 sintered glass crucible which has been dried to constant weight at 130 to 150°. Wash the precipitate well with cold water and dry at 130 to 150° to constant weight.

DETERMINATION OF AMODIAQUINE HYDROCHLORIDE AS AMODIAQUINE BASE

This method depends upon the precipitation of amodiaquine base which is formed as a filterable precipitate when the salt is decomposed in aqueous solution with dilute ammonia.

$$C_{20}H_{22}ON_3Cl, 2HCl, 2H_2O + 2NH_3 \rightarrow C_{20}H_{22}ON_3 \downarrow + 2NH_4Cl + 2H_2O$$

Amodiaquine hydrochloride contains water of crystallisation, and the percentage base is calculated with reference to the substance dried over phosphorus pentoxide at a pressure not exceeding 5 mm of mercury. The drying is carried out on one portion of the sample, and the assay on a separate portion.

Method. Accurately weigh the sample (about 0.3 g) into a 100 ml beaker provided with a stirring rod and clock-glass cover. Dissolve in 50 ml of Water and add dilute ammonia solution, with continuous stirring until the solution is just alkaline. Allow to stand for 30 minutes, and filter through a No. 4 sintered glass crucible, previously dried to constant weight at 105°. Complete the transfer of the precipitate and wash with water, until the washings no longer give a reaction for chloride (test with silver nitrate solution). Dry the residue to constant weight at 105°.

DETERMINATION OF THIAMINE HYDROCHLORIDE AS THIAMINE SILICOTUNGSTATE

Thiamine Hydrochloride. *Determination of the percentage of* $C_{12}H_{17}ON_4SCl,HCl$ *calculated with reference to the substance dried at* 105°

This method depends upon the precipitation of insoluble thiamine silicotungstate which is formed on the addition of a solution of silicotungstic acid to a slightly acidified solution of the sample. The precipitating reagent is a complex silicate, $SiO_2,12WO_3,xH_2O$ of somewhat variable composition in respect of

the degree of hydration. The quality of the reagent, however, is controlled by the requirement that it should contain $\not< 85$ per cent of WO_3 and $\not< 1.85$ per cent of SiO_2. Thiamine silicotungstate, on the other hand, is of constant composition:

$$2C_{12}H_{17}ON_4SCl,HCl + [SiO_2,12WO_3] + 6H_2O \rightarrow$$
$$(C_{12}H_{17}ON_4SCl)_2,[SiO_2(OH)_2,12WO_3],4H_2O$$

$\therefore$ 674.6 g $C_{12}H_{17}ON_4SCl,HCl \equiv 3480$ g of residue

$\therefore$ 0.1936 g $C_{12}H_{17}ON_4SCl,HCl \equiv 1$ g of residue

Method. Accurately weigh the sample (about 0.05 g) and dissolve it in Water (50 ml) in a 250 ml beaker provided with stirring rod and clock-glass cover. Add hydrochloric acid (2 ml) (Note 1), heat to boiling, and add silicotungstic acid solution (10 per cent; 4 ml) in a rapid stream of drops (Note 2). Boil the solution for 2 minutes (Note 3) and filter through a No. 4 sintered glass crucible, previously dried to constant weight at 105°. Wash with a boiling mixture of hydrochloric acid (1 volume) and Water (19 volumes) (Note 4), then with Water (10 ml) and finally with 2 portions of 5 ml each of acetone. Dry the residue to constant weight at 105°.

Notes on the determination of Thiamine Hydrochloride as thiamine silicotungstate

Note 1. Excess of hydrochloric acid is necessary to produce a readily filterable precipitate.

Note 2. The rate of addition does not influence the result, if the thiamine hydrochloride is reasonably pure, but slow addition in the presence of appreciable impurity may give poor results.

Note 3. Boiling for less than 2 minutes gives low results. Longer periods of boiling do not affect the result.

Note 4. The volume of wash liquid should be restricted to about 50 ml. Prolonged washing gives low results.

DETERMINATION OF BENZYLPENICILLIN

Benzylpenicillin may be determined gravimetrically by quantitative precipitation as the 1-ethylbenzylpiperidinium salt. The precipitation is carried out with 1-ethylpiperidine after the sodium or potassium salt of benzylpenicillin has been converted with phosphoric acid to the parent acid and the latter extracted into amyl alcohol.

Method. Dissolve the sample (0.12 g) accurately weighed in ice-cold Water (5 ml) and cool in ice (Note 1). Add amyl acetate (5 ml), previously saturated with 1-ethylpiperidinium benzylpenicillin at room temperature, cooled in ice and filtered. Add ice-cold phosphoric acid (20 per cent v/v; 0.5 ml), stopper, shake immediately for 15 seconds, and centrifuge for 30 seconds. Remove the aqueous layer completely by pipette. Add freshly-ignited and powdered anhydrous sodium sulphate (0.5 g), stir, and cool in ice (5 minutes). Centrifuge (30 seconds), cool again in ice (5 minutes). Pipette 3 ml of the supernatant liquid into an accurately-weighed centrifuge tube. Add ice-cold dry acetone (3 ml) previously saturated with 1-ethylpiperidinium benzylpenicillin at room temperature, cooled in ice and filtered. Add 1-ethylpiperidine amyl acetate solution (1.5 ml; Note 2), stir, stopper the tube and cool in ice for 2 hours. Centrifuge (1 minute), break the surface so that all crystalline particles are covered by liquid, and again centrifuge (1 minute). Decant the supernatant liquid, wash the precipitate with ice-cold dry acetone in amyl acetate (1:1; 2 ml) previously saturated with 1-ethylpiperidinium benzylpenicillin, and again centrifuge (1.5 minutes). Decant the supernatant liquid, wash twice with solvent ether (2 ml), centrifuging (1.5 minutes) and decanting each time. Dry to constant weight under vacuum at room temperature.

Note 1. All solutions should be ice-cold.

Note 2. 1-Ethylpiperidine amyl acetate solution is prepared from 1-ethylpiperidine (1 ml) and amyl acetate (8 ml), saturated at room temperature with 1-ethylpiperidinium benzylpenicillin, cooled in ice, and filtered.

DETERMINATION OF HALIDES

Potassium Chloride. *Determination of the percentage of* Cl

The determination depends upon the reaction expressed by the following equations:

$$Cl^- + Ag^+ \rightarrow AgCl$$
$$\therefore \quad 35.46 \text{ g } Cl^- \equiv 143.34 \text{ g AgCl}$$
$$\therefore \quad 0.2474 \text{ g } Cl^- \equiv 1 \text{ g AgCl}$$

Excess of silver nitrate solution is added to the chloride solution containing nitric acid, the precipitate of AgCl collected on a sintered glass or Gooch crucible, washed with very dilute nitric acid solution to remove soluble substances and then dried at 130 to 150° to constant weight (as AgCl). The following procedure applies to the determination of chloride in a soluble chloride containing no interfering substances (Note 1).

Method. Accurately weigh the sample (about 0.2 to 0.3 g) (sodium or potassium chloride: Note 2) into a 400 ml beaker provided with a stirring rod and clock-glass cover. Add Water (150 ml), stir until dissolved, and add concentrated nitric acid (0.5 ml) (Note 3). *Carry out the rest of the experiment in subdued light* (Note 4). To the solution, add *0.1N* silver nitrate solution slowly with constant stirring until a slight excess is present (Note 5). The precipitate coagulates when the silver and chloride ions are in approximately equivalent amount if the silver nitrate is added slowly with constant stirring. Allow the precipitate to settle and add a few drops of silver nitrate solution—no further precipitation should occur if sufficient reagent has been added. Heat the suspension nearly to boiling, stirring vigorously until the precipitate coagulates completely and the supernatant liquid is clear (about 2 or 3 minutes required). Add 2 drops of silver nitrate solution to check that no further precipitation occurs. Set the beaker aside (covered with the clock-glass) for 1 to 2 hours in the dark (Notes 4 and 6). In the meantime dry a No. 3 sintered glass crucible (or a Gooch crucible) to constant weight (±0.0002 g) at 130 to 150°. Washing with Water and then ethanol (Note 7), with gentle suction prior to drying in the oven, is a quick method. This can also be used when the silver chloride has been collected in the filter crucible. Decant the clear cold supernatant solution through the sintered glass (or Gooch) crucible using gentle suction. Wash the precipitate in the beaker two or three times by decantation with cold *0.01N* nitric acid solution (Note 8; approx. 1 to 2 ml of concentrated nitric acid to 1 litre of Water), and then transfer to the crucible. Remove any small film of silver chloride particles from the beaker by means of a policeman. *Check that all the precipitate has been removed from the beaker.* Wash the precipitate in the crucible with small portions of *0.01N* nitric acid, until a few ml of the last washing collected in a test tube give no turbidity on the addition of 2 drops of *0.1N* hydrochloric acid (Note 9). Finally, wash the precipitate with two small portions of Water to remove most of the nitric acid and then with a few portions of ethanol added from a teat pipette. Heat the crucible and its contents at 130 to 150° for 45 minutes, cool in a desiccator and weigh. Repeat the heating and cooling until a constant weight (±0.0002 g) is attained.

Notes on the determination of chloride as silver chloride

Note 1. Anions which yield silver salts insoluble in dilute nitric acid solutions, e.g. bromide, iodide, cyanide, sulphide, must be absent. The presence of heavy metal salts may also interfere.

Note 2. The weight of solid, or volume of solutions used, should contain approximately 0.1 g Cl.

Note 3. The nitric acid is present to prevent the precipitation of silver salts which are insoluble in neutral solution, e.g. silver carbonate and phosphate.

Note 4. Silver chloride is sensitive to light; decomposition to silver and chlorine occurs, the silver remaining colloidally dispersed in the precipitate and imparting to it a purplish colour.

Note 5. A slight excess of silver nitrate reduces the solubility of silver chloride (common ion effect). Under the conditions of the precipitation, little occlusion of silver or other ions occurs.

Note 6. The heating, stirring and standing converts any colloidal precipitate to completely coagulated particles which can be readily filtered. Since silver chloride is more soluble in hot aqueous solution than in cold solution, the liquid is allowed to cool to room temperature before filtering.

Note 7. Silver chloride is insoluble in ethanol; the latter removes much of the water and, being more volatile than water, assists in the drying process.

Note 8. If water is used, silver chloride may become colloidal again and pass through the filter. The wash solution should contain an electrolyte to prevent the peptisation; nitric acid is suitable because any acid adhering to the precipitate is volatilised during the drying process.

Note 9. This indicates that all the silver nitrate (and therefore presumably all water-soluble salts) has been washed out of the precipitate.

Cognate Determinations

Ethyl Iodide. The ethyl iodide is refluxed with ethanolic silver nitrate to give quantitative precipitation of silver iodide, ethanol being used as solvent to bring the reactants into a homogeneous system.

$$C_2H_5I + AgNO_3 \rightarrow C_2H_5.NO_3 + AgI \downarrow$$

4-Nitrobenzyl Chloride. Determined as for Ethyl Iodide.

DETERMINATION OF CHOLESTEROL AS CHOLESTEROL-DIGITONIN COMPLEX

This method depends upon the fact that all 3β-hydroxysterols, of which cholesterol is an example, form an insoluble molecular addition complex with the steroidal saponin, digitonin. The complexes have very low solubilities, and are stable and strongly crystalline.

Cholesterol

Wool Alcohols. *Determination of the percentage of Cholesterol*

$$\underset{\text{Cholesterol}}{C_{27}H_{46}O} + \underset{\text{Digitonin}}{C_{56}H_{92}O_{29}} \rightarrow \underset{\text{Complex}}{C_{83}H_{138}O_{30}} \downarrow$$

∴ 386.3 g Cholesterol ≡ 1616 g of complex
∴ 0.239 g Cholesterol ≡ 1 g of complex

Method. Accurately weigh the sample (about 0.1 g) into a 100 ml conical flask and dissolve in ethanol (90 per cent; 12 ml). Cork and allow to stand at room temperature for twelve hours, filter through a Gooch crucible, and wash with ethanol (90 per cent; 5 ml). To the filtrate and washings, add a solution of digitonin (0.5 per cent w/v; 40 ml) in ethanol (90 per cent), and warm to 60° to ensure that complex formation is complete. Filter through a prepared Gooch crucible, previously dried to constant weight at 105°. Wash the precipitate with ethanol (90 per cent), acetone and hot carbon tetrachloride, allow to drain, and dry to constant weight at 105°.

DETERMINATION OF GOLD SALTS OF ORGANIC ACIDS AS METALLIC GOLD

The method depends upon the oxidation of organic matter to volatile products by digestion with sulphuric acid. Oxidation occurs at the expense of the sulphuric acid which is reduced to sulphur dioxide and hydrogen sulphide. The latter reduces the gold salts to metallic gold which is precipitated. The subsequent addition of nitric acid and digestion ensures the complete decomposition of organic matter and oxidation of inorganic substances such as sulphur (from thiomalic acid) which might otherwise be weighed as gold.

Sodium Auriothiomalate. *Determination of the percentage of* Au

Sodium Auriothiomalate is of undefined constitution and, moreover, contains a small but variable proportion of moisture. The percentage of gold is calculated with reference to the substance dried over phosphorus pentoxide at a pressure not exceeding 5 mm of mercury.

Method. Accurately weigh about 0.2 g of sample into a 100 ml Kjeldahl flask. Add concentrated sulphuric acid (10 ml) and heat gently to boiling (fume cupboard). Boil gently until the liquid is clear and pale yellow in colour. Cool, add nitric acid (1 ml) dropwise, and then boil again for 1 hour. Cool, dilute with Water (70 ml), boil for five minutes and filter through a Whatman No. 42 filter paper. Wash with hot water, allow to drain, and dry the paper and precipitate, still in the funnel, in an oven at 120°. Transfer the paper and precipitate to an ignited and weighed silica crucible. Heat carefully to burn off the paper and then ignite to constant weight.

Cognate Determinations

Gold Chloride. $(HAuCl_4,3H_2O)$
An aqueous solution of the sample is treated with sodium hydroxide to precipitate auric hydroxide which undergoes a mutual oxidation-reduction with hydrogen peroxide; metallic gold is deposited quantitatively:

$$2HAuCl_4 + 8NaOH \rightarrow 2Au(OH)_3 + 8NaCl + 2H_2O$$
$$2Au(OH)_3 + 3H_2O_2 \rightarrow 2Au + 6H_2O + 3O_2$$

Sodium Auriothiomalate Injection

DETERMINATION OF HISTAMINE AS HISTAMINE NITRANILIC ACID COMPLEX

Histamine Acid Phosphate. *Determination of the percentage of* $C_5H_9N_3,2H_3PO_4$, *calculated with reference to the substance dried to constant weight at 105°*

This method depends upon the precipitation of insoluble histamine nitranilic acid complex, which is formed on the addition of a solution of nitranilic acid to one of the histamine salts.

$$C_5H_9N_3,2H_3PO_4 + C_6H_2N_2O_8 \rightarrow C_5H_9N_3.C_6H_2N_2O_8 \downarrow + 2H_3PO_4$$
$$\therefore \quad 307.1 \text{ g } C_5H_9N_3,2H_3PO_4 \equiv 341.3 \text{ g } C_5H_9N_3.C_6H_2N_2O_8$$
$$\therefore \quad 0.9001 \text{ g } C_5H_9N_3,2H_3PO_4 \equiv 1 \text{ g } C_5H_9N_3.C_6H_2N_2O_8$$

Method. Accurately weigh the sample (about 0.15 g) into a 250 ml beaker with a stirring rod and clock-glass cover. Add Water (10 ml) and dissolve the sample. Add nitranilic acid solution (3.5 per cent w/v in 95 per cent ethanol; 10 ml), stir and allow to stand for 15 minutes. Add ethanol (95 per cent; 10 ml), stand in an ice-bath for 3 hours and filter through a No. 3 sintered glass crucible, previously dried to constant weight at 130°. Complete the transfer of the precipitate and wash with four quantities, each of 5 ml of ethanol (95 per cent), and finally with ether (10 ml). Dry to constant weight at 130°.

Determine the loss in weight on drying a separate portion of the sample at 105° and complete the calculation.

Cognate Determination

Histamine Acid Phosphate Injection

DETERMINATION OF IRON SALTS AS FERRIC OXIDE (Fe$_2$O$_3$)

The determination depends upon the reaction expressed by the following equation:

$$2Fe^{3+} + 6OH^- \rightarrow 2Fe(OH)_3 \downarrow$$
$$2Fe(OH)_3 \rightarrow Fe_2O_3 + 3H_2O$$
$$\therefore \quad 2 \times 55.85 \text{ g Fe} \equiv 159.7 \text{ g Fe}_2O_3$$
$$\therefore \quad 0.6994 \text{ g Fe} \equiv 1 \text{ g Fe}_2O_3$$

The iron is first converted to the ferric state by boiling the solution with nitric acid. A slight excess of ammonium hydroxide solution is then added to the solution to precipitate ferric hydroxide which is filtered off on filter paper, washed, ignited to ferric oxide and weighed. Other metals such as aluminium or trivalent chromium, which precipitate as hydroxides under these conditions, must not be present. Similarly, various ions, e.g. phosphate or arsenate, which give precipitates with ferric ions in weakly basic media, must be absent.

Ferrous Ammonium Sulphate. *Determination of the percentage of* Fe^{2+}

Method. Accurately weigh the sample (about 1 g) (Note 1) into a 400 ml beaker provided with stirring rod and cover glass. Add Water (20 ml) hydrochloric acid (3 ml) and heat to dissolve the sample. Add nitric acid (2 ml) and boil gently for 5 minutes to oxidise the ferrous iron to the ferric state (Note 2). When oxidation is complete, dilute to about 200 ml, heat nearly to boiling and add Dilute Ammonia Solution in a slow stream from a small beaker, until a slight excess is indicated by the odour of the escaping steam after blowing away the fumes (Note 3); stir constantly during the addition. Add a pulp of filter paper obtained by shaking one or two Whatman accelerators or a quarter of an ashless tablet with Water (10 ml) in a corked test tube (Note 4). Cover the beaker with the clock-glass to prevent loss due to spurting, heat the contents to boiling, boil gently for about 1 minute, remove the beaker from the source of heat, allow the precipitate to settle slightly and rinse the under surface of the clock-glass into the beaker. *Immediately* (Note 5) decant the supernatant liquid through the Whatman No. 41 or 541 filter paper fitted well in a funnel (see page 122) leaving as much as possible of the precipitate in the beaker (Note 6). Wash the precipitate by decantation using 4×100 ml portions of hot 1 per cent ammonium nitrate solution (Note 7). Transfer the precipitate to the filter using hot ammonium nitrate solution from a washbottle with a fine jet. Use a policeman to remove the traces of precipitate adhering to the walls of the beaker. *Immediately* (Note 8) wash the precipitate on the filter with hot ammonium nitrate solution (1 per cent) until a few drops of the last washings acidified with nitric acid give no opalescence with silver nitrate solution (Note 9).

While filtration is proceeding, ignite a crucible (and lid if required—its use is not essential) at red heat, cool in a desiccator for 20 minutes, weigh and repeat the heating and cooling until constant weight is attained. Allow the filter paper and contents to drain in the funnel, transfer to an oven and leave until most of the moisture has been removed but do not dry completely. Carefully fold the filter paper around the precipitate to form a small bundle, transfer to the weighed crucible and dry over a *small* flame (Note 10). If the lid of the crucible is used place it on the crucible inclined on the pipe-clay triangle so that it rests partly on the crucible and partly on the triangle. Char the paper gradually but do not allow it to ignite. Should the paper ignite stop heating and cover the crucible with a lid. Burn off the carbon at as low a temperature as possible and under good oxidising conditions (Note 11). When all the carbon has been burnt off, ignite the precipitate at red heat (maximum temperature of Bunsen burner) for 15 to 20 minutes, taking care to exclude the burner gases from the interior of the crucible (Note 12). Allow the crucible to cool slightly and then place in a desiccator for 30 minutes, weigh and repeat the ignition and cooling until constant weight has been attained.

Notes on the determination of iron as ferric oxide

Note 1. Sufficient sample is required to give a residue of about 0.20 to 0.25 g but not more because of the bulky nature of the precipitate.

Note 2. Boiling for five minutes is usually sufficient to complete the oxidation of the iron. However, it is advisable to apply the following test for the completion of the reaction; transfer one drop of the solution to a clean test tube by means of the stirring rod and dilute with about 1 ml of distilled water and mix. Transfer one drop of the diluted solution to a white tile and add a drop of freshly prepared (approx. 0.01 per cent) potassium ferricyanide solution. If a blue colour is produced, ferrous iron is still present; add nitric acid (1 ml) to the main solution and boil for a further three minutes. The solution remaining in the test tube must be rinsed into the beaker.

Note 3. Precipitation at or near the boiling point, and heating the solution at 100° for a short time before filtration, gives better coagulation of the precipitate.

Note 4. The addition of filter paper pulp is advantageous though not essential. The pulp tends to prevent the gelatinous ferric hydroxide precipitate from blocking the filter paper pores during the filtration.

Note 5. The mixture should not be boiled for more than the stated time or allowed to stand for more than one or two minutes before decantation of the supernatant liquid because the precipitate tends to become slimy.

Note 6. A filter paper method is required because the gelatinous precipitate would block a Gooch or sintered glass crucible. Suction must not be applied.

Note 7. The electrolyte solution prevents peptisation of the precipitate during the washing procedure. The bulky precipitate is washed more readily by decantation than by washing after transference to the filter.

Note 8. The precipitate must not be allowed to dry and crack before washing is complete. Otherwise, these cracks and channels in the gelatinous mass preclude the thorough washing of the whole of the precipitate.

Note 9. Complete absence of chloride in the last washing indicates that the washing process is complete. Chloride ions must not be left in the precipitate because their presence would result in loss of iron as volatile ferric chloride in the ignition process. Chloride ions are added initially because their presence can be detected readily. It follows that if they have been removed from the precipitate as indicated by the test, then other contaminating ions have also been completely removed from the precipitate. Ammonium nitrate is decomposed to volatile products and is removed during the ignition process.

Note 10. A *small* flame is necessary because too much heat would lead to the sudden expulsion of steam and possible loss of some of the precipitate.

Note 11. This precaution is necessary to avoid partial reduction of the ferric oxide to Fe_3O_4 or even metallic Fe during the burning off of the filter paper and charred material.

Note 12. The inner portion of the Bunsen flame possesses reducing properties and will reduce the ferric oxide.

DETERMINATION OF LEAD AS LEAD CHROMATE (PbCrO₄)

The determination depends upon the reaction expressed by the following equation:

$$Pb^{2+} + CrO_4^{2-} \rightarrow PbCrO_4 \downarrow$$
$$\therefore \quad 207.2 \text{ g Pb} \equiv 323.2 \text{ g PbCrO}_4$$
$$\therefore \quad 0.6411 \text{ g Pb} \equiv 1 \text{ g PbCrO}_4$$

The lead chromate is filtered off, washed with water and dried at 120°. This method of determining lead is of limited use because of the general insolubility of metal chromates.

Lead Nitrate. *Determination of the percentage of* $Pb(NO_3)_2$

$$Pb(NO_3)_2 \equiv PbCrO_4$$
$$\therefore \quad 331.2 \text{ g Pb(NO}_3)_2 \equiv 323.2 \text{ g PbCrO}_4$$
$$\therefore \quad 1.025 \text{ g Pb(NO}_3)_2 \equiv 1 \text{ g PbCrO}_4$$

Method. Dry a No. 4 sintered glass crucible (or a Gooch crucible) at 120° to constant weight. Dissolve about 0.3 g of sample, accurately weighed, in Water (150 ml) containing acetic acid (5 ml) in a 400 ml beaker equipped with a stirring rod and clock-glass. Heat just to the boiling point and add solution of potassium chromate (5 ml) dropwise from a pipette with stirring. Boil gently for 10 minutes, remove the flame and allow the precipitate to settle; the supernatant liquid should be coloured slightly yellow—if this is not so add more solution of potassium chromate. Filter through the prepared Gooch or sintered glass crucible, wash with hot water by decantation; transfer the precipitate to the crucible removing any traces of solid adhering to the beaker with the aid of a policeman. Wash the precipitate with hot Water until the washings are colourless and then dry to constant weight at 120°.

DETERMINATION OF NICKEL

The determination depends upon the fact that nickel is precipitated as a complex when an alkaline solution of dimethylglyoxime (H.Dmg) is added followed by a slight excess of ammonia solution to a hot slightly acidic solution of a nickel salt:

$$Ni^{2+} + 2H.Dmg \rightarrow Ni(Dmg)_2 + 2H^+$$
$$\therefore \quad 58.71 \text{ g Ni} \equiv 288.9 \text{ g } C_8H_{14}O_4N_4Ni$$
$$\therefore \quad 0.2032 \text{ g Ni} \equiv 1 \text{ g } C_8H_{14}O_4N_4Ni$$

The precipitate of nickel dimethylglyoxime is washed and dried at 100° to 120°.

Because dimethylglyoxime is almost insoluble in water, it is added as a 1 per cent ethanolic solution. A large excess of reagent should be avoided because precipitation of dimethylglyoxime may occur; a very large excess of the reagent may result in an increase in the solubility of nickel dimethylglyoxime in the mixture because of the ethanol in the reagent.

Nickel Ammonium Sulphate. *Determination of the percentage of* Ni^{2+}

Method. Dissolve about 0.3 to 0.4 g, accurately weighed, in Water (100 ml) in a 400 ml beaker provided with clock-glass cover and stirring rod. Add dilute hydrochloric acid (10 ml) and dilute to 200 ml with Water. Heat to between 70° and 80°, add a slight excess (30 to 35 ml) of the dimethylglyoxime reagent (1 per cent) followed immediately by the dropwise addition of dilute ammonia solution, with constant stirring until no further precipitation of the red nickel dimethylglyoxime complex occurs, and a slight excess of ammonia is present (Note 1). Place the beaker on a boiling water-bath, and as soon as the precipitate has settled, test for complete precipitation by addition of a little more reagent and a little more dilute ammonia solution (Note 2). Leave on a water-bath for 30 minutes, remove and allow to cool for 1 hour. In the meantime, heat a sintered glass or Gooch crucible at 110° to 120° to constant weight. Filter the *cold* solution through the crucible, wash the precipitate with *cold* Water by decantation, transfer the precipitate to the crucible with the aid of a jet of cold Water and a glass rod (Note 3). Remove the last traces of precipitate adhering to the beaker by means of a policeman. Wash the precipitate with cold Water until free from chloride, and dry at 100° to 120° for 45 minutes. Cool and weigh, then reheat and cool to constant weight.

Notes on the determination of nickel as nickel dimethylglyoxime complex

Note 1. Precipitation by adding ammonia to a hot weakly acidic solution containing the reagent gives a more readily filterable precipitate than obtained by direct precipitation from an ammoniacal solution.

Note 2. A check for complete precipitation is essential since it is rather difficult to ensure that sufficient ammonia has been added at the precipitation stage, owing to the bulky flocculent nature of the precipitate.

Note 3. The mixture must be thoroughly cooled and cold water used as the wash liquid because the precipitate is appreciably soluble in hot water.

DETERMINATION OF SULPHATE AS BARIUM SULPHATE (BaSO₄)

The determination depends upon the reaction expressed by the following equation:

$$SO_4^{2-} + Ba^{2+} \rightarrow BaSO_4$$
$$\therefore \quad 96.07 \text{ g } SO_4^{2-} \equiv 233.43 \text{ g } BaSO_4$$
$$\therefore \quad 0.4115 \text{ g } SO_4^{2-} \equiv 1 \text{ g } BaSO_4$$

The solution containing sulphate ions is acidified with hydrochloric acid, heated almost to boiling point, and a slight excess of barium chloride solution is added. The precipitated barium sulphate is filtered off, washed with water and ignited.

Potassium Sulphate. *Determination of the percentage of* K$_2$SO$_4$

Method. Accurately weigh the sample (about 0.3 g) (Note 1), into a 400 ml beaker provided with a stirring rod and clock-glass cover. Add Water (250 ml) and concentrated hydrochloric acid (1 ml) (Note 2), and heat nearly to boiling point. Dilute 20 ml of barium chloride solution (5 per cent) to approx. 100 ml with Water in a small beaker. Heat to boiling and add to the sulphate solution slowly with constant stirring (Note 3), testing from time to time for complete precipitation after allowing the precipiate to settle slightly, until a slight excess of barium chloride solution has been added. Place the covered beaker in a water-bath almost at the boiling point for 1 hour (Note 4). The volume of the solution must not fall below 200 ml, Water being added to maintain the volume if necessary. Allow to cool slightly (10 minutes) and then continue the determination by one of the following methods.

(*a*) During the precipitation, 'prepare' a Gooch crucible (Note 5). Heat to red heat inside a nickel crucible (a porcelain one may also be used), allow to cool slightly in the air, transfer to a desiccator using crucible tongs which have been heated in a Bunsen flame (Note 6), cool and weigh. Reheat, cool and weigh until constant weight is attained. Decant the clear warm supernatant liquid through the Gooch crucible. Test the filtrate for complete precipitation by adding a few drops of barium chloride solution. Wash the precipitate three times by decantation with small portions of warm Water, transfer the precipitate to the crucible with the aid of a jet of hot Water from a wash bottle. Use a policeman to remove any precipitate adhering to the beaker or the glass rod. Wash with small portions of hot Water until the final washings give no precipitate with a few drops of silver nitrate solution (Note 7). Wash the precipitate with a few portions of ethanol added from a teat pipette (Note 8). Heat the Gooch crucible surrounded by a nickel (or porcelain) crucible to red heat, cool in a desiccator for 30 minutes and weigh. Repeat the heating and cooling until constant weight is attained.

(*b*) During the precipitation procedure heat a silica crucible to redness for 10 minutes, cool in a desiccator for 30 minutes and weigh. Repeat until constant weight is attained. If only a porcelain crucible is available, heat inside an outer jacket crucible, allow to cool slightly and handle with hot crucible tongs. (Note 5) Decant the hot clear supernatant liquid through an ashless filter paper (Whatman No. 540). Do not add more liquid than will come to within 10 mm of the top of the paper. Wash the precipitate by decantation with a few portions of hot Water and then transfer to the filter with the aid of a jet of hot Water from the wash bottle. Use a policeman to loosen any precipitate adhering to the beaker or glass rod. Wash the filter paper and its contents with small portions of hot Water, directing the jet near the top of the filter paper. Allow to drain well between each washing (Note 9). Continue the washing until a few ml of the final washings give no precipitate with 2 drops of silver nitrate solution. Allow to drain well, cover with a filter paper, and place in an oven at 100° until most of the moisture is removed. Fold the moist paper around the precipitate and place in the weighed crucible. Heat the crucible *gently* on a pipe clay triangle a few cm above a small flame. Place the crucible in a jacket crucible if it is made of porcelain. Gradually increase the heat until the paper chars and volatile matter is expelled. Do not allow the paper to burst into flame because mechanical loss of precipitate may occur. A crucible lid is not necessary if the above precautions are taken. When charring is complete, raise the temperature of the crucible to a dull redness and burn off the carbon with a free acess of air. Do not allow the reducing Bunsen flame to come into contact with the inside of the crucible. When the tarry carbon residue has been burnt off, the residue should be white. Ignite for 15 minutes at a dull red heat. If the precipitate is slightly discoloured, *cool* and add 5 drops of a mixture of strong sulphuric acid and ethanol (1 : 1), heat gently at first and then more strongly at a dull red heat for 15 minutes (fume cupboard). Allow the crucible to cool slightly in the air, transfer to a desiccator with crucible tongs, cool for 30 minutes and weigh. Repeat the heating and cooling until constant weight (±0.0002 g) is attained.

(c) The following method is suitable for obtaining quick results, which are about 0.1 to 0.2 per cent above the theoretical value because of occluded water in the precipitate.

Prepare a Gooch crucible, wash well with Water, and then with ethanol added from a teat pipette. Dry in an oven at 120° to constant weight. Decant the solution through the crucible, transfer the precipitate and wash until free from chloride as described under Method (a). Finally wash the precipitate well with ethanol added from a teat pipette, and then with absolute ethanol. Dry to constant weight in an oven at 120°.

Notes on the determination of sulphate as barium sulphate

Note 1. A weight of sample to contain 0.05 g of sulphur is required.

Note 2. Hydrochloric acid is added for the following reasons:

(a) It prevents precipitation of the barium salts as carbonate and phosphates which are insoluble in neutral and alkaline solutions.

(b) The co-precipitation of barium hydroxide is prevented.

(c) It promotes formation of a coarse, easily filterable precipitate of barium sulphate. The solubility of barium sulphate is increased in the presence of hydrochloric acid, but under the conditions of acidity described, and in the presence of barium chloride the solubility of barium sulphate is negligible.

Note 3. There is a tendency to co-precipitation with the barium sulphate precipitate; precipitation using dilute solutions near the boiling point reduces this tendency.

Note 4. The digestion period allows time for complete precipitation and ageing of the precipitate to make it more readily filterable.

Note 5. A silica Gooch is preferable to a porcelain one in this determination. Alternatively, a No. 4 sintered silica (Vitrosil filtering crucible) may be used. A No. 4 sintered glass crucible should not be used because, although retaining the barium sulphate precipitate, it cannot be heated above 450°.

Note 6. Heating the crucible tongs prior to touching the hot crucible helps to avoid cracking the crucible if it is made of porcelain.

Note 7. This indicates that barium chloride (and therefore presumably water-soluble salts) have been washed out of the precipitate.

Note 8. This treatment removes most of the water.

Note 9. It is important that all soluble salts are washed from the filter paper as well as from the precipitate.

Cognate Determination

Sodium Sulphate contains not less than 99.0 per cent of Na_2SO_4 determined on the substance dried to constant weight at 105°.

$$Na_2SO_4, 10H_2O \equiv Na_2SO_4 \equiv BaSO_4$$
$$\therefore \quad 142.0 \text{ g } Na_2SO_4 \equiv 233.4 \text{ g } BaSO_4$$
$$\therefore \quad 0.6086 \text{ g } Na_2SO_4 \equiv 1 \text{ g } BaSO_4$$

Weigh accurately about 0.25 g of the dried sodium sulphate sample from the 'loss on drying' determination (see p. 21 for this procedure) into a 400 ml beaker and proceed as described above for potassium sulphate.

DETERMINATION OF PROGUANIL AS PROGUANIL-CUPRIC COMPLEX

Proguanil Tablets. *Determination of the percentage of* $C_{11}H_{16}N_5Cl,HCl$

This method depends on the precipitation of insoluble proguanil-cupric complex, which is formed on the addition of ammoniacal cupric chloride solution to a solution of proguanil hydrochloride.

$$Cl-\langle\ \rangle-NH.CNH.C.NH.CH(CH_3)_2$$
$$\overset{NH}{\underset{NH}{\|\ \|}}$$

$$2C_{11}H_{16}N_5Cl,HCl + CuCl_2 \xrightarrow{NH_3} (C_{11}H_{15}N_5Cl)_2Cu + 4HCl$$

$\therefore$ 580.2 g $C_{11}H_{16}N_5Cl,HCl \equiv 568.9$ g $(C_{11}H_{15}N_5Cl)_2Cu$

$\therefore$ 1.020 g $C_{11}H_{16}N_5Cl,HCl \equiv 1$ g $(C_{11}H_{15}N_5Cl)_2Cu$

Method. Accurately weigh the sample (about 0.6 g) into a 250 ml beaker with stirring rod and clock-glass cover. Add Water (50 ml) and warm to dissolve. Cool in iced water until the temperature of the solution is down to 10° and then add ammoniacal cupric chloride solution, stirring continuously, until the solution acquires a permanent deep-blue colour. Allow the solution to stand for 90 minutes and filter through a No. 4 sintered glass crucible previously dried to constant weight at 130°. Complete the transfer of the precipitate, wash first with a mixture of dilute solution of ammonia (1 volume) and Water (5 volumes) and then with cold water until the washings are colourless indicating the complete removal of soluble copper salts. Dry the precipitate to constant weight at 130°.

Determine the loss in weight on drying a separate portion of the sample at 105°, and complete the calculation.

13 Solvent Extraction Methods

Solvent extraction methods are used principally for the determination of alkaloids in crude drugs, galenicals, injections and tablets. Such methods are also applicable to the control of synthetic bases, their salts and preparations, and to a number of non-basic substances which may be readily isolated by extraction with an organic solvent.

The methods may be classified according to the process used after extraction, viz:

        (*a*) Gravimetric
        (*b*) Volumetric
        (*c*) Spectrophotometric and Colorimetric

The Gravimetric Method

This is limited in its application to the determination of alkaloids in those drugs and galenicals which contain relatively high concentrations of alkaloids (over 0.5 per cent). At lower alkaloid concentrations, accuracy can only be achieved by the use of excessively large quantities of material, which, apart from being uneconomic, can also lead to manipulative difficulties. The method is suitable only where a moderately simple isolation technique yields the alkaloid in a pure state. In practice this ideal is seldom achieved, and the manipulative technique necessary to obtain a pure product is often troublesome and tedious. A further objection to the method is the need to dry the extracted alkaloids to constant weight. The removal of water or other solvent may involve prolonged heating, and many alkaloids are unstable under these conditions.

The Volumetric Method

The volumetric method offers the advantage that it is applicable when the alkaloidal residue still contains some impurities; these must be neutral and must not be contaminated with large quantities of coloured impurities which would mask a visual end point. A back-titration method is always used because of the insolubility of alkaloids and organic bases in water. Care is necessary if good results are to be obtained. The forward titration is always small and the use of a dilute volumetric solution to give a larger forward titration defeats its own object since the end point is less clearly defined. For example, *0.02N* solutions are used for solanaceous alkaloids because of the small percentage of alkaloid present but the end points are noticeably flat and difficult to judge accurately. Alkaloids are, for the most part, weak bases, hence methyl red is the indicator of choice. Phenolic alkaloids such as cephaeline in ipecacuanha, and morphine in opium, which are amphoteric, tend to exert a buffering action and give a flat end point. This buffer-like action is caused by competition between the phenolic hydroxyl group of the alkaloid and the indicator for the sodium hydroxide. Certain substances, for example Levallorphan Tartrate and Pethidine Injection, are determined by non-aqueous titration.

Spectrophotometric and Colorimetric Methods

Spectrophotometric methods are frequently quicker and simpler than volumetric and gravimetric methods, and can usually be applied with greater accuracy to smaller quantities. Not all substances amenable to solvent extraction methods exhibit the absorption characteristics suitable for spectrophotometric determination. Chlorpheniramine Tablets, Bisacodyl Tablets, Levorphanol Tablets, and many similar substances are assayed by solvent extraction and measurement of the ultraviolet absorption at the wavelength of maximum absorption of the active principle. A number of steroids, exemplified by cortisone acetate in Cortisone Injection, are isolated by extraction, and converted to a coloured complex, the absorption of which can be measured at specified wavelengths of maximum absorption in the visual spectrum. Colorimetric assays, employing comparison methods, are also specified in a few special cases where other methods are inapplicable. For example, in the determination of morphine in Camphorated Opium Tincture, the quantity of alkaloid is too small to be determined by volumetric or gravimetric methods. A specific colour reaction between morphine and sodium nitrite is used and the colour produced compared with the standard colour. Since it is often difficult to exclude traces of colouring matter from the alkaloid isolated in determinations of this type it may be necessary to compare colours which are not of exactly the same shade. The eye is not very sensitive to the fine distinctions of colour, shade, and depth of colour and accurate comparisons are difficult. Much more satisfactory results can be obtained with the use of a photoelectric colorimeter or absorptiometer, when measurements may be made at selected wavelengths. The principles underlying these determinations are discussed fully in the appropriate Chapter in Part 2.

DETERMINATION OF THE SALTS OF ALKALOIDS AND OTHER BASES

Since these substances contain little extraneous matter, the extraction process is less complicated than that used for crude drugs and galenicals. The following general principles are involved:

(1) Most alkaloids are sparingly soluble in water, but readily soluble in organic solvents such as ether and chloroform. Phenolic alkaloids, e.g. morphine, are exceptions in that they are relatively insoluble in both water and organic solvents; they can, however, be extracted from water by mixed organic solvents such as chloroform-ethanol.

(2) Alkaloidal salts, such as hydrochlorides, sulphates, etc., are water-soluble but insoluble in organic solvents.

(3) Alkaloids and their salts are readily interconvertible by reaction with either acid or alkali. Ammonia is preferable to sodium hydroxide as alkali as it is volatile and any excess can be readily removed during the final drying of the alkaloidal residue.

(4) Most alkaloids and all alkaloidal salts are non-volatile.

The scheme of extraction on p. 288 can be evolved by taking advantage of these properties.

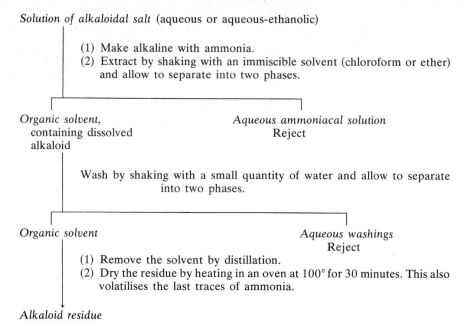

Solution of alkaloidal salt (aqueous or aqueous-ethanolic)

  (1) Make alkaline with ammonia.
  (2) Extract by shaking with an immiscible solvent (chloroform or ether)
      and allow to separate into two phases.

Organic solvent,   *Aqueous ammoniacal solution*
  containing dissolved   Reject
  alkaloid

  Wash by shaking with a small quantity of water and allow to separate
      into two phases.

Organic solvent   *Aqueous washings*
                 Reject
  (1) Remove the solvent by distillation.
  (2) Dry the residue by heating in an oven at 100° for 30 minutes. This also
      volatilises the last traces of ammonia.

Alkaloid residue

Distribution Law and Partition Coefficients

When a substance is shaken with two immiscible solvents, it distributes itself
between them in a constant ratio which is characteristic of the substance and
the solvents employed at a given temperature. That is,

$$\frac{\text{Concentration in solvent A}}{\text{Concentration in solvent B}} = \frac{C_A}{C_B} = K$$

The constant K is known as the partition coefficient. It varies with temperature.

The law applies strictly only to dilute solutions, for completely immiscible
solvents, and for solutes which have the same molecular weight in both solvents
and which do not affect the miscibility of the solvents. For concentrated
solutions, the limiting case occurs when the two solvents are shaken with excess
solute. Two saturated solutions are obtained and the partition coefficient is
then equal to the ratio of the solubilities of the solute in the two solvents. This
provides a rough approximation from which approximate distribution coefficients can be calculated. This is not strictly valid at lesser concentrations, but
provides a useful guide where accurate distribution coefficients are not available; moreover, it provides for conditions frequently met in practice insofar as
the solubility in one or other of the solvents is usually low.

In practice, the amount of material extracted depends on the amount of
material present, the partition coefficient and the volumes of the two solvents.
The efficiency of the process can be increased by increasing the volume of the
extracting solvent and by increasing the number of extractions. This is
illustrated by the following examples of the distribution between chloroform
and water of 1 g of a solid of partition coefficient, $K = 9$.

Example 1. *Extraction using equal volumes of solvents*
Consider the partition of 1 g of solid between 20 ml volumes of chloroform and water.

$$\frac{\text{Concentration in chloroform}}{\text{Concentration in water}} = K = \frac{9}{1}$$

then, if x g dissolves in the chloroform

$$\frac{x/20}{(1-x)/20} = \frac{9}{1}$$

$$\therefore \quad \frac{x}{(1-x)} = 9$$

$$\therefore \quad x = 0.9$$

Hence, after the first extraction, one-tenth of the substance (0.1 g) remains in the aqueous phase. A second extraction with the same volume of solvent will reduce the weight of substance remaining in the aqueous phase to one-hundredth (0.01 g) of that originally present.

Example 2. *Extraction with unequal volumes of solvents*
Consider the partition of 1 g of solid between 40 ml of chloroform and 20 ml of water. Again,

$$\frac{\text{Concentration in chloroform}}{\text{Concentration in water}} = \frac{9}{1}$$

then if x g dissolves in the chloroform

$$\frac{x/40}{(1-x)/20} = \frac{9}{1}$$

$$\therefore \quad x/2 = 9(1-x)$$

$$\therefore \quad x = 18/19$$

Hence, after the first extraction, one-nineteenth of the substance (0.0526 g) remains in the aqueous phase. A second extraction with a further 40 ml of chloroform will reduce the weight of substance remaining in the aqueous phase to 1/361 of that originally present.

Choice of Solvent

The choice of solvent depends on the solubility characteristics of the substance to be extracted, characteristics on which the partition coefficient also depends.

Chloroform has the advantage that it is non-inflammable, and heavier than water. The latter property facilitates separation of the organic phase. Ether on the other hand is both inflammable and lighter than water, so that not only are the separation techniques more complicated, but care must also be exercised in the subsequent removal of the solvent. Mixed solvents are sometimes necessary. Morphine is insoluble in both ether and chloroform and is best extracted with a mixture of chloroform and ethanol (or isopropanol).

Extraction with Chloroform

Procaine Penicillin. *Determination of the percentage of* $C_{13}H_{20}N_2O_2$
Chloroform is used as the extracting solvent, and since it is heavier than water, only two separators are required. For convenience of description these two separators are designated A and B respectively. Before commencing the extraction check that the taps are properly greased and that there is no possibility of leakage from either taps or stoppers.

Method. Accurately weigh the sample (0.1 g) into separator A. Add Water (20 ml) and dissolve. Make alkaline by the addition of excess of sodium carbonate solution (5 ml) (Note 1). Add chloroform (30 ml), stopper the separator and shake gently. Invert the separator and cautiously open the tap to release the pressure. Close the tap and shake vigorously for at least two minutes. Return the separator to an upright position in the stand, when the contents will separate rapidly into two layers, of which the chloroform is the lower (Note 2). Remove the stopper, washing its surface with a few drops of chloroform (teat pipette). Carefully open the tap and run off the whole of the chloroform solution into separator B without allowing any aqueous layer to follow. It is most important to make a clean separation between the two phases, and any residual emulsion should be allowed to break.

Whilst the stem of separator A is still inside the neck of separator B, pour about 1 ml of chloroform into A, allow the chloroform to sink straight to the bottom, and then run into B as before (Note 3). Finally wash the outside of the stem of A with a few drops of chloroform (teat pipette) allowing the washings to run directly into B (Note 4). Wash the chloroform solution in separator B by shaking vigorously for 2 minutes with Water (20 ml). Allow to separate and run off the lower chloroform layer into a clean dry alkaloidal flask (150 ml). *Wash through* with chloroform, running the washings into the flask. Retain the aqueous liquor in separator B as this is required for washing successive chloroform extracts.

Repeat the extraction of the aqueous layer in A by shaking with a further 20 ml of chloroform. Allow to separate into two phases and run off the lower layer into B. *Wash through* with chloroform. Shake the chloroform solution now in B with the aqueous wash liquor, already present, as described above. Allow to separate and run off into the same alkaloidal flask.

Complete the extraction of the base from the alkaline solution in A, by shaking with two further 10 ml portions of chloroform. Transfer each extract in turn to B, wash and run into the alkaloidal flask. Test the remaining alkaline liquor in A as follows to ensure that the base has been completely extracted. Shake with 2–3 ml of chloroform and run off the latter into a test tube. Add an equal volume of dilute sulphuric acid, heat on a boiling water-bath until all the chloroform has volatilised. Add a few drops of Mayer's reagent (K_2HgI_4) when there should be no opalescence or precipitate, indicating the absence of base. If base is still present, continue the extraction with further 10 ml portions of chloroform as before until the extraction is complete.

Fit the alkaloidal flask containing the chloroform solution for distillation (Fig. 29), and heat in a water-bath until most of the solvent has been recovered (Note 5). Do not on any account allow solid matter (base) to separate from solution. Add ethanol (95 per cent; 5 ml) and complete the evaporation to dryness (Note 6) on a water-bath, gently rotating the flask to give a thin layer of base.

Dry the base in an oven at 100° for half an hour (Note 7). Care should be taken to place the flask on the shelf of the oven in such a position as to avoid any possibility of charring which may occur in close proximity to the heating elements. Cover the top of the flask to prevent specks of dirt from falling into the flask.

Dissolve the residue in 1.2 ml of chloroform (Note 8). Pipette into the flask 20 ml of 0.01N H₂SO₄ and warm gently in a water-bath to remove the chloroform. Cool to room temperature and back-titrate the exess sulphuric acid with 0.01N NaOH using methyl red as indicator.

$$C_{13}H_{20}N_2O_2 \equiv H \equiv 1000 \text{ ml } N$$
$$\therefore \quad 236.3 \text{ g } C_{13}H_{20}N_2O_2 \equiv 10\ 000 \text{ ml } 0.1N$$
$$\therefore \quad 0.00236 \text{ g } C_{13}H_{20}N_2O_2 \equiv 1 \text{ ml } 0.01N$$

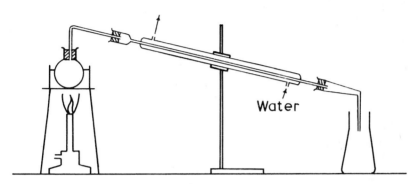

Water

Fig. 29. Chloroform recovery apparatus

Notes on the Determination of Procaine in Procaine Penicillin

Note. 1. This decomposes Procaine Penicillin liberating procaine base (white precipitate). Sodium carbonate is only used in practice when ammonia will not do, since, whereas traces of the latter can always be removed from the residue by heat, the former is non-volatile. Traces of sodium carbonate remaining in the residue due to insufficient washing could vitiate the result of the assay. In this case Na₂CO₃ forms soluble sodium penicillinate.

Note 2. Procaine base, which is very much more soluble in chloroform than in water, becomes distributed between the two phases in proportion to its solubility in each of the two solvents. A larger volume of solvent is used at this stage than is stipulated in the official process, so that any unavoidable loss of solvent results in the smallest possible loss of base.

Note 3. This small volume of chloroform displaces any of the procaine solution from the tap and stem of A.

Note 4. Chloroform does not run cleanly away from the end of the separator stem, so that small quantities of the base are deposited on the outside of the stem, owing to creeping and subsequent evaporation of the solvent.

The operations covered by Notes 3 and 4 comprise a *wash through* technique which should become standard procedure in all manipulations involving transfer from one separating funnel to another.

Note 5. Chloroform is expensive and therefore should always be recovered.

Note 6. The addition and evaporation of ethanol helps to remove the last traces of chloroform and water.

Note 7. This removes traces of volatile bases, which are liable to be present.

Note 8. Alcohol (95 per cent) may be used instead of chloroform. This is less satisfactory since it is not removed prior to titration and therefore tends to flatten the end point. Moreover chloroform dissolves the base more readily, especially if it is contaminated with traces of grease from the separator taps.

Cognate Determinations

The following substances are extracted with chloroform as described above, but the residue is determined volumetrically by titration in non-aqueous media (for details of titration procedures see Chapter 7).

Benztropine Mesylate
Carbenoxolone Sodium
Dextromethorphan Hydrobromide
Dextropropoxyphene Capsules
Dextropropoxyphene Napsylate
Pethidine Hydrochloride
Pethidine Injection
Phenindamine Tartrate

The following substances are determined by extraction with chloroform as described above, but the residue is determined gravimetrically using the procedure described under Phenobarbitone Sodium (p. 305).

Hydroxychloroquine Sulphate
Mepacrine Hydrochloride
Mepacrine Tablets
Methohexitone Injection
Quinine Dihydrochloride Injection
Thiopentone Sodium. This is a mixture of the monosodium derivative of 5-ethyl-5-(1-methylbutyl)thiobarbituric acid and exsiccated sodium carbonate and is assayed first for sodium, and then on the residual solution from this estimation for 5-ethyl-5-(1-methylbutyl)thiobarbituric acid.
Thiopentone Injection

The following is extracted with chloroform or dichloromethane and the residue determined by measurement of the ultraviolet absorption.
Cortisone Injection

The following is extracted with $CHCl_3$ and determined colorimetrically.
Betamethasone Sodium Phosphate Injection

Extraction with Chloroform-Alcohol Mixtures

Mixed solvents are not often used in the determination of salts and simple preparations. They are useful, however, for the extraction of amphoteric bases such as morphine and nalorphine from their salts and preparations. The use of a mixed solvent is also often very valuable in the extraction of biological materials and of ion-pair complexes. One per cent of amyl alcohol or octanol in chloroform materially assists recovery of drugs and metabolites from plasma and counteracts adsorption of coloured complexes on glass surfaces.

Morphine Sulphate Tablets. *Determination of the weight of* $(C_{17}H_{19}NO_3)_2,H_2SO_4,5H_2O$ *per tablet of average weight*

Method. Weigh and powder 20 tablets. To an accurately weighed quantity of powder, equivalent to about 0.1 g of morphine sulphate in a separator, add Water (25 ml), and N sodium hydroxide (5 ml). Mix well, add ammonium sulphate (1 g), shake to dissolve and add ethanol (95 per cent, 20 ml). Extract with a mixture of chloroform/ethanol 95 per cent (3/1; 40, 20, 20 and 20 ml), wash each extract with water (5 ml) and filter through a small plug of cotton wool. Evaporate the solvent and dissolve the residue in *0.05N* hydrochloric acid (10 ml). Boil and cool the solution, add Water (15 ml) and titrate the excess of acid with *0.05N* sodium hydroxide, using methyl red as indicator.

$$0.01897 \text{ g } (C_{17}H_{19}NO_3)_2,H_2SO_4,5H_2O \equiv 1 \text{ ml } 0.05N \text{ HCl}$$

Cognate Determinations

Nalorphine Hydrobromide. Chloroform-isopropanol extraction.
Nalorphine Injection. Chloroform-isopropanol extraction.

Extraction with Ether

Chloroquinine Phosphate. *Determination of the percentage* $C_{18}H_{26}ClN_3,2H_3PO_4$ *calculated with reference to the substance dried to constant weight at* 105°

Since ether is less dense than water, the actual separation technique is slightly more complex than that using chloroform. Three separators are required and these are designated A, B and C in the following description.

Method. Accurately weigh the sample (0.7 g approx.) introducing it via a funnel directly into separator A containing Water (30 ml) and mix well to dissolve. Add sodium hydroxide solution (3 ml) (Note 1) and ether (50 ml), stopper and shake to dissolve the precipitated chloroquine; continue shaking for two minutes (N.B. release pressure). Allow the solutions to separate. Remove the stopper carefully, wash it and the neck of the separator with about 1 ml of ether from a teat pipette, allowing the washings to run into the separator (Note 2). Run off the lower aqueous solution into separator B. *Wash through* the tap and stem of A with 2–3 ml of Water, running this off into B.

Repeat the extraction of the aqueous liquor now in separator B, shaking with a further 20 ml of ether. Allow to separate as before, remove the stopper, and wash the latter and the neck of the separator with 1 ml of ether. Run off the aqueous liquor into separator C, *washing through* the tap and stem with 2–3 ml of Water. Combine the ether extracts by running the second extract into A, washing the tap and stem with 2–3 ml of ether.

Continue the extraction of the aqueous solution now in separator C, with a further 20 ml of ether. Allow to separate and run off the aqueous layer into B. Transfer the ether layer to separator A. Complete the extraction of the aqueous solution now in separator B with a final 20 ml of ether. Allow to separate and run the aqueous layer into C. Transfer the ether layer to A. Four extractions should be sufficient in this and most other cases, but if necessary a fifth or even more extractions should be made to ensure complete extraction of the base. Wash the combined ether extracts with successive quantities of water (10 ml) until the washings are no longer alkaline to titan yellow paper (Note 3). Shake the combined washings with ether (25 ml) and add the ether to the combined ether extracts. Evaporate the ether to reduce volume (2–3 ml) (Note 4) and add *0.1N* HCl (50 ml). Warm to dissolve the base, cool and back-titrate the acid with *0.1N* NaOH using bromocresol green as indicator (Note 5). Carry out a blank determination. The difference between the titrations is the amount of *0.1N* HCl required by the chloroquine phosphate.

$$0.02579 \text{ g } C_{18}H_{26}ClN_3,2H_3PO_4 = 1 \text{ ml } 0.1N \text{ HCl}$$

Notes on the Determination of Chloroquine Phosphate

Note 1. Basification of the solution liberates the free base which is insoluble in water.

Note 2. Ether readily evaporates about the stopper and the neck of the separator with the possibility of losses.

Note 3. The washings remove the strong base, sodium hydroxide, which would interfere with the titration later. Titan yellow changes colour over the range pH 12–13 and so is not affected by the weak base chloroquine.

Note 4. Ether is inflammable, and distillation must be carried out well away from any naked flames.

Note 5. The titration is strong base/strong acid but is carried out in the presence of the hydrochloride of a weak base. An indicator on the acid side of pH 7 is therefore chosen.

Loss on drying. Chloroquine Phosphate may contain up to 1.5 per cent of volatile matter and allowance is made for this in the assay by calculating the percentage of $C_{18}H_{26}ClN_3,2H_3PO_4$ with reference to the sample dried to constant weight at 105°. A loss on drying is therefore determined on a separate portion of the sample.

Cognate Determinations

Chloroquine Phosphate Injection

Chloroquine Sulphate

Chloroquine Sulphate Injection

Dimenhydrinate. Determination of diphenhydramine.

Dimenhydrinate Injection

Emetine Hydrochloride. Emetine is precipitated from the sample by the addition of sodium hydroxide, which retains the phenolic alkaloid cephaëline in the aqueous phase. The emetine is extracted by shaking out with successive volumes of solvent ether. The combined ether solutions are washed repeatedly with water until, after being re-extracted with ether, the washings are neutral to litmus. The emetine is re-extracted with a known volume of standard acid and then with water. The excess acid is back-titrated with standard sodium hydroxide solution using methyl red as indicator.

Emetine and Bismuth Iodide. Sodium hydroxide is used to decompose the complex, liberate emetine and to precipitate bismuth as hydroxide. This stage is complete when all red colour has disappeared. The assay is then completed in the same way as for Emetine Hydrochloride.

Emetine Injection

Lignocaine and Adrenaline Injection. Determination of lignocaine hydrochloride.

Lignocaine Hydrochloride Injection

Mephentermine Injection

Mephentermine Sulphate

Methadone Injection

The following are extracted with ether and determined volumetrically by non-aqueous titration.

Levallorphan Tartrate

The following are extracted with ether and determined by ultraviolet absorption:

Levallorphan Injection

Levorphanol Injection

DETERMINATION OF ALKALOIDS IN CRUDE DRUGS AND GALENICALS

Separation of Non-Alkaloidal Matter

In addition to their active principles, crude drugs and galenical preparations obtained from them contain large quantities of extraneous vegetable matter, which must first be removed if the alkaloid is to be extracted in a pure state. The non-alkaloidal matter present includes substances of many different chemical types, which can be classified as follows.

Water-soluble matter, such as sugars, glycosides, starches, proteins, gums, mucilages, tannins and saponins. These substances are present in both crude drugs and in galenicals. Although soluble in ethanol and water (the solvents most frequently used in galenical preparations) these substances are completely insoluble in chloroform and ether. They are easily removed during the course of the assay process since they are retained in the aqueous phase during extraction of the alkaloid from alkaline solution by shaking with ether or chloroform.

Resins, fats, oils and colouring matters. These are all readily soluble in organic solvents, but insoluble in acid, neutral, and, in most instances, in alkaline aqueous solutions. They may be removed by a preliminary extraction of an acidified aqueous solution of the galenical with ether or chloroform. The alkaloids readily form water-soluble salts in acid solution, and are retained in the aqueous phase whilst the unwanted material is removed in the organic solvent.

Organic acids, although often insoluble in water, all form water-soluble ammonium salts, so that in the final extraction of the alkaloid base from ammoniacal solution the acid is retained in the latter solvent, whilst the alkaloid passes into the organic phase.

Organic Bases Other than the Required Alkaloid

Since the solubility properties of such bases are similar to those of the alkaloids, their separation from the alkaloids often presents considerable difficulty, and special methods are usually adopted to fit a particular case as in the following examples.

(*a*) *The solanaceous drugs.* In addition to the active alkaloids these contain a number of volatile bases which can be removed from the alkaloidal residue by heating in an oven under specified conditions.

(*b*) *Nux Vomica.* This contains principally two alkaloids, brucine and strychnine, though only the latter is determined in the official assay process. Both alkaloids are isolated together by solvent extraction, and brucine can be converted to an alkali-soluble product by oxidation with nitric and nitrous acids. The resulting solution is made alkaline with sodium hydroxide and strychnine extracted with chloroform (see also p. 298).

(*c*) *Opium.* This contains a mixture of alkaloids of which the most important is the phenolic base morphine. The bulk of the remaining alkaloids are non-phenolic. Opium is assayed on its content of phenolic alkaloids (calculated as morphine) which are separated from the non-phenolic group by conversion to water-soluble calcium salts in a preliminary extraction with calcium hydroxide.

(d) *Ipecacuanha*. This also contains phenolic and non-phenolic alkaloids. These are determined together to give a total alkaloid content by a direct extraction process.

The General Method for Liquid Galenicals

(A) Dilute the liquid extract with water and acidify to convert the alkaloids to their water-soluble salts. Extract the solution by shaking with chloroform. Allow to separate into two phases. Repeat the chloroform extraction procedure several times and combine the extracts.

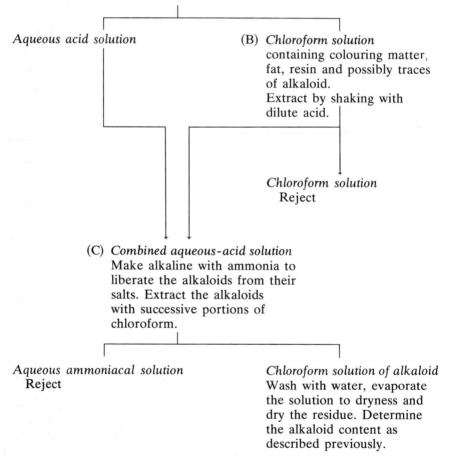

Aqueous acid solution

(B) *Chloroform solution* containing colouring matter, fat, resin and possibly traces of alkaloid. Extract by shaking with dilute acid.

Chloroform solution Reject

(C) *Combined aqueous-acid solution* Make alkaline with ammonia to liberate the alkaloids from their salts. Extract the alkaloids with successive portions of chloroform.

Aqueous ammoniacal solution Reject

Chloroform solution of alkaloid Wash with water, evaporate the solution to dryness and dry the residue. Determine the alkaloid content as described previously.

Note. Vigorous and continuous shaking is necessary at each extraction to ensure attainment of equilibrium in the distribution of alkaloid between two immiscible solvents. Unfortunately the presence of vegetable extractives renders the formation of emulsions likely and in these cases gentle shaking for longer periods reduces the likelihood of their formation. Compatible with the time available, it is essential to allow the two phases to separate as completely as possible. Ethanol may on occasion be used to break up or prevent troublesome emulsions (Hyoscyamus), though it should be used judiciously to avoid upsetting the partition ratio of alkaloid and other substances between the two phases. In some cases the application of gentle heat is effective (i.e. under running hot water) though great care should be exercised if the solvent is readily volatile and inflammable (ether). Coarse emulsions can be broken by scratching with a piece of wire at the emulsified interface.

The General Method for Crude Drugs and Dry Extracts

Preliminary treatment with a suitable organic solvent is used to exact the alkaloid, the solvent being so chosen that only the minimum of other unwanted vegetable material is also extracted. This part of the process resembles the preparation of a liquid galenical, the main steps in the process being as follows.

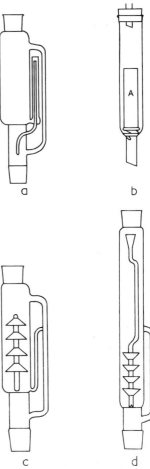

(a) Moisten with the solvent and allow to stand so that penetration of the tissues by the solvent can occur.

(b) Make alkaline by moistening with a small quantity of ammonia and macerate for half to one hour.

(c) Extract the alkaloid either by percolation or, where suitable, by a process of continuous extraction with an organic solvent. There are various means of carrying out a continuous extraction and a Soxhlet apparatus (Fig.30a) or the pharmacopoeial system (Fig. 30b) are two common methods for powders. Continuous liquid-liquid extraction is also used and extractors for use with solvents heavier and lighter than water are illustrated in Figs. 30c and 30d respectively. The advantage of continuous extraction is that only a small volume of solvent is necessary and tedious evaporation of large volumes of solvent can be avoided. It is also a much more rapid extraction process than percolation. The principal objection to the continuous process is that certain alkaloids cannot withstand the combined effect of heat and alkali. This applies especially to the solanaceous alkaloids.

For liquid/solid extraction the sample is held in a paper thimble (Soxhlet apparatus) or in the tube A (pharmacopeial system) fitted with a base of calico or other suitable fabric to retain the powder.

Fig. 30. Continuous extraction apparatus

For liquid/liquid extraction two modes are possible, one using a solvent heavier than water (downward displacement, Fig. 30c) and one using a solvent lighter than water (upward displacement, Fig. 30d). Extraction by downward displacement involves condensation of the refluxing solvent which falls through the liquid solution or suspension to the bottom of the extractor and back to the flask holding the boiling solvent. To assist rapid extraction the condensed solvent passes over baffles where the droplets are reduced in size on passing through the solution to be extracted. Upward displacement involves collecting the condensate in such a way as to convey it to the bottom of the solution to be extracted. The solvent rises through the solution, being broken up by the baffles and returns to the flask holding the boiling solvent. The extract obtained in this way contains a considerable quantity of non-alkaloidal matter and requires further purification.

(d) Treat the liquid extract from (c) with successive portions of dilute acid to convert the alkaloids to water-soluble salts. The acid solution may then be treated by the process described in the *General Method for Liquid Galenicals* although for Nux Vomica the solution is examined directly.

Nux Vomica

The principal alkaloidal constituents of Nux Vomica are brucine and strychnine which are extracted together in the assay. Formerly brucine was converted by reaction with nitrous and nitric acids to a red alkali-soluble nitro compound which was separated from strychnine. Use is now made of ultraviolet absorption to determine strychnine in the presence of brucine, an example of the analysis of a two-component mixture as described in Part 2.

Nux Vomica. *Determination of the percentage of strychnine*

Method. Mix about 0.4 g of sample in fine powder accurately weighed with ethanol (70 per cent, 2 ml) (Note 1). Add dilute ammonia solution (5 ml) and mix thoroughly. Transfer the powder to a 60 ml ground glass downward displacement liquid-liquid extractor (Fig. 30c) fitted with four baffle discs on the distributor, with the aid of water (25 ml) and chloroform (20 ml) (Note 2). The flask on the extractor apparatus (Fig. 30c) should be about 100 ml volume.

Attach a reflux condensor and boil under reflux for 4 hours with occasional swirling of the contents. Cool and transfer the chloroform extract to a separator and extract with N H_2SO_4 (20, 20, 20, 20 ml). Wash the combined acid extracts with chloroform (10 ml) and reject the lower chloroform layer. Filter the acid solution through a small plug of cotton wool into a shallow dish and warm, with stirring, to remove traces of chloroform. Cool and dilute to 100 ml with N H_2SO_4. Mix well and dilute 20 ml to 100 ml with N H_2SO_4.

Measure the extinction in a 1 cm cell at 262 nm and 300 nm. Calculate the percentage of strychnine from the formula

per cent strychnine $= 5(0.321a - 0.467b)/w$ where a is the extinction

at 262 nm, b is the extinction at 300 nm and w is the weight of powder taken.

Note 1. The addition of ethanol (70 per cent) is to assist thorough wetting by aqueous solvent.

Note 2. In order to avoid carry-over of suspended powder into the reflux flask the extractor must contain chloroform (60 ml) before adding the suspension. Care should be taken to avoid leaving powder on the distributor.

Cognate Determinations

Nux Vomica Liquid Extract
Nux Vomica Tincture

Solanaceous Alkaloids

Belladonna Tincture. *Determination of the percentage w/v of alkaloids calculated as hyoscyamine*

Method. The tinctures are concentrated by evaporation on a water-bath and then submitted to a preliminary process.

Make alkaline with ammonia and extract the alkaloids with chloroform

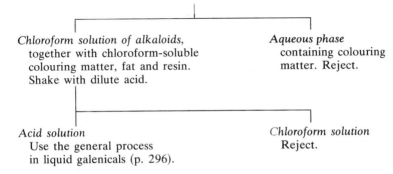

Chloroform solution of alkaloids,
together with chloroform-soluble
colouring matter, fat and resin.
Shake with dilute acid.

Aqueous phase
containing colouring
matter. Reject.

Acid solution
Use the general process
in liquid galenicals (p. 296).

Chloroform solution
Reject.

Cognate Determinations

Stramonium Tincture
Hyoscyamus Tincture. The tincture is concentrated to one twenty-fifth of its volume by careful evaporation. A low temperature of evaporation is necessary to prevent decomposition of the hyoscyamine. The subsequent procedure is the same as that described for Belladonna Tincture.

Belladonna, Dry Extract

In addition to desiccated extracted substance, the preparation contains a proportion of the corresponding powdered leaf which has been added to adjust the alkaloid content of the preparation to the required standard figure. This added leaf may or may not contain alkaloids. On the assumption that it does, the early stages of the extractions are suitably modified to ensure extraction of the alkaloid from the leaf tissue.

Belladonna Dry Extract is macerated with ethanol (50 per cent) for 30 minutes. After making alkaline with ammonia to liberate the alkaloids from their salts, the solution is extracted with successive portions of chloroform, the latter being filtered after separation to remove pieces of leaf tissue. This preliminary process is completed by extraction of the alkaloids from the chloroform into dilute acid in the presence of ethanol to reduce emulsification.

Belladonna Herb; Prepared Belladonna Herb; Hyoscyamus; Stramonium

These are drugs which consist essentially of vegetable tissue and the first step in the determination of their alkaloid content is the isolation of these alkaloids. Following the General Method for Crude Drugs (page 297) the sample (10 g) in fine powder is moistened with ammonia solution to liberate the alkaloids from their salts and macerated with a mixture of ether and ethanol for four hours. The mixture is transferred to a percolator and the extraction continued by percolation first with ether-chloroform (3:1) and afterwards with ether. Concentrate the percolate to 50 ml and dilute with ether to reduce the density below that of water. The alkaloids are extracted from the percolate by shaking with dilute acid and this solution is submitted to the General Method for Liquid Galenicals (page 296).

Ipecacuanha

This drug contains a number of alkaloids of which the most important are emetine and cephaëline. Earlier Pharmacopoeias required both phenolic and non-phenolic alkaloids to be determined, but now only total alkaloids are determined, these being calculated as emetine.

Ipecacuanha. *Determination of total alkaloids, calculated as emetine*

The general method for crude drugs is used, but with modifications in the later stages of the process. The alkaloids are extracted by a combination of maceration and percolation with solvent ether after the addition of ammonia to liberate the alkaloids from their salts. The percolate is then evaporated and the residue extracted first with 20 ml of N HCl and then with a mixture of 3 volumes of $0.1N$ HCl and one volume of ethanol (95 per cent); the latter reduces emulsification. The mixed acid solutions are washed with successive volumes of 10. 5 and 5 ml of chloroform to remove colouring matter, each chloroform solution being washed with the same 20 ml of $0.1N$ HCl. The combined acid solutions are made alkaline with ammonia and extracted with chloroform in the usual way. The alkaloidal residue, after moistening with 5 ml of ethanol (95 per cent) and evaporating, is dried at 80° for 5 minutes in a current of air. Further heating causes the residue, which should be golden yellow in colour, to darken considerably. The alkaloidal residue is determined volumetrically by back-titration procedure. The end point is indistinct since not only may it be masked by the colour of the solution, but the presence of phenolic alkaloids has a definite buffering effect. The end point is improved by dilution with water and also by the use of more than the usual amount of indicator (methyl red 10–15 drops).

Cognate Determination

Prepared Ipecacuanha. Some modification is necessary owing to the possible presence of lactose which may be used as a diluent. A mixture of strong ammonia solution and ethanol is used, instead of dilute ammonia solution, to liberate the alkaloids from their salts prior to the preliminary extraction process. In this way the water content of the medium is reduced and much of the lactose can be filtered out with the vegetable debris. Proceed as under Ipecacuanha.

Ipecacuanha Liquid Extract. *Determination of the percentage w/v of total alkaloids, calculated as emetine*

Total alkaloids are determined by the General Method for Liquid Galenicals (p. 296).

(*a*) Particular care should be exercised to remove as much colour as possible in this initial chloroform extraction.

(*b*) A mixture of acid and ethanol is used to wash the chloroform. Ethanol reduces emulsification, which can also be modified if only moderate shaking is used.

(*c*) At this stage 5 ml of dilute ammonia solution is sufficient to make the solution alkaline. Avoid the use of any large excess of ammonia and use only moderate shaking to avoid emulsions.

The alkaloids are determined volumetrically and the notes under Ipecacuanha should be consulted.

Cognate Determination

Ipecacuanha Tincture is determined in the same way as the Liquid Extract, using a larger volume, but without any preliminary concentration, since the latter would cause considerable darkening of colour with consequent difficulties at the end point.

Determination of Non-Phenolic Alkaloids

These are no longer specifically determined in the official assay processes, but may be determined in the following way. After the extraction and volumetric determination of total alkaloids transfer the solution to a separator with dilute acid, and make alkaline with sodium hydroxide solution. Phenolic alkaloids form water-soluble sodium salts, whilst the non-phenolic alkaloids are precipitated and separated by extraction with ether. Titrate in the usual way, the alkaloidal residue remaining after evaporation of the ether.

Opium

Although several alkaloids are present in opium and its preparations only the principal one, morphine, is determined. Morphine is insoluble in chloroform and most other organic solvents, but its isolation is facilitated by the fact that the molecule is phenolic and hence soluble in solutions made alkaline with sodium hydroxide.

Opium. *Determination of the percentage w/v of morphine, calculated as anhydrous morphine*

Method. Triturate the sample (8 g) with Water (30 ml) and calcium hydroxide (2 g) in a mortar. This liberates the alkaloids from their salts, and converts morphine and narcotine into their water-soluble calcium salts. Transfer the mixture (grease the lip of the mortar) to a tared flask and make up to exactly 90 g with Water. Shake occasionally during half an hour and filter through a dry Buchner funnel into a dry Buchner flask. Transfer exactly 52 ml of the filtrate to a clean dry conical flask, and add ammonium chloride (2 g), ether (25 ml) and 90 per cent ethanol (5 ml). The function of the ammonium chloride is to decompose the calcium salts of the alkaloids.

$$(RO)_2Ca + 2NH_4Cl \rightarrow 2ROH + 2NH_3 + CaCl_2$$

Ether acts as a solvent for narcotine and the ethanol facilitates the crystallisation of the morphine which separates from solution. Cork the flask, shake for 5 minutes and occasionally during half an hour, so that the total shaking time is about 15 minutes. Allow to stand overnight. Decant the ethereal solution as completely as possible through a funnel with a tight plug of cotton wool retaining the crystals in the flask as far as possible. Wash the contents of the flask with a further quantity of ether (10 ml) and decant through the filter. Wash the filter with solvent ether (5 ml) added slowly and in small quantities. Pour the aqueous liquor from the flask onto the filter and wash the flask and filter with morphinated water (a saturated solution of morphine in chloroform water), until the filtrate is free from chloride. Displace the air in the flask gently with air from a bellows to ensure removal of ammonia fumes. Transfer the contents of the funnel back to the flask with water from a wash bottle. Add $0.1N$ HCl (25 ml), boil gently to remove CO_2 and back-titrate the excess acid with $0.1N$ NaOH using methyl red as indicator. The end point is flat, due to the buffering effect of the phenolic hydroxyl group of morphine, and is indicated by the last shade of orange.

A correction of $+0.052$ g must be added to the calculated amount of anhydrous morphine, due to the solubility of morphine in the ether-water mixture. The correction as well as the general process is empirical and both depend upon 52 ml of the filtrate corresponding to exactly 5 g of the sample taken; these calculations are based on the average sample of morphine containing some definite percentage of water. It is assumed therefore that all are average samples.

Cognate Determinations

Powdered Opium
Opium Tincture only contains about 1 per cent of morphine and is therefore concentrated by evaporation to dryness before proceeding as for Opium.

Camphorated Opium Tincture. *Determination of the percentage w/v of morphine, calculated as anhydrous morphine*

The percentage of morphine is very low (0.5 per cent w/v) and a colorimetric method is adopted, depending upon the fact that nitrous acid reacts with morphine to convert it into an intensely yellow nitroso derivative. The colour is intensified and becomes darker on the addition of ammonia.

Method. Evaporate the sample (5 ml) to dryness on a water-bath and extract the residue with calcium hydroxide solution (10 ml). Filter into a separator and wash the evaporating dish and the filter with a further quantity of calcium chloride solution (10 ml). Add ammonium sulphate (0.1 g) to liberate the morphine and extract twice with alcohol-free chloroform (10 ml) to remove camphor and anise oil. Wash the chloroform extracts with the same 10 ml of Water, and reject the chloroform. Add ethanol (30 ml) and chloroform (30 ml) to the combined aqueous layer and washings and shake. Separate the lower chloroform-ethanol solution of morphine and wash by shaking with 95 per cent ethanol (5 ml) and Water (10 ml). Repeat the extraction with two further quantities (45 ml) of chloroform (2)-ethanol(1), washing with the same liquid as before. Evaporate the combined chloroform-ethanol solution of morphine to dryness, dissolve the residue in N HCl (10 ml) and dilute with water to 50 ml.

To 10 ml of the solution (representing 1 ml of sample), add freshly prepared 1 per cent w/v sodium nitrite (8 ml) allow to stand for 15 minutes and add dilute ammonia solution (12 ml). Compare the yellowish-brown colour which is produced with that obtained under the same conditions from standard amounts of morphine solution (see Part 2).

Colchicum

The determination of colchicine presents certain difficulties since unlike most other alkaloids, it is non-basic, the nitrogen being present in an acylamino group. Colchicine is soluble in both water and chloroform but insoluble in ether. Thus colchicine may be readily extracted into chloroform from both acid and alkaline solutions. The alkaloid being non-basic cannot be titrated, and therefore is determined gravimetrically.

Colchicum Corm. *Determination of the percentage of alkaloids*

Method. Macerate the sample (20 g in coarse powder), with 95 per cent ethanol (30 ml) warming on a water-bath for about 15 minutes. Transfer the mixture to a continuous extraction apparatus, and extract with ethanol (90 per cent). Cool, set aside for half an hour, when fatty matter is precipitated. Filter and wash the precipitate with ethanol until the alkaloid is completely extracted. Evaporate the combined filtrate to dryness on a water-bath, and wash the residue into a separator first using 20 per cent w/v aqueous sodium sulphate (5 ml) and then ether (50 ml). The function of the sodium sulphate is threefold, to prevent emulsification, to coagulate resinous material, and to render the aqueous phase less soluble in ether. Shake the separator when the ether dissolves resin and any residual fat. Repeat the extraction of resinous material with a further quantity of ether, and wash the ether solutions with further quantities of sodium sulphate solution. Combine the aqueous solutions and warm on a water-bath to remove traces of ether; transfer to a graduated flask (50 ml), add a small quantity of talc and make up to volume with sodium sulphate solution. Shake well and set aside for one hour, with occasional shaking, to clarify the solution. Filter and take 40 ml (equivalent to 16 g of colchicum), wash with ether again to remove foreign matter. Extract with chloroform in the presence of sodium hydroxide, which forms water-soluble derivatives with any remaining resinous matter. Repeat the extraction with chloroform until all the alkaloid has been isolated; wash the chloroform solution with dilute aqueous alkali and filter through a double filter paper into a tared flask. Distil off the chloroform, removing the last traces by repeated evaporation with ethanol; colchicine forms an addition compound with chloroform which is not readily decomposed by heat at 100°. Dry over phosphorus pentoxide under reduced pressure for 3 hours and weigh the residue. This residue is impure. Extract with water to separate colchicine which is water-soluble. Filter and dissolve the insoluble residue (resin) on the filter in ethanol. Combine the ethanol solution with the remainder of the water-insoluble material in the tared flask. Dry over phosphorus pentoxide under reduced pressure for 3 hours and weigh the residue. Subtract the weight of this residue from the total weight of impure alkaloid. The difference in weight is equivalent to colchicine.

Cognate Determinations

Colchicum Liquid Extract; Colchicum Tincture. Evaporate the sample to dryness and transfer the residue to a separator with sodium sulphate solution and ether. Proceed as for the determination of colchicine in Colchicum Corm.

Ergot

Ergot is no longer official in the British Pharmacopoeia though salts of pure ergot alkaloids are included. The alkaloids are present in ergot in small quantities, and hence are expensive. For this reason a colorimetric method is adopted. This depends on the blue colour obtained in acid solution on the addition of *p*-dimethylaminobenzaldehyde in sulphuric acid containing a trace of ferric chloride. For the sake of completeness, the preliminary processes used for ergot are described.

Ergot. *Determination of the percentage of total water-soluble alkaloids*

Method. Defat the powdered sample by continuous extraction with light petroleum (b.p. 40–50°). Remove the powder and carefully dry it in a porcelain dish at a temperature not exceeding 40°.

Total alkaloids. Transfer the defatted sample (2.5 g) to a 200 ml flask and add anaesthetic ether (100 ml). Insert the stopper, shake well and set aside for 10 minutes. Anaesthetic ether is specified to avoid the possibility of reaction with ether peroxides. Add 1 per cent ammonium hydroxide (20 ml) and shake vigorously every 5 minutes during half an hour. Add powdered tragacanth (1.5 g) and shake until the ether separates. Decant exactly 50 ml of the clear ethereal solution (equivalent to 1.25 g of ergot) and transfer to a separator. Extract with successive portions of 15, 15, 10 and 10 ml of 1 per cent w/v aqueous tartaric acid, and make up the acid extracts to exactly 60 ml.

Take 3 ml of the above solution with 6 ml of the *p*-dimethylaminobenzaldehyde reagent, allow to stand for 5 minutes and compare the colour obtained with that produced by a standard amount of ergometrine maleate (3 ml of freshly prepared 0.002 per cent w/v ergometrine maleate in 1 per cent w/v tartaric acid). Compare the colour intensities in a suitable colorimeter (see Chapter 1).

Dimethylaminobenzaldehyde reagent. Dissolve *p*-dimethylaminobenzaldehyde (0.125 g) in a cooled mixture of sulphuric acid (65 ml) and water (35 ml). Add 5 per cent w/v aqueous ferric chloride solution (0.1 ml).

Water-soluble alkaloids may be determined in the same tartaric acid solution as follows. Pipette 25 ml of the tartaric acid solution into a separator, make alkaline with ammonia, and extract with successive volumes of 40, 30, 30 and 20 ml of anaesthetic ether. Mix the ethereal solutions and wash with five successive portions of water to remove water-soluble alkaloids. Re-extract the washed ethereal solution with successive portions of 10, 5, 5 and 5 ml of 1 per cent w/v tartaric acid solution. Mix the acid solutions, remove dissolved ether as before and dilute to 30 ml with Water. Compare the colour obtained on addition of *p*-dimethylaminobenzaldehyde reagent with a standard. Calculate the percentage of water-soluble alkaloids by difference.

DETERMINATION OF THE SALTS OF ACIDS

The general principles of the determinations are similar to those for bases (p. 287) except that the solubilities are reversed at certain points. Thus the salts are water-soluble but are insoluble in organic solvents, as for alkaloidal salts. However, addition of mineral acid to an aqueous solution of the salt liberates the free acid which is extractable by organic solvents. This type of assay relates, in the main, to barbiturate-type salts.

Extraction with Ether

Phenobarbitone Sodium. *Determination of the percentage of* $C_{12}H_{11}N_2NaO_3$ *calculated with reference to the substance dried to constant weight at* 130°

The method follows closely that described for Chloroquine Phosphate (page 293).

Method. Heat a clean alkaloidal flask (100–150 ml) for half an hour in an oven at 105°. Transfer to a desiccator and allow to cool for exactly 30 minutes, and weigh (Note 1). Reheat for 15 minutes, cool in the desiccator (30 minutes) and weigh. Repeat until the weight of the flask is constant.

Accurately weigh the sample (0.5 g approx.) introducing it via a funnel directly into separator A containing Water; dissolve in Water (total volume 50 ml). Add dilute hydrochloric acid (10 ml) (Note 2), and ether (50 ml), stopper and shake to extract the precipitated phenobarbitone; continue shaking for two minutes (N.B. release pressure). Continue as described for Chloroquine Phosphate from 'Allow the solutions . . . to ensure complete extraction'.

Wash the combined ether solutions in A by shaking with 5 ml of Water. Allow to separate, run off the water and wash the latter with ether (5 ml). Reject the water. Repeat the washing with a further 5 ml of Water, rewashing with the same 5 ml of ether. Combine the ether solutions. Transfer to the dried and weighed alkaloidal flask. Fit up the flask for distillation and evaporate to dryness on a hot water-bath (Note 4). Cover the flask with a piece of paper (Note 5) and dry on the oven shelf (Note 6) to constant weight at 105°.

Notes on the determination of Phenobarbitone Sodium

Note 1. The flask must be cooled under desiccating conditions to prevent the deposition of moisture. Because of its size the flask must be cooled for 30 minutes to ensure that it is quite cold before weighing. A warm vessel will set up air currents inside the balance case, making accurate weighing impossible.

Note 2. Acidification of the solution causes the precipitation of phenobarbitone which is insoluble in water. Hydrochloric acid is used as it is completely volatile, and traces contaminating the residue would be readily removed in the drying process.

Note 3. Ether readily evaporates about the stopper and the neck of the separator with the possibility of losses.

Note 4. Ether is inflammable, and distillation must be carried out well away from any naked flames.

Note 5. This is to avoid possible contamination from the oven.

Note 6. Placing the flask centrally on the shelf avoids all possibility of charring the residue due to overheating in close proximity to the oven elements.

Loss on drying. Samples of Phenobarbitone Sodium may take up a small but variable amount of atmospheric moisture. Allowance is made for this in the assay, by calculating the percentage of $C_{12}H_{11}N_2NaO_3$ with reference to the sample dried to constant weight at 130°. A loss on drying is therefore determined on a separate portion of the sample.

Cognate Determinations

Amylobarbitone Sodium
Pentobarbitone Sodium

DETERMINATION OF ACIDS IN CRUDE DRUGS AND GALENCIALS

Extraction with Ether

Balsams

Balsams are exudates from the trunks of trees, and are composed of resins (both acidic and neutral), alcohols, esters of balsamic acids and the free balsamic acids themselves (benzoic and cinnamic acids).

Determination of total balsamic acids. Reflux the sample (1 to 1.5 g) with *0.5N* ethanolic KOH for one hour to hydrolyse the esters. Remove the ethanol by distillation and warm with Water (50 ml) until the residue is evenly diffused. Under these conditions the potassium salts of all acids (benzoic, cinnamic and resin acids) dissolve and the suspension consists of neutral insoluble resins and high molecular weight alcohols. Addition of magnesium sulphate precipitates the resin acids as salts and assists in the subsequent filtration. Filter and wash the residue with hot Water (20 ml). Acidify the mixed filtrate and washings to liberate the free acids and extract with four successive volumes each of 40 ml of ether. Reject the aqueous solution. Shake the mixed ethereal solutions with successive volumes (20, 20, 10, 10 and 10 ml) of 5 per cent aqueous sodium bicarbonate, separating and washing each aqueous solution with the same 20 ml of solvent ether. Acidify the mixed alkaline solutions with hydrochloric acid to liberate the acids, and extract with successive portions of chloroform. Dry the chloroform extract by filtering through a plug of cotton wool supporting a layer of anhydrous sodium sulphate. Remove the chloroform using a current of air, ceasing the process as soon as the last trace of chloroform has been removed to avoid volatilisation of the aromatic acids. Dissolve the residue in neutral 95 per cent ethanol (10 ml) and titrate with *0.1N* NaOH, using phenol red as indicator.

1 ml *0.1N* NaOH ≡ 0.01482 g of total balsamic acids, calculated as cinnamic acid

Tolu Balsam contains between 35 and 50 per cent of *total balsamic acids,* calculated with reference to the dry ethanol-soluble matter.

Cognate Determinations

Benzoin Tincture B.P.C. using 30 ml
Benzoin Tincture Compound B.P.C. using 10 ml

Male Fern

Filicin, the principal constituent of Male Fern, is a mixture of acidic substances, chiefly filmarone and filicic acid, which can be separated from non-acidic material and weighed.

Male Fern Extract. *Determination of the percentage w/w of Filicin*

Method. Accurately weigh the sample (1 g approx.) in a small beaker and wash into a separating funnel with 40 ml of ether. If film remains in the beaker, use some of the barium hydroxide solution specified below to complete the transference. Add 50 ml (exactly) of 2 per cent barium hydroxide solution, shake vigorously for 5 minutes and allow to separate. Barium salts of the acidic substances are formed and remain in the aqueous phase, whilst oily substances and colouring matter dissolve in the ether. Much material remains insoluble in either phase and sticks to the sides of the separator. Filter the aqueous liquid, wash with ethereal layer with water (2×5 ml) and filter the washings through the same filter. Acidify the combined aqueous liquids with hydrochloric acid to liberate acidic substances and extract with successive volumes of 30, 20, 15 and 10 ml of chloroform. Bulk the chloroform extracts, dry with anhydrous sodium sulphate (2 g) and filter, washing the sodium sulphate with chloroform (2×5 ml). Evaporate to dryness and dry the residual filicin to constant weight at 100°.

Soft Soap. *Determination of the percentage of fatty acids*

Method. Dissolve a sample of the soap (30 g) in water (100 ml), transfer the solution to a separator, and acidify with dilute sulphuric acid to precipitate the fatty acids. Extract the latter by shaking with three successive portions of 70 ml of solvent ether. Wash the ether solutions with successive portions of Water until the latter are free from mineral acid. Evaporate the ether solution in a tared flask and dry the residue to constant weight at 80°.

The fatty acids obtained in the assay process are further examined to ensure that they comply with certain characteristics.
Acid value $\not> 205$.
Iodine Value (iodine monochloride method) $\not< 83$.
Solidifying Point $\not> 31°$.
Limit Test for Resin Shake 0.5 ml with 2 ml of acetic anhydride until a clear solution is obtained. Cool to 15.5° and transfer one drop to a white porcelain tile. Mix with one drop of 50 per cent H_2SO_4 when there should be no transient violet colour.

Extraction with Buffer Solution

Cochineal

Cochineal is a colouring agent and to maintain a reasonable standard in this respect, a value is obtained by extraction of the carminic acid with a buffer solution of pH 8.0, which gives a stable colour for measurement of the extinction (Part 2).

Method. Weigh exactly 0.5 g of cochineal in moderately fine powder and add solution of standard pH 8.0 (60 ml). Heat on a water-bath for 30 minutes, cool and dilute to 100 ml with more buffer solution. Filter and dilute 5 ml of filtrate to 100 ml with the pH 8.0 buffer. Measure the extinction of the solution at the maximum at about 530 nm. A limit of $\not< 0.25$ in a 1 cm cell is imposed.

DETERMINATION OF NEUTRAL SUBSTANCES

Unsaponifiable Matter

Unsaponifiable matter is determined in the Vitamin D containing oils Cod-liver Oil and Halibut-liver Oil, and also in Emulsifying Wax where it forms not less than 88.0 per cent calculated with reference to the anhydrous substance and consists almost entirely of the major component cetostearyl alcohol. By contrast, in oils, the unsaponifiable matter is the small non-glyceride fraction consisting chiefly of steroids. Unsaponifiable matter is calculated as the percentage w/w of ether-soluble material remaining after complete saponification. Glycerides are saponified to water-soluble soaps and glycerol, and the unsaponifiable matter is the small fraction of the original material which is then soluble in ether. There is a tendency for the soaps to hydrolyse in solution, and traces of fatty acid so formed may dissolve out in the ether. Except in the examination of Emulsifying Wax the residue is always titrated with alkali to ensure that large quantities of such acid are not present.

Method. Accurately weigh between 2 and 2.5 g of oil and reflux for 1 hour with $0.5N$ ethanolic potassium hydroxide. Transfer the contents of the flask to a separator with Water, and whilst still warm extract vigorously with three successive portions of 50 ml of ether. Take great care to release the pressure in the separator. Separate the ethereal solution and mix with 20 ml of Water. Very gently rotate the separator for a few minutes. Avoid violent shaking. Allow to separate and run off the aqueous layer. Continue washing, shaking vigorously with two successive portions of 20 ml of Water. Shake vigorously with $0.5N$ aqueous KOH (20 ml) and wash again with 20 ml of Water. Repeat the dual alkali-water washing procedure twice more. Wash with successive portions of Water until the latter are no longer alkaline to phenolphthalein. Evaporate the ether solution in a tared flask. Evaporate the residue with a small volume of acetone (3 ml). Dry to constant weight at a temperature not exceeding 80°, and weigh. Dissolve the residue in freshly boiled and neutralised ethanol (95 per cent) and titrate with $0.1N$ alcoholic NaOH. If a titre of more than 0.1 ml is obtained reject the results and repeat the determination.

14 Medicaments in Formulations

The determination of single compounds in simple formulated products (dosage forms) has been discussed in earlier chapters, but the methods described therein are restricted to those which apply to both the parent drug, and with minimal modification, to its formulation. Frequently, however, the vehicle or basis of the formulation or additional active ingredients may interfere with the normal assay process. Separation methods or analytical methods which differentiate one substance from another are, therefore, required in order to determine the actual composition of the formulation.

In this Chapter methods are described for the analyses of a number of more complex formulations requiring special treatment. Some of the formulations considered are drawn from official compendia. Others are proprietary preparations. Since the proprietary preparations considered may not always be readily available overseas, an indication is given of likely vehicles or bases so that suitable mixtures can be prepared for instructional purposes. Many of the methods are based on basic procedures and instrumental techniques described elsewhere in either this volume or Part 2 of the book, to which readers should refer.

The standards demanded of preparations include not only identification of the active ingredients and their control but also control of factors affecting the safety in use, stability and, in certain cases, the bioavailability of the medicine. When drugs which are difficultly-soluble in water are incorporated into tablets the latter may comply with the official disintegration test but additional tests are necessary to indicate that the tablets are likely to be satisfactory clinically. Therefore, in spite of possible limitations, dissolution tests are used to study the rate of solution of such drugs in formulation. Dissolution tests are essential if satisfactory manufacturing standards are to be maintained for sustained or delayed release preparations. Again, to ensure adequate penetration of drugs such as Isoprenaline Sulphate into the respiratory tract when formulated as a suspension in aerosol inhalations, the particle size of the compound should be controlled so that it is not greater than about 8 μm. These and similar controls are highlighted in the sections which follow.

AEROSOL INHALATIONS

Dose

Aerosols supply unit doses and contain the medicament in suspension, or solution, under pressure in a mixture of inert propellants which, for anti-asthmatic sprays, is usually one of dichlorodifluoromethane and dichlorotetrafluoroethane. They may contain a surface-active agent, stabilising agents and other adjuvants. Correct unit dose from the pressurised container is dependent upon satisfactory functioning of the special metering valve as well as on the correct composition of the mixture or solution. The limits for the amount of active ingredient stated on the label to be available to the patient are wide, viz. 75–125 per cent, because of the nature of the preparation and its method of use. Thus, some of the dose delivered by the metering valve is retained in the oral adapter. The amount available to the patient is, therefore, the difference between the amount delivered by the metering valve and the amount retained in the oral adapter.

Isoprenaline Aerosol Inhalation. *Determination of the content of* $C_{11}H_{17}NO_3,\frac{1}{2}H_2SO_4$ *per spray*

The determination is based upon the formation of a colour when phenols are treated with iron salts under standard conditions (Part 2, Chapter 9) and the result obtained is an average value for 10 sprays.

Determination of the Amount of Isoprenaline Sulphate Delivered by Metering Valve

Method. Remove the oral adapter from the pressurised container, gently shake the container, actuate the valve several times, wash the valve stem with methanol and discard the washings (Note 1). Replace the oral adapter, invert the assembled unit, and place in a small beaker containing carbon tetrachloride (25 ml) and *0.01N* H_2SO_4 (25 ml) ensuring that the mouth of the adapter is completely immersed below the carbon tetrachloride layer (Note 2). Fire 10 sprays by pressing on the base of the container, gently shaking the container and contents of the beaker between each spray. Remove the aerosol unit from the beaker, wash it carefully with *0.01N* H_2SO_4 (10 ml) and transfer the contents of beaker and washings to a separator. Shake thoroughly (5 minutes), allow to separate and transfer the aqueous liquid to a 50 ml volumetric flask. Extract the carbon tetrachloride layer with *0.01N* H_2SO_4 (5 ml portions) and add the extracts to the 50 ml flask. Adjust to volume with more acid and mix well.

To 20 ml of extract in a 25 ml volumetric flask, add ferrous sulphate-citrate solution (0.3 ml) and aminoacetate buffer solution (3 ml), mix, allow to stand for 15 minutes and dilute to volume with water. Measure the extinction of the solution in 2 cm cells at about 530 nm using a blank of reagents. By reference to a previously prepared calibration curve calculate the average amount of isoprenaline sulphate delivered per spray.

Determination of the Amount of Isoprenaline Sulphate Retained in the Oral Adapter

Method. Remove the adapter, rinse with methanol and dry. Gently shake the pressurised container, actuate the valve several times, wash the valve stem with methanol and allow to dry. Replace the oral adapter and fire 20 sprays carefully into the air, with gentle shaking between each spray. Remove the oral adapter and immerse it in a mixture of carbon tetrachloride (25 ml) and *0.01N* H_2SO_4 (25 ml). Swirl the contents of the beaker, remove the adapter and continue as described above from the words 'wash it carefully with *0.01N* H_2SO_4,' but using a 4 cm cell for measurement of extinction. Calculate the average amount of isoprenaline sulphate retained per spray. Hence calculate the average content of isoprenaline sulphate per spray.

Note 1. This is a clean-up procedure and check to ensure smooth-working of metering valve.

Note 2. Complete immersion of the adapter in an absorbing liquid avoids loss of medicament as the spray emerges. Depending on the formulation, some aerosol inhalations give turbid or opalescent solutions on spraying directly into aqueous acid; the carbon tetrachloride dissolves this fatty material.

Uniformity of Dose. *Determination of the uniformity of dose from Isoprenaline Aerosol Inhalation*

The content of isoprenaline sulphate as determined above is an average value and for uniformity of dose each spray must be monitored in some way. Strictly, it is the metering valve that is being checked and as it may be required to deliver up to 400 sprays it is useful to apply a test for uniformity of dose at three stages, viz. when the container is full, when half-full and towards the end of its nominal number of sprays. The effect of sedimentation or caking of medicament or valve clogging may well be shown up in inaccurate dosage.

Method A. Remove the adapter from the pressurised container and determine the loss in weight of the pressurised container after each spray for a total of 10 sprays. Calculate the coefficient of variation for the results which refer, in the main, to loss of propellant.

Method B. Remove the adapter from the pressurised container, shake the container and invert it in a small beaker containing $0.01N$ H_2SO_4 (10 ml). Fire one spray below the surface of the liquid by pressing on the base of the container. Mix, filter if necessary through a sinter-glass filter (Porosity 3) and measure the extinction of the filtrate in a 2 cm cell. Calculate the content of isoprenaline sulphate using an $E_{1cm}^{1\%}$ value of 100 for isoprenaline sulphate. Repeat the determination for a total of 10 results and calculate the coefficient of variation. Compare the result with that obtained under A above as an indication of the relative precision of the two methods.

Identification of Medicament

The direct firing of spray into an aqueous or ethanolic system is used to obtain a solution on which tests for identity can be carried out. They take the form of chemical tests for particular groups or characteristic reactions e.g. the blue colour produced by ergot alkaloids with p-dimethylaminobenzaldehyde solution. In addition, thin-layer or paper chromatographic tests are applied (Part 2).

Particle Size

The medicament is intended to exert its action in the bronchioles and alveoli of the respiratory system. It is essential, therefore, that adequate penetration into the lungs takes place on inhalation. For this to occur the particle size of the emitted solution, or suspended particles in suspensions should be in the range 2–8 μm. The particle size must not be too small, however, as loss of medicament may occur on breathing out. Particle size is therefore an important part of the specification for aerosol inhalations even although difficulties occur in its determination. These have been emphasised by Kirk who has designed a simple system to simulate the upper part of the respiratory tract (Fig. 31). The tubing is lined with 3 per cent agar gel to reproduce the wet conditions of the lung.

With the vacuum pump set to give a flow of 16 l./min the aerosol adapter is fitted to the mouthpiece 'A' with a suitable gasket and a definite number of sprays fired into the apparatus. The fine particles collect on the Millipore filter which is removed for quantitative determination of medicament. The results are expressed as percentage of dose emitted from the valve to reach the filter or as percentage of dose leaving the mouth-piece to reach the filter. These particular doses are determined as described above. The results are related to particle size but the true value of the apparatus lies in the ability to compare aerosol inhalations for penetrative efficiency.

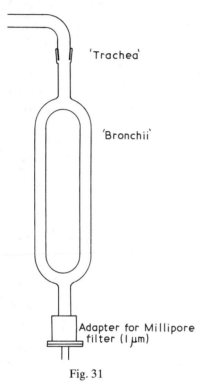

Fig. 31

Isoprenaline Aerosol Inhalation. *Determination of the particle size of the suspended isoprenaline sulphate*

The main difficulty associated with many methods is that of representative sampling and this is particularly so in the following.

Method A. With the oral adapter in position fire one spray onto a clean dry microscope slide held about 5 cm from the end of the mouthpiece of the adapter and perpendicular to the direction of the spray. Rinse the slide carefully with carbon tetrachloride (2 ml), allow to dry and examine the residue under a microscope. Most of the individual particles have a diameter not greater than 5 μm; no individual particle has a length, measured along its longest axis, greater than 20 μm.

Method B. Place several slides at the bottom of a tank of about 10 l. capacity and fire 6 sprays into the top of the tank. Cover immediately and allow to stand overnight for the particles to settle. Measure the size of the particles deposited on the slides as in A above.

Method C. Place several microscope slides at the bottom of a tin of dimensions approximately 25 cm side, in which a hole is cut in the bottom. Fire several sprays upwards through the hole and proceed as in B.

Cognate Determinations

Ergotamine Aerosol Inhalation. The determination is carried out by using 1 per cent tartaric acid solution instead of *0.01N* H$_2$SO$_4$ and measuring the extinction of the solution at 317 nm.

Isoprenaline Aerosol Inhalation Strong

Orciprenaline Aerosol Inhalation. The sprays are fired into *0.01N* HCl and any fatty material is extracted with ether. The extinction of the solution at 276 nm is used to determine orciprenaline.

Salbutamol Aerosol Inhalation. The sprays are fired into dehydrated alcohol and a colour is developed on an aliquot portion by oxidation with alkaline potassium ferricyanide in the presence of dimethyl-*p*-phenylenediamine. The coloured product is extracted into chloroform and the extinction determined at about 605 nm.

salbutamol

CAPSULES

Capsules consist of a medicament, in some cases mixed with a suitable diluent and stabiliser, enclosed in a shell which may be hard or soft. The former is composed mainly of gelatin and the latter of gelatin with varying proportions of glycerol and/or sorbitol solution to give the required degree of softness.

Disintegration Test

Unless otherwise specified, capsules are required to disintegrate in not more than fifteen minutes when determined in the prescribed apparatus, which is identical with that for tablets (p. 341) except that the tube of the basket is made of glass. Five capsules are used and disintegration is complete when no solid particle which would not pass through the gauze remains above it. The nature of the shell may give rise to aggregation in the test and if this occurs five capsules may be examined individually. The longest time taken by one of the capsules is taken to be the disintegration time. In the test, the use of a guided disc (see p. 342) is not permitted.

Uniformity of Weight

Capsules vary in weight and a test for Uniformity of Weight is applied. It is not always practicable to separate the contents quantitatively, since these may consist not only of active ingredient, but also fillers. The contents may also be present as a paste with a suitable liquid (usually an oil). In order to get a uniform test, which is generally applicable, it is simpler to remove the contents and weigh the shells. Two methods are used, one for capsules containing dry contents (Method A) and the other for those containing a liquid or paste (Method B). Method A involves weighing a capsule, opening and removing the contents without loss of shell material and then weighing the shell material. The difference represents the weight of contents. This is repeated on a further nineteen capsules and the average weight of contents is calculated. The weight of contents of each capsule should not deviate from the average weight of contents by a percentage greater than that shown in column A

Table 27. Uniformity of weight of capsule contents

Average weight	Percentage deviation A	B
120 mg or less	±10	±20
More than 120 mg	±7.5	±15

(Table 27) except that for two capsules the weight may deviate by not more than the percentage shown in column B.

Method B requires that the contents be expressed as much as possible and the remaining oil adhering to the shells be removed by washing with ether. After allowing the ether to evaporate, the individual shells are weighed. Ten capsules are required for this method and the requirement is that the content of each capsule does not differ from the average weight of contents by more than ±7.5 per cent, except that for one capsule the weight of contents may differ by not more than ±15 per cent.

Determination of Active Ingredients in Capsules

Standard analytical procedures are used to assay the active ingredients in the mixed contents of the capsules taken for the test for Uniformity of Weight. Solvent extraction procedures are used where simpler methods are unsuitable, as for example on Amylobarbitone Sodium Capsules, Phenytoin Capsules and on Pentobarbitone Capsules in which starch is frequently used as a diluent.

Examination of the mixed capsule contents is satisfactory for dry powders and oily solutions but if the contents consist of an oily or pasty suspension of medicament the mixed sample may not truly represent the bulk because of (a) sedimentation of powder or crystals and (b) absorption of water-soluble drugs by the soft gelatin capsule. In such cases it is better to digest the capsules with $0.1N$ hydrochloric acid either on a water-bath or in an autoclave to obtain complete recovery of water-soluble drugs, e.g. Nicotinamide, Riboflavine and Thiamine Hydrochloride in Vitamins Capsules.

Cough Capsules. *Determination of Promethazine Hydrochloride, Ephedrine Hydrochloride and Noscapine in Capsules*

The formulation for one capsule is stated to be

Promethazine Hydrochloride	3.6 mg
Ephedrine Hydrochloride	7.2 mg
Noscapine	12.5 mg

The capsule contents are an oil made opaque by suspended crystals as shown by microscopic examination under low power. A solution containing all three components is, therefore, prepared by digestion with acid.

From this point there is no single method which may be described as the best, as a number of possibilities exist. Some of these may be eliminated very quickly e.g. solvent extraction methods for separation of three bases are likely to prove very difficult if not impossible. Gas-liquid chromatography offers a solution but the differing volatility of the three bases requires differing column conditions for each e.g. on OV17 (3 per cent, 1 m) at 240° the bases emerge at 0.5 min (Ephedrine), 11 minutes (Promethazine) and 120 minutes (Noscapine). Thus, if a single column is to be used, temperature programming may offer a solution. Another possible approach via a separation method is by thin-layer chromatography followed by solvent extraction of the bases from the plate and determination of each by a suitable method e.g. ultraviolet absorption. Ephedrine, however, is weakly absorbing (Part 2) and some means of increasing the absorption is preferable. This leads naturally to considering the properties of the individual components and the possibility of their determination by reactions characteristic of each. Thus promethazine and other phenothiazine compounds give stable orange to red colours with oxidising agents in acid/ethanol solution and with palladous chloride in acid solution (Compare Promethazine Elixir B.P.C.). Ephedrine is oxidised to benzaldehyde by sodium periodate (compare Ephedrine Elixir B.P.C.). Noscapine absorbs radiation at 314 nm but promethazine absorbs to the same extent at about 310 nm. The following method is based on these properties.

Method.
Digest 5 capsules with *0.1N* hydrochloric acid (50 ml) at 50°C until the capsules are dissolved. Add sodium chloride (Note 1), cool, extract the oily solvent with ether (2×50 ml) and wash the bulked ether layers with saturated sodium chloride solution (2.5 ml). Add the aqueous washings to the acid layer and basify with ammonia. Extract immediately with ether (3×50 ml), bulk the ether extracts and wash once with saturated sodium chloride (5 ml containing 1 drop of ammonia solution). Reject the aqueous solutions and extract the bases from the ether layer with *0.1N* sulphuric acid (50, 20 ml). Transfer the acid extracts to a 100 ml volumetric flask, and wash the ether layer with Water (20 ml). Add the washing to the flask, adjust to volume with Water and mix well. This is the solution to be examined.

Determination of Promethazine Hydrochloride. To the solution to be examined (10 ml), add palladous chloride solution (20 ml) and dilute to 50 ml with water. Dilute the palladous chloride solution (20 ml) to 50 ml with water and use as a blank. Measure the absorbance in 2 cm cells at about 472 nm and calculate the concentration of promethazine hydrochloride using the value of 110 for the $E_{1cm}^{1\%}$ of promethazine hydrochloride in this colour reaction. Hence calculate the amount of promethazine hydrochloride in the contents of one capsule.

Determination of Noscapine. Dilute the solution to be examined (10 ml) to 100 ml with Water and measure the absorbance at 310 nm at which wavelength both promethazine hydrochloride and noscapine have the same $E_{1cm}^{1\%}$ of about 87 (Note 2), whereas ephedrine hydrochloride has zero absorption. Calculate the total concentration of promethazine hydrochloride and noscapine per capsule and subtract the value of promethazine hydrochloride found by direct determination above to give the amount of noscapine per capsule.

Determination of Ephedrine Hydrochloride. Dilute the solution to be examined (5 ml) to 100 ml with water. To the dilution (5 ml), in a separator, add saturated sodium bicarbonate solution (1.5 ml) and sodium periodate solution (5 per cent, 3 ml). Allow to stand 15 minutes, add *0.1N* sulphuric acid (10 ml) and cyclohexane (spectroscopic grade, 15 ml). Shake well for 3 minutes, allow to separate and reject the lower aqueous phase. Wash the cyclohexane layer with a further 10 ml of acid. If necessary, filter the cyclohexane layer through a plug of cotton wool and record the absorption curve (Note 3) in 2 cm cells over the region 275–225 nm using as blank the cyclohexane layer obtained in the same procedure but using Water (5 ml) instead of the solution to be examined. Calculate the amount of ephedrine hydrochloride per capsule by means of a calibration curve obtained by treating known amounts of ephedrine hydrochloride in the same way.

Note 1. The presence of gelatin gives rise to emulsions and the addition of sodium chloride is a common method of ensuring separation of the phases in a reasonable time.

Note 2. If a recording instrument is used, the wavelength and $E_{1cm}^{1\%}$ values should be determined by examination of standard solutions.

Note 3. It is essential to record the absorption curve to confirm that it corresponds exactly, in relative proportions, to those obtained with the standard solutions. Only in this way can some confidence be obtained in the result which, if calculated from a measurement at a single wavelength (about 241 nm), could be subject to error because of irrelevant absorption. This is very likely if insufficient acid is used after the oxidation procedure and traces of bases appear in the cyclohexane layer. It is for this reason that a second acid wash is introduced.

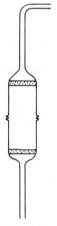

Fig. 32. Flow-through cell for determination of release rates

Sustained-release Capsules

The method of testing sustained-release capsules involves the determination of the rate of release of the active ingredient. Although, ideally, *in vivo* methods should be employed, some measure of the reproducibility of the manufacturing process and release characteristics of batches of capsules can be obtained by *in vitro* methods. If the result of such tests show good correlation with those of *in vivo* methods the quality control analyst has some confidence in using the *in vitro* method routinely.

Many variations in the apparatus exist and although that described for tablets (page 342) can be used, the nature of the formulation sometimes leads to difficulties. For example, the capsule contents may consist of small pellets coated with waxy or fatty materials which cause aggregation in the test. The rotating basket is better replaced by a flow-through system (Fig. 32) leading to a spectrophotometer if the active ingredient absorbs ultraviolet radiation.

The two sintered-glass filters are held together by a paraffin wax film (parafilm) and the capsule is placed on the porosity 1 sinter. A more uniform flow through the system is obtained than occurs when a single sinter is suspended in the dissolution medium, particularly if a small sinter is used as it soon becomes blocked. The system must be leak-proof, and a rapid circulation of solvent is necessary to ensure satisfactory dissolution profiles.

CREAMS

Creams consist of water, fatty materials and emulsifying agents which yield either water-miscible (oil-in-water type) or oily (water-in-oil type) emulsions. The active ingredients may be in solution in the basis or present as a suspension, in which case, particle size of the medicament may be important, as for example in Hydrocortisone Cream. Control of microbial contamination of creams (and other pharmaceutical preparations), although important, is outside the scope of this chapter but physical characters such as appearance, uniformity, viscosity, pH and checks to detect cracking are important aspects of quality control.

Water is not normally determined but it is an instructive exercise to compare the water content of Cetrimide Cream B.P.C. and that of a proprietary preparation using the Karl Fischer method (Chapter 11). The exercise can be extended to characterisation of the emulsifying agent used in the proprietary preparation by means of infrared spectrophotometry.

For many creams the medicament may be determined without prior separation, as the basis does not interfere with the assay process, e.g. Cetrimide Cream, and Salicylic Acid and Sulphur Cream. The correct choice of solvent may also avoid a lengthy separation procedure e.g. Dimethicone Cream using toluene, Hydrocortisone Cream using dehydrated ethanol, Mexenone Cream using methanol (simple filtration) and Proflavine Cream using chloroform/alcohol as solvent. The end-method of analysis in all these examples is spectrophotometry. Frequently, however, extraction and separation methods cannot be avoided, particularly where the concentration of medicament is small as in certain steroidal preparations.

Dimethicone Cream. *Determination of Cetrimide and Dimethicone*

The formula of the B.P.C. cream is as follows:

Dimethicone 350	100 g
Cetrimide	5 g
Chlorocresol	1 g
Cetostearyl Alcohol	50 g
Liquid Paraffin	400 g
Purified Water, freshly boiled and cooled	444 g

Cetrimide is determined without difficulty by the method for Cetrimide Emulsifying Ointment (p. 268) because the basis for the cream does not interfere. Dimethicone, however, is determined by an infrared absorption method, of which there are few examples in official compendia. It must therefore be separated from the water present in the cream. Silicones have strong infrared absorption bands but two, centred at about 1050 and 800 cm^{-1} respectively, are so broad that they are not satisfactory for quantitative analysis. Of the remaining three absorption bands, only that at about 1270 cm^{-1} is suitable because the other constituents of the cream, cetostearyl alcohol and liquid paraffin, and the solvent interfere with absorption in the region of the other two bands at about 2900 and 1400 cm^{-1}. It would be difficult to compensate for these accurately and their choice would, moreover, be incorrect because they are not specific for silicones.

Method. Mix the cream (about 5 g accurately weighed) with anhydrous sodium sulphate (10 g) in a 50 ml beaker until a uniform mixture is obtained (Note 1). Extract the mixture with slightly warm toluene (6×5 ml), filtering each extract through a porosity 3 sintered glass filter into a receiver. Transfer the filtrate to a 50 ml volumetric flask, rinse the receiver with toluene, add to the volumetric flask and make up to volume with toluene.

Record the infrared absorption curve in 0.1 mm cells over the region 1350 to 1200 cm^{-1} using as blank a 4 per cent w/w solution of liquid paraffin in toluene (Note 2). Determine the extinction due to dimethicone by the base-line technique (Part 2). Calculate the concentration of dimethicone by reference to a calibration curve prepared by examining solutions containing 0.2, 0.4, 0.6, 0.8 and 1.0 per cent w/v of dimethicone in 4 per cent w/w liquid paraffin in toluene.

Note 1. The anhydrous sodium sulphate abstracts the water as the decahydrate leaving a readily filtered organic layer (compare Triamcinolone Cream, below).

Note 2. A 4 per cent w/w solution of liquid paraffin in toluene is to allow for both liquid paraffin and cetostearyl alcohol (alkane absorption) in the sample solution. The actual interference at 1270 cm^{-1} is, however, quite small.

Triamcinolone Cream (0.1 per cent.) *Determination of Triamcinolone Acetonide*

The cream is water-miscible and extraction procedures are necessary to obtain the acetonide in chloroform solution. The determination is based upon the reaction of α,β-unsaturated steroidal ketones with isoniazid to yield

yellow products with an absorption maximum at about 415 nm. This is a particularly useful reaction in that it enables determinations, previously made difficult by the nature of the base or vehicle, to be carried out simply as in Progesterone Injection.

Reagent. Isoniazid 0.5 g in methanol (500 ml) containing hydrochloric acid (0.63 ml).

Method. Transfer the cream (about 2 g accurately weighed) to a 50 ml glass stoppered flask, add chloroform (30 ml) and anhydrous sodium sulphate (0.5 g). Shake thoroughly (10 minutes) and filter the chloroform through a phase separating paper (Note). Wash the flask and filter with chloroform (10 ml), transfer the filtrate and washings quantitatively to a 50 ml volumetric flask and adjust to volume with chloroform.

Transfer 5 ml of the solution to a 25 ml volumetric flask and add isoniazid solution (10 ml). Warm at 55° for 45 minutes, cool and adjust to volume with chloroform. Measure the extinction at about 415 nm in a 1 cm cell using as blank 5 ml of chloroform treated in the same way as 5 ml of solution. The standard solution (5 ml) used for comparison is 0.005 per cent triamcinolone acetonide in chloroform treated as described above.

Calculate the concentration of triamcinolone acetonide by reference to the two extinctions.

Note. Where a mixture consists of organic solvent and solid matter, the use of a sintered gláss crucible is satisfactory for filtration. If, however, droplets or a layer of water is present phase separating paper is the system of choice.

Whatman No. 1 PS is a filter paper impregnated with a stable silicone to make it water repellant. The method of use is as follows.

(1) Fold the paper in the normal way and place it in a conical filter funnel.
(2) Pour the mixed phases directly into the paper. It is not normally necessary to allow the phases to separate cleanly before pouring.
(3) Allow the solvent phase to filter completely through the paper.
(4) If required, wash the retained aqueous phase with a small volume of clean solvent. This is normally necessary when separating lighter than water solvents in order to clean the meniscus of the aqueous phase.

DO NOT

Allow the aqueous phase to remain a long time in the funnel after phase separation, or it will begin to seep through. This is caused by evaporation of the solvent from the pores of the paper, creating a local vacuum, which in turn draws the water through.

EMULSIONS

Pharmaceutically a distinction is drawn between emulsions which are oil-in-water preparations for internal use and creams and applications which are intended for external use.

Liquid Paraffin and Phenolphthalein Emulsion. *Determination of Liquid Paraffin and Phenolphthalein*

The determination of liquid paraffin is based upon solvent extraction and that for phenolphthalein on the formation of a red colour with alkali. The conditions for development of the colour are critical and for a discussion of these the original paper by Allen, Gartside and Johnson (1962) should be consulted.

Method

Liquid Paraffin. Weigh accurately about 5 g of the well shaken emulsion into a separator, add Water (10 ml) and extract with two successive portions each consisting of a mixture of ethanol (95 per cent, 10 ml), light petroleum (b.p. 40–60°, 15 ml) and ether (15 ml). Complete the extraction with a mixture of light petroleum (15 ml) and ether (15 ml). Bulk the extracts and wash with $0.5N$ NaOH (15 ml) and water (15 ml). Transfer the solvent extract to a tared flask quantitatively and evaporate the solvent. Add acetone (5 ml) and again evaporate; repeat the addition of acetone and evaporation until a clear water-free residue is obtained. Dry at 105° for 15 minutes, cool and weigh. Calculate the percentage w/w of liquid paraffin.

Phenolphthalein. Weigh accurately about 3 g of the well-mixed emulsion into a centrifuge tube and add about 1–2 g of Filtercel or cellulose powder if the Filtercel is not available. Mix into a paste and gradually add ethanol (95 per cent, 20 ml) with constant stirring and mixing. Centrifuge, decant the supernatant liquid into a 100 ml volumetric flask and complete the extraction by washing the residue with ethanol (95 per cent, 2×20 ml). Add the washings to the flask and dilute to volume with ethanol (95 per cent). Evaporate the solution (5 ml) to dryness in a small beaker (Note 1) and dissolve the residue in glycine buffer solution (Note 2). Transfer to a 100 ml volumetric flask and rinse the beaker with more buffer solution, adding the washings to the flask. Dilute to volume with buffer solution, mix and measure the extinction at 555 nm in a 1 cm cell. Complete the measurement within 10 minutes of adding the buffer solution (Note 3) and calculate the percentage of phenolphthalein using an $E_{1cm}^{1\%}$ value of 1055 for phenolphthalein in glycine buffer solution.

Note 1. It is essential to remove all traces of ethanol otherwise a reduction in the intensity of the pink colour occurs.

Note 2. Maximum colour develops over a narrow range of pH and the buffer solution should be adjusted if necessary to pH 11.1. It is made by adding *0.1N* NaOH (100 ml) to a solution (100 ml) containing aminoacetic acid (0.75 g) and sodium chloride (0.58 g).

Note 3. The colour is stable up to about 15 minutes and decreases slowly thereafter.

EYE-DROPS

Eye-drops are sterile preparations and those that are aqueous (other than single-dose forms) contain a bactericide and fungicide. Of the preservatives, only benzalkonium chloride and chlorhexidine acetate may interfere in the determination of the active ingredient when the assay involves the use of sodium tetraphenylborate. Eleven of the twenty six preparations in the British Pharmaceutical Codex use this method and three procedures are specified.

Method 1 is used where no interference is to be expected e.g. with strong solutions (1 per cent or more) containing phenylmercuric acetate as preservative. The normal method outlined on page 270 is used.

Method 2 is specified where benzalkonium chloride is present as preservative. A determination is carried out at pH 10 when only the preservative is titratable (compare page 267).

Method 3 is specified for those preparations containing chlorhexidine acetate. The modification consists in precipitating the chlorhexidine as sulphate by addition of sodium sulphate. After allowing time for precipitation to occur the assay is continued as described for Method 1.

Many of the preparations can be assayed by direct measurement of ultraviolet absorption on a diluted aliquot portion of the sample. This procedure is adopted for Chloramphenicol Eye-drops, Proxymetacaine Eye-drops and Prednisolone Eye-drops. Several preparations are of interest in illustrating important analytical aspects.

Phenylephrine Eye-drops. *Determination of the percentage of* $C_9H_{14}ClNO_2$

The determination is based upon measurement of the deep red colour given by many phenols with 4-aminophenazone in slightly alkaline conditions on the addition of potassium ferricyanide. The reaction is given by the following equation.

The oxidation product of phenols is often soluble in chloroform and an extraction procedure increases the sensitivity. Otherwise dilution with water is possible as in the present example. Ferricyanide reacts with 4-aminophenazone to give a red colour in neutral and slightly acid solutions. The order of addition of reagents is, therefore, important. The ferricyanide solution should be added last to the sample solution containing the 4-aminophenazone and sodium bicarbonate. Although the red colour formed in acid solution with the two reagents disappears on making alkaline it may well be that some change has occurred which would explain the variable results observed by Ensor for salicylamide in formulations. He suggested that ferricyanide be added last and no difficulties were then experienced.

INJECTIONS

In maintaining the quality of injections, consideration must be given not only to the amount of active ingredient present but also to factors affecting the stability and safety in use of the formulation. Specification for containers, closures and particulate matter in solutions therefore become important in addition to factors affecting dosage and its determination e.g. volume or weight in the containers, and the presence or absence of bactericides, which may influence choice of assay.

Limit Test for Particulate Matter

The Pharmacopoeia requires that solutions for injection should not contain particles of foreign matter that can readily be observed on visual inspection using ordinary and plane-polarised light. In particular, containers of 500 ml capacity or more should be examined individually for the presence of undesirable particles. This inspection, however, is subjective and a more discerning test designed to limit the number of particles of particular sizes is applied solely to large volume solutions (500 ml or more per container) such as intravenous fluids made from various sugars.

Apparatus. No particular apparatus is mandatory so long as it is capable of counting the numbers of particles having equivalent sphere diameters equal to or greater than 2 μm and equal to or greater than 5 μm. A Coulter counter is satisfactory (Part 2), and as it is based upon electrical resistance, a filtered sodium chloride solution should be added to the injection if necessary e.g. for non-conducting dextrose injections.

Sampling. Contamination with particles on opening large-volume containers is not likely to be great but if ampoules containing 1 to 2 ml of injection are examined some precautions are necessary. Somerville and Gibson referring to a paper by Emerot and Dahlinder adopted a procedure of immersing the ampoules in a water bath at 70° followed by wiping off to reduce drastically misleading contamination before opening. Heating the ampoules to 70° produced a positive internal pressure, which reduced the possibility of airborne contamination.

Standard. As only large-volume injections are examined officially the standard for these is that the mean counts of particles per 1.0 ml do not exceed

1000 equal to or greater than 2 μm and 100 equal to or greater than 5 μm, the mean being determined from the examination of five containers. If only one container is available the standard is relaxed to 2000 and 100 respectively.

Volume in Single-dose Containers

In order to permit withdrawal of the nominal volume from the container an excess (overfill) of the injection must be present. Requirements for total volume are specified in Table 28.

When the volume of injection does not exceed 2.0 ml the average volume, based on examination of ten containers, deviates by not more than 5 per cent from the requirements for total volume and in no case does the volume deviate by more than 10 per cent from the requirements.

For volumes greater than 2.0 ml the requirements are that the total volume in each container is not less than the nominal volume and does not exceed the requirements for total volume by more than 5 per cent.

Table 28. Overfill volumes for injection

	Nominal volume (ml)			Excess volume (ml) Mobile injections	Viscous injections
		Not exceeding	0.5	0.10	0.12
Exceeding	0.5	Not exceeding	1.0	0.10	0.15
Exceeding	1.0	Not exceeding	2.0	0.15	0.25
Exceeding	2.0	Not exceeding	5.0	0.30	0.50
Exceeding	5.0	Not exceeding	10.0	0.50	0.70
Exceeding	10.0	Not exceeding	20.0	0.60	0.90
Exceeding	20.0	Not exceeding		2 per cent	3 per cent

Determination of Volume of Injection Solution

A suitable method for determining the volume is based on that described in the B.P.C. for ampoules of solvent supplied with ampoules of dry sterile powder for preparing an injection immediately before use.

Method. Examine eleven ampoules to check that the volume of injection is approximately the same in each. Select a syringe of volume not greater than twice the nominal volume of the injection and fitted with a needle. Draw up into the syringe some of the injection from one of the ampoules and discharge it from the syringe with the needle pointing upwards to remove air bubbles. Use this prepared syringe to transfer the injection from the remaining ten ampoules to a measuring cylinder of capacity not greater than twice the volume to be measured. Measure the individual and total volumes and calculate the average volume in each ampoule. Reserve the solution for assay purposes.

Uniformity of Weight

This requirement applies to those injections prepared by dissolving the contents of a sealed container before use. Two methods are specified. Method A for use when the contents consist wholly of one substance and Method B for use when a mixture with inert material or another active ingredient is present.

Method A. Remove any paper labels attached to the container, wash the outside with water and dry carefully. Open the container, weigh all parts and remove the contents. Wash with Water and then with ethanol (95 per cent), dry at 105°, cool and weigh. The difference between the weights represents the weight of contents. Repeat with a further nine containers. The requirements are that the weight of contents of each container do not deviate from the weight stated on the label by a greater percentage than that shown in column A (Table 29) except that in one container the weight may deviate by not more than that shown in column B.

Method B. Determine the weight of contents as described for Method A but mixing the contents of the sealed containers for determination of the active ingredient or ingredients if a mixture.

The proportionate amount should be within the specification in the monograph except that in one container the amount may lie within a range which allows not more than twice the tolerance permitted in the monograph.

Table 29

Weight stated on label	Percentage deviation	
	A	B
120 mg or less	±10	±15
More than 120 mg and less than 300 mg	±7.5	±12.5
300 mg or more	±5	±10

Neostigmine Injection. *Determination of Neostigmine Methylsulphate and Chlorocresol in an injection containing 0.5 mg/ml and 0.1 per cent respectively*

There is no official method for the determination and suggested procedures involving hydrolysis and colour reactions for the resulting phenol appear unnecessarily long. It is not possible to extract the neostigmine with organic solvents because it is a quaternary ammonium compound. This is an advantage in that phenolic bactericides are readily removed from a slightly acidified solution by extraction with ether. Neostigmine has a significant, albeit low intensity absorption at about 270 nm and this then serves as a simple means of its determination. Chlorocresol contributes to the absorption at about 270 nm and neostigmine contributes a small amount to the chlorocresol absorption. It is simpler to separate the components than to use the method described for Pethidine Injection (Part 2).

Neostigmine. Dilute the injection (5 ml) with *0.1N* H_2SO_4 (5 ml) and Water (5 ml). Extract with ether (3×25 ml), bulk the ether extracts and wash with Water (5 ml), adding the washing to the reserved aqueous liquid. Dilute the aqueous liquid to 25 ml and record the absorption curve using water as blank and 4 cm cells. Calculate the mg/ml of neostigmine methylsulphate in the injection using a standard neostigmine methylsulphate.

Chlorocresol. Dilute the injection (5 ml) with *0.1N* H_2SO_4 (5 ml) and Water to 50 ml. Record the absorption curve using the *solution obtained for neostigmine methylsulphate* (*diluted 10 ml to 20 ml*) as blank in 1 cm cells (Note). Calculate the percentage of chlorocresol by reference to a standard.

Note. This solution compensates for the small contribution made by neostigmine to the chlorocresol absorption. This contribution may be observed if the neostigmine blank is replaced by water and the absorption curve recorded on the same chart paper. A similar experiment may be carried out with Pethidine Injection in which case the curves should be superimposable over the region of the chlorocresol absorption. If they are not it indicates that the extraction technique of the analyst must be improved.

Mannitol Injection. *Determination of the percentage of* $C_6H_{14}O_6$

The principle underlying the assay is oxidation of mannitol to formic acid and formaldehyde by sodium periodate followed by determination of excess periodate. The latter determination uses a different end-method of analysis from that described under Glycerol Suppositories (page 338), and the reactions taking place overall are as follows.

$$\begin{array}{l} \text{CH}_2\text{OH} \\ | \\ \text{HO}-\text{C}-\text{H} \\ | \\ \text{HO}-\text{C}-\text{H} \\ | \\ \text{H}-\text{C}-\text{OH} \\ | \\ \text{H}-\text{C}-\text{OH} \\ | \\ \text{CH}_2\text{OH} \end{array} + 5\text{NaIO}_4 \longrightarrow 2\text{H.CHO} + 4\text{H.COOH} + 5\text{NaIO}_3 + \text{H}_2\text{O}$$

$$\text{NaIO}_4 + 2\text{HI} + \text{H}_2\text{O} \xrightarrow[\text{solution}]{\text{NaHCO}_3} \text{NaIO}_3 + \text{I}_2 + 2\text{KOH}$$

$$\underset{\text{(excess)}}{\text{Na}_3\text{AsO}_3} + \text{I}_2 + \text{H}_2\text{O} \longrightarrow \text{Na}_3\text{AsO}_4 + 2\text{HI}$$

$$\downarrow \text{NaHCO}_3$$

$$\text{NaI} + \text{H}_2\text{O} + \text{CO}_2$$

$$182.2 \text{ g } C_6H_{14}O_6 \equiv 5\text{NaIO}_4 \equiv 5\text{Na}_3\text{AsO}_3 \equiv 10\text{I} \equiv 10\text{H}$$
$$18.22 \text{ g } C_6H_{14}O_6 \equiv 1000 \, N$$
$$1.822 \text{ g } C_6H_{14}O_6 \equiv 1000 \, 0.1N$$
$$0.001822 \text{ g } C_6H_{14}O_6 \equiv 1 \text{ ml } 0.1N$$

Method. Dilute a volume of the injection containing about 0.4 g of mannitol to 100 ml with Water. Transfer 10 ml of the dilution to a stoppered flask, add sodium periodate solution (2.14 per cent, 20 ml) (Note 1) and dilute sulphuric acid (2 ml). Heat on a water-bath for 15 minutes (Note 2). Cool, add sodium bicarbonate (3 g) and 0.2N Na$_3$AsO$_3$ (25 ml) and mix. Add potassium iodide solution (20 per cent, 5 ml), allow to stand for 15 minutes (Note 3) and titrate with 0.1N iodine to the first trace of yellow colour. Repeat the operation without the mannitol and calculate the percentage of mannitol from the difference in titrations.

Note 1. The volume of periodate solution must be accurately measured.

Note 2. Loosen the stopper of the flask to release any pressure.

Note 3. The liberated iodine immediately oxidises the arsenite present in the presence of bicarbonate but no reaction occurs between the iodide and iodate under neutral or slightly alkaline conditions. In the blank determination all the periodate is available to liberate iodine from potassium iodide and hence more of the arsenite is oxidised than occurs when sample is present. The iodine titre for the blank is, therefore, less than that for the sample.

Cognate Determination

Sorbitol Injection. For strong solutions Greenwood has shown that refractometry (Part 2) is a rapid and convenient method.

Procyclidine Injection B.P.C. *Determination of Procyclidine Hydrochloride*

The principle underlying the method is the formation of an indicator base complex which is soluble in chloroform at pH 5 used in the extraction procedure. A similar method is adopted for Procyclidine Tablets (page 268) but in order to increase the sensitivity the indicator is further extracted into 0.1N NaOH. For a fuller account of the factors involved in the formation and breakdown of the complexes see page 266.

Many bases will give this reaction and the procedure is therefore no more specific than a determination by measurement of ultraviolet absorption which is here used as an identity test.

Method. Dilute the injection to give a solution containing 0.002 per cent of Procyclidine Hydrochloride. Transfer 10 ml to a separating funnel, add chloroform (25 ml, by pipette), a solution of sodium phosphate 0.2 per cent, sodium dihydrogen phosphate 4.3 per cent (*10M*) and bromocresol purple solution (0.1 per cent in *0.01N* NaOH; 10 ml). Shake thoroughly and allow to separate. Transfer the yellow chloroform layer to a stoppered flask containing anhydrous sodium sulphate (Note), mix well and measure the extinction in 1 cm cells at 407 nm. Repeat the procedure using a standard solution of procyclidine hydrochloride (0.002 per cent) and calculate the percentage of procyclidine hydrochloride in the injection.

Vitamins B and C Injection for Intravenous Use, B.P.C. *Determination of Pyridoxine Hydrochloride*

Ampoule 1 contains Thiamine Hydrochloride, Pyridoxine Hydrochloride, Riboflavine and Nicotinamide. Ampoule 2 contains Ascorbic Acid.

The injection is prepared immediately before use by mixing the contents of ampoules 1 and 2. The nicotinamide may be present in ampoule 2 rather than ampoule 1 but the general procedure is in no way affected. In spite of the various components, each is determined by properties peculiar to itself and no separation procedure is involved.

Thiamine Hydrochloride: precipitation as a silicotungstate (Chapter 12).

Riboflavine: absorption in the visible region of the spectrum (Part 2).

Nicotinamide: by ammonia distillation (Chapter 5).

Ascorbic Acid: by iodine oxidation (Chapter 8).

Pyridoxine Hydrochloride: interference in the direct determination of pyridoxine by ultraviolet absorption is caused by thiamine hydrochloride, nicotinamide and riboflavine depending upon the wavelength chosen. However, by a correct choice of buffer solutions a ΔE value (Part 2) may be obtained at 328 nm characteristic of pyridoxine alone. The spectral changes that occur when using solutions of pH 2-7 are shown in Fig. 33.

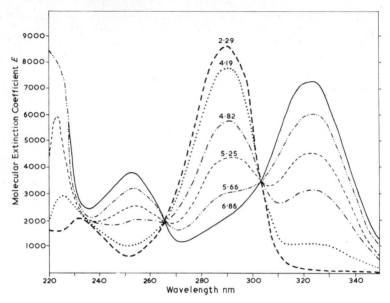

Fig. 33. Absorption spectra of pyridoxine at pH values less than 7

Method. Dilute the injection with water to give a 0.01 per cent solution of Pyridoxine Hydrochloride. Dilute the solution (10 ml) to 100 ml with glycerinated phosphate buffer pH 7.0 (Solution A) and glycerinated *0.1N* hydrochloric acid (Solution B). Measure the extinction in 1 cm cells at the maximum at about 328 nm of Solution A using Solution B as blank.

Calculate the percentage pyridoxine hydrochloride using a value of 340 for the $E_{1cm}^{1\%}$ of pyridoxine hydrochloride under these conditions.

LOZENGES

The active ingredient is incorporated in a flavoured basis consisting (in Codex preparations) mainly of sucrose. Although the British Pharmaceutical Codex describes two methods of preparation, viz. by moulding or compression, the latter may be used for all the lozenges described therein. Standards for uniformity of weight of compressed lozenges are those applicable to tablets.

Amphotericin Lozenges B.P.C. *Determination of Amphotericin A in the Amphotericin used in the lozenges*

The parent drug Amphotericin B.P. is required to contain not more than 15.0 per cent of Amphotericin A. The method is of interest because it consists of a spectrophotometric determination of a mixture of two components as described in Part 2 (Chapter 9). The method is complicated, however, in that one of the two standards used, Amphotericin B (A.S.) contains a small percentage F of Amphotericin A. Also, in order to calculate the appropriate $E_{1cm}^{1\%}$ values, S_1 and S_2, for the lozenges, the total content of Amphotericin must be known. Fortunately, a microbiological assay can be used for this purpose, and the formula for calculating the content of Amphotericin A in the Amphotericin used becomes:

$$F + \frac{100(BS_2 - bS_1)}{(aB - Ab)}$$

where

$A = E_{1cm}^{1\%}$ of Amphotericin A (A.S.) at 282 nm

$a = E_{1cm}^{1\%}$ of Amphotericin A (A.S.) at 304 nm

$B = E_{1cm}^{1\%}$ of Amphotericin B (A.S.) at 282 nm

$b = E_{1cm}^{1\%}$ of Amphotericin B (A.S.) at 304 nm

$S_1 = E_{1cm}^{1\%}$ of the sample at 282 nm*

$S_2 = E_{1cm}^{1\%}$ of the sample at 304 nm*

$F =$ declared content of Amphotericin A in Amphotericin B A.S.

The quality control of the lozenges consists of general appearance, uniformity of diameter, uniformity of weight, content of active ingredient and a limit on a likely impurity.

Method. Powder the lozenges and accurately weigh a quantity equivalent to about 50 mg of amphotericin into a 100 ml volumetric flask. Add dimethyl sulphoxide (20 ml) (Note), mix well and dilute to 100 ml with dehydrated methanol. Filter and dilute the filtrate (10 ml) to 50 ml with dehydrated methanol. Measure the extinction in 1 cm cells at 282 and 304 nm using a blank of 0.8 per cent v/v dimethyl sulphoxide in dehydrated methanol. In the same way determine the extinction of 0.008 per cent Amphotericin B (A.S.) and 0.0008 per cent Amphotericin A in 0.8 per cent v/v dimethyl sulphoxide in dehydrated methanol.

Calculate the appropriate $E_{1cm}^{1\%}$ values, and substitute in the equation to determine the percentage of Amphotericin A.

Note. Dimethyl sulphoxide is a very useful solvent for the initial penetration and solution of constituents of even crude drugs.

Throat Lozenges. *Determination of 2,4-Dichlorobenzyl Alcohol and Amyl-m-cresol*

The proprietary lozenge is red, pleasantly flavoured and about 2.5 g in weight. The approximate content of active ingredients is stated to be:-

2,4-dichlorobenzyl alcohol	1.2 mg
amyl-*m*-cresol	0.6 mg

In principle the two compounds should be readily separated from the sugar basis by solvent extraction with ether under slightly acid conditions. In turn, the phenolic compound should be extractable with sodium hydroxide from ethereal solution in accordance with the principles described in Chapter 13. In practice, however, this latter separation does not occur and a possible ultraviolet spectrophotometric method of assay for each component is not feasible.

The relatively small quantities involved therefore suggest a gas-chromatographic procedure. Versamide 900 is frequently used for phenols

---

*with respect to the content of Amphotericin determined by microbiological assay.

but polypropylene glycol adipate (Reoplex 400), 15 per cent on Chromosorb W 80–100 mesh is also very useful. Both components are separated on this column (1 m) at 190°, and for an internal standard chloroxylenol is best. Typical retention times are

2,4-dichlorobenzyl alcohol	10 minutes
amyl-*m*-cresol	12 minutes
chloroxylenol	14 minutes

Method. Weigh and powder five lozenges. Dissolve a weight of powder equivalent to one lozenge in Water (15 ml), add dilute hydrochloric acid (2 drops), sodium chloride (2 g) and internal standard solution (0.02 per cent, 5 ml). Extract with ether (3× 20 ml), bulk the ether extracts, wash with saturated sodium chloride solution (5 ml) and dry with anhydrous sodium sulphate (2 g). Decant most of the ether (Note) and evaporate to low volume (1–2 ml) using a little ethanol if necessary to obtain a clear solution free from traces of suspended water. Inject about 1 μl into the column and compare the result with that obtained for a mixture of the components (amyl-*m*-cresol, 0.012 per cent, 5 ml; 2,4-dichlorobenzyl alcohol, 0.024 per cent, 5 ml) treated in the same way.

Calculate the content of each component in the normal way (Part 2).

Note. The presence of internal standard makes quantitative recovery at this point unnecessary.

Extension. (*a*) Investigate by polarimetric and other means the nature of the sugar basis.

(*b*) An odour of menthol is detected on preparing the solution used for the assay. Confirm its presence by GLC.

MIXTURES

The complexity of the formulation of many liquid preparations intended for oral use makes separation methods of paramount importance in the determination and identification of active ingredients. Sample handling involves weighing rather than pipetting and determination of weight per ml is therefore involved in many determinations.

Elixirs

The satisfactory presentation of unpalatable medicaments requires the presence of flavourings and sweetening agents in the formulation. The general appearance of the preparation is similar to a syrup but sucrose need not necessarily be present in the vehicle and the British Pharmaceutical Codex tends to draw a distinction between Syrups and Elixirs.

The advent of potent yet unstable compounds may require a special type of formulation in which the dry ingredients, in the form of granules or powder, are mixed with a specified quantity of water before issue. Quality control of the preparation includes inspection on reconstitution to ensure a satisfactory product. For derivatives of penicillanic acid, specifications also include a stability test on the regenerated elixir or mixture.

Phenethicillin Elixir. *Determination of Phenethicillin*

The elixir is freshly prepared by adding a specified quantity of water to the dry powder. The determination is based upon the formation of a hydroxamic acid on opening of the β-lactam ring in the presence of hydroxylamine, and the formation of a characteristic red colour which such acids give with a ferric salt.

The reaction depends on the presence of the intact β-lactam ring so that inactive hydrolysis products will not, therefore, be determined; the stability test is based upon this. As control of pH is important in maintaining stability a limit of 5.2–6.2 is imposed on the regenerated elixir.

Method. Weigh accurately about 3 ml of the freshly prepared elixir (Note 1), dilute to 50 ml with Water and transfer 2 ml of the solution to each of two test-tubes.

To one tube add $2N$ NaOH (1 ml), allow to stand exactly 45 minutes and add $2N$ H_2SO_4 (1 ml). To the second tube add Water (2 ml).

Add hydroxylamine hydrochloride solution (6 ml) (Note 2) to each tube and allow to stand for 40 minutes. Add ferric reagent (2 ml) (Note 3) to each tube and allow the colour to develop (20 minutes). Measure the extinction of the solution in the second tube in a 1 cm cell at 490 nm using the solution in the first tube as blank. Calculate the concentration w/w of phenethicillin by reference to a calibration curve prepared with phenethicillin potassium (A.S.). Determine the weight per ml of the elixir and convert the per cent w/w to terms of w/v. The specification requires 95.0–120.0 per cent of the stated amount as determined by the method above.

Stability Test. Reserve a portion of the prepared elixir, store at $15° \pm 1°$ for 7 days and repeat the determination. The result should not be less than 90 per cent of that found for the freshly prepared elixir.

Note 1. 3 ml contains the equivalent of 75 mg of phenethicillin if the preparation is stated to contain 125 mg in 5 ml. For stronger solutions the quantity should be correspondingly reduced.

Note 2. The reagent is freshly prepared by mixing equal volumes of hydroxylamine hydrochloride solution (35.76 per cent in Water) and a solution containing sodium hydroxide (17.3 g) and sodium acetate (3.18 g) in 100 ml. The mixture is adjusted to pH 7.0 by addition of the alkaline solution or hydrochloric acid, as appropriate, and diluted with 3 volumes of ethanol (90 per cent).

Note 3. Ferric ammonium sulphate (30 g) dissolved in a mixture of Water (70 ml) and sulphuric acid (9.3 ml) and diluted to 100 ml with Water.

Cognate Determinations

Ampicillin Mixture
Cloxacillin Elixir
Phenoxymethylpenicillin Elixir. The method uses absorption of iodine as a measure of phenoxymethylpenicillin.
Propicillin Elixir

Bronchial Spasm Relaxant. *Determination of Ephedrine Hydrochloride, Caffeine, Sodium Salicylate and Sodium Iodide*

The composition of the elixir is stated to be as follows:

Ephedrine Hydrochloride	22.89 mg
Caffeine and Sodium Salicylate	178.35 mg
Sodium Iodide	57 mg
Tincture Belladonna	0.52 ml
Vehicle (red) to	5 ml

The elixir is not viscous so that a suitable vehicle could be made from diluted glycerol with a red dye added to simulate the colour of the preparation. Quality control would include determination of weight per ml, a check on colour from batch to batch, clarity and identity tests for active ingredients as well as quantitative determinations. The latter are based mainly upon separation procedures already described.

Ephedrine Hydrochloride. Pipette the sample (5 ml) (Note 1) into a 50 ml separator and add Water (5 ml) and dilute hydrochloric acid (0.5 ml). Extract the caffeine and salicylic acid with chloroform (4 × 30 ml), washing each extract with Water (10 ml). Reject the chloroform (Note 2). Add ammonia solution to the combined aqueous liquids and extract with ether (4 × 30 ml) (Note 3). Bulk the ether extracts and wash with saturated sodium chloride solution (2 × 5 ml), washing each extract with ether (20 ml). Reject the aqueous extracts and add the ether washing to the bulk. Extract the ether with $0.1N$ H_2SO_4 (20, 20, 15 ml), bulk the acid extracts, warm gently to remove ether, cool and dilute to 100 ml with $0.1N$ H_2SO_4. Record the absorption curve over the region 220–300 nm in 4 cm cells using $0.1N$ H_2SO_4 as blank. On the same chart record a standard curve for 0.023 per cent ephedrine hydrochloride in $0.1N$ H_2SO_4. Calculate the concentration of ephedrine hydrochloride using a base-line technique if necessary (Note 4).

Caffeine and Sodium Salicylate. Pipette 5 ml accurately (see above) into a separator containing Water (10 ml) and 20 per cent NaOH (Bench Reagent, 2 ml). Extract the caffeine and ephedrine with chloroform (30, 30, 20, 10 ml), washing each chloroform extract with the same volume (5 ml) of Water in a second separator. Reserve the aqueous liquors. Bulk the chloroform extracts and dilute to 100 ml with chloroform. Mix well and dilute 5 ml to 250 ml with chloroform. Mix and record the absorption curve over the range 250–320 nm. Calculate the percentage of *caffeine* using data obtained from pure caffeine dried at 80° for 2 hours.

Acidify the reserved aqueous liquors with dilute HCl and extract the salicylic acid with ether (40, 30, 30, 20 ml). Wash each ether extract with the same 10 ml of Water and bulk the extracts. Extract the salicylic acid with $0.1N$ NaOH (30, 20, 10 ml) and dilute the extracts to 100 ml with $0.1N$ NaOH. Dilute 5 ml of the solution to 100 ml with $0.1N$ NaOH and record the adsorption curve.

Calculate the percentage of sodium salicylate using data obtained from pure salicylic acid (about 0.003 per cent in $0.1N$ NaOH) as a convenient standard.

Sodium Iodide. Dilute the sample (5 ml) with Water (100 ml) and add acetate buffer (pH 5.0, 75 ml). Titrate the solution with $0.02N$ $AgNO_3$ using a glass electrode as reference and a clean silver electrode as indicator electrode. Record the titration curve (Note 5), determine the end-point in the normal manner (Part 2) and calculate the percentage of sodium iodide.

Note 1. The sample is sufficiently mobile for a pipette to be used.

Note 2. It is possible to extract the salicylic acid from the chloroform solution with alkali and use the aqueous solution for the determination of sodium salicylate. The chloroform still contains caffeine.

Note 3. It is a useful precaution when spectrophotometric end-methods of analysis are used, to wash organic solvents with acid and alkali before use.

Note 4. The absorption curves serve to check the purity of the final extract and enable small corrections to be carried out. It must be emphasised that a standard solution should be examined under the same conditions on the same chart paper for precise and accurate results to be obtained.

Note 5. It is worthwhile to continue the titration beyond the first obvious end-point (iodide) until a second less obvious one is covered. The difference between the first and second end-points corresponds to the chloride present. Calculate the percentage of ephedrine hydrochloride based upon this titration (about 5 ml). Although this is a simple convenient check for the analyst in the quality control laboratory of the actual manufacturer it is not so for a public analyst. The latter would not know if the ingredients in the vehicle contained chloride as impurity and could not therefore, make a suitable allowance in the rather small titration. The presence of iodide in this context is less likely and the result, correspondingly, more reliable.

A glass electrode is convenient in this titration although a mercury/mercurous sulphate half-cell with saturated potassium sulphate can also be used.

Solutions

Cough Mixture. *Determination of Ephedrine Hydrochloride, Codeine Phosphate and Promethazine Hydrochloride*

The mixture is made to the following formula:

Ephedrine Hydrochloride	0.144 g
Codeine Phosphate	0.180 g
Promethazine Hydrochloride	0.072 g
Liquid Glucose	70 g
Burnt Sugar	q.s.
Water to	100 ml

The principles underlying the determination of ephedrine and promethazine are described under Capsules (Cough) (p. 314). Codeine can be determined by spectrophotometry and, in so doing, an indication of possible oxidation of the promethazine in the mixture is obtainable. In the event that the complete absorption curve indicates the presence of promethazine sulphoxide a modified method is possible to allow for this impurity. The method is based on the fact that promethazine hydrochloride is soluble in chloroform whereas the sulphoxide hydrochloride and codeine phosphate are not.

Method. Prepare an acid extract of the mixture (25 ml) as described on page 315 but omitting the digestion procedure. Determine the ephedrine and promethazine on aliquots of the solution.

Codeine. (i) Dilute the acid extract (10 ml) to 100 ml with *0.1N* H_2SO_4 and record the absorption curve over the region 260–400 nm in a 4 cm cell. If the curve appears as in Fig. 34a the codeine (and promethazine) are determined by making use of the calculation for a two-component mixture (Note 1) in which one component interferes with the absorption of the other (Note 2). If, however, the trace appears as in Fig. 34b or an indication of irrelevant absorption at about 330 nm is observed, the presence of promethazine sulphoxide must be suspected and a simple correction procedure is not possible for the three-component mixture. Proceed as in (ii).

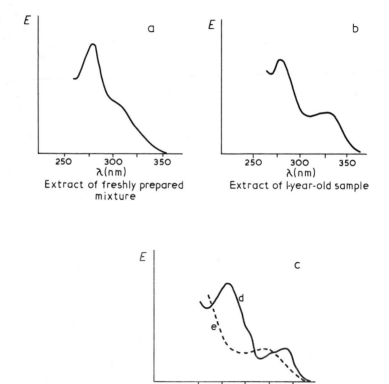

Extract of freshly prepared mixture

Extract of 1-year-old sample

d – codeine + promethazine sulphoxide
e – promethazine hydrochloride extract

Fig. 34. Effect of promethazine sulphoxide on absorption curves of extracts of cough mixture

(ii) Weigh accurately about 2.4 g of sample and transfer to a separator with the aid of Water (6 ml in all). Add dilute hydrochloric acid (1 drop) and extract the promethazine hydrochloride with chloroform (30, 20, 20, 20 ml) washing each extract with the same 5 ml of Water. Transfer each extract to a 100 ml volumetric flask, add ammonia solution (2 drops) (Note 3) and dilute to 100 ml with chloroform. Shake well and add anhydrous sodium sulphate (2 g). Shake to give a clear solution (Note 4).

Add the acid washing to the extracted sample and make alkaline with ammonia. Extract with chloroform as described above. Record the absorption curve in 4 cm cells using chloroform as blank over the region 270–400 nm. Calculate the concentration of codeine after allowing for the contribution of the promethazine sulphoxide at the codeine absorption maximum at about 285 nm (Note 5). Determine the weight per ml of the mixture and calculate the concentration in terms of w/v.

Note 1. Standard solutions of codeine phosphate and promethazine hydrochloride should be examined at the same time as the sample on the same chart.

Note 2. Strictly, it is a three-component mixture but ephedrine does not absorb radiation at 285 nm.

Note 3. To convert salt (λ_{max} 303 nm) to base (λ_{max} 310 nm).

Note 4. This solution may be examined for promethazine and will give a result free from the effect of sulphoxide. However, if commercial preparations are examined flavouring material may contribute to the absorption and an analyst must therefore consider this aspect in assessing his results.

Note 5. The absorption curve of promethazine sulphoxide is required to calculate the correction at 285 nm based upon an extinction at 340–345 nm. The absorption curve for this purpose is readily obtained by oxidising about 3–4 mg of promethazine hydrochloride in glacial acetic acid (0.5 ml) and water (2 ml) with hydrogen peroxide (100 vol., 0.15 ml). Warm on a water-bath (3 min) and remove any unchanged promethazine with chloroform. Basify the aqueous liquid and extract the sulphoxide with chloroform for examination. A typical ratio E_{285}/E_{345} is 1.57 and a trace for the solutions obtained from an old sample of mixture is shown in Fig. 34c.

Cognate Determination

Determine the active ingredients in the following preparation:

Pseudoephedrine Hydrochloride	30 mg
Triprolidine Hydrochloride	1.25 mg
Codeine Phosphate	10 mg
Colour	q.s.
Syrup base to	5 ml

Hint. Triprolidine contains a pyridine nucleus, a characteristic feature of which is that the ultraviolet absorption of an acid solution is greater than that of an alkaline solution at a particular wavelength.

Mouthwash. *Determination of chloroxylenol in a preparation stated to contain 1.02 per cent w/v*

The preparation appears similar, except in content of chloroxylenol, to that of Chloroxylenol Solution B.P.C. Gas-liquid chromatography is the method of choice using the conditions described under Lozenges (page 326) but with chlorocresol as internal standard.

Method. Dilute the sample (5 ml) with methanolic chlorocresol solution (1 per cent, 5 ml) and sufficient methanol to produce 25 ml of dilution. Dilute the sample (5 ml) to

25 ml with methanol. Inject 1 μl volume into the gas-chromatography apparatus and calculate the percentage of chloroxylenol by the method of peak height ratios. Use a standard solution of chloroxylenol 1.0 per cent (5 ml) diluted with chlorocresol solution (1 per cent, 5 ml) and methanol to 25 ml.

Adopt the standard method of 3 injections of each solution and make sure that no interference from constituents of the preparation occurs at the position of the internal standard peak.

Suspensions

Chloramphenicol Palmitate Mixture. *Determination of the percentage of Chloramphenicol and the content of Polymorph A*

Chloramphenicol Palmitate is known to exist in at least two principal forms. When administered as an oral suspension, as in this mixture, one of the forms referred to as Polymorph A in the British Pharmaceutical Codex is virtually inactive. The two forms can be distinguished by their infrared absorption spectra and a limit on the amount of Polymorph A is imposed by means of this technique. The quality control therefore consists of a visual check for uniformity of the suspension, an assay for total chloramphenicol, a limit test and weight per ml.

Polymorph A. Mix the suspension (about 20 ml) with Water (20 ml) and centrifuge for 15 minutes (ideally at about 18 000 rpm). Remove the supernatant liquid, which contains flavouring and preservative agents, and re-suspend the residue in Water by mixing thoroughly with a small quantity first and diluting with more Water. Centrifuge, and repeat the washing and centrifuging twice more. Suspend the solid in Water (10 ml), filter by suction and dry the residue first by pressing between filter paper and then in an oven at 70° for $\frac{1}{2}$ hour.

Prepare a liquid paraffin mull (Part 2) of the dry residue so as to obtain an absorption band equivalent to about 20–30 per cent transmittance at about 12.3 μm. To achieve this intensity of absorption a concentrated mull is required.

Prepare in a similar manner the absorption curves for standards of Chloramphenicol Palmitate (A.S.) containing 20 per cent and 10 per cent of chloramphenicol palmitate (Polymorph A (A.S.)). The curve obtained for the 20 per cent mixture is used to determine the exact wavelengths of minimum and maximum absorptions at about 11.3, 12.65 and 11.65, 11.86 μm respectively. Using a base line technique (Part 2) obtain the corrected extinctions at about 11.65 and 11.86 μm and calculate the ratio $E_{11.65}/E_{11.86}$ for sample and 10 per cent mixture. The extinction ratio for the sample should be less than that for the 10 per cent mixture.

Chloramphenicol. Mix the sample and weigh about 12 g accurately. Dilute to 1000 ml with Water and mix thoroughly to disperse the suspension. Allow to stand (10 min) and mix thoroughly once more. Immediately pipette 5 ml of the suspension into a 100 ml volumetric flask and dilute to the mark with ethanol (95 per cent). Measure the extinction in a 1 cm cell at about 271 nm (maximum) and calculate the percentage w/w of chloramphenicol palmitate in the mixture. Convert to terms of per cent w/v by means of the weight per ml of the preparation and to chloramphenicol with the factor 0.575.

Primidone Mixture. *Determination of the percentage of* $C_{12}H_{14}N_2O_2$

The determination is made difficult by normal extraction methods because of the poor solubility of primidone in water and organic solvents. It is, however, sufficiently soluble in methanol to allow simple dilution of the preparation and examination by gas-liquid chromatography. The method described in the British Pharmaceutical Codex does not involve the use of an internal standard. A number of injections of sample and standard must, therefore, be made in order to assess the precision of the injection technique.

Method. Shake the mixture thoroughly and to an accurately weighed quantity equivalent to about 0.4 g of primidone add methanol (75 ml) and mix whilst warming on a water-bath (5 min). Cool and dilute to 100 ml with methanol, shake well and allow precipitated solids to settle out.

Inject 2 μl onto a GLC column (Note) at 270° and repeat the injection after elution of the primidone. Make a further 6 injections in the same way and assess the precision of the results. Continue the assay by injecting solutions of primidone (0.375, 0.400 and 0.425 per cent in methanol) in duplicate, each duplicate standard being followed by an injection of sample. Calculate the concentration (w/w) of primidone by reference to a calibration curve obtained with the standards. Determine the weight per ml of the sample and calculate the concentration w/v.

Note. A 1.5 m column of XE60 silicone gum rubber 2 per cent on Gas Chrom Q (60–80 mesh) is suitable.

OINTMENTS

Ointments are generally semi-solid greasy preparations and distinct from creams in which water forms a significant proportion of the whole. The necessarily chemical inertness of an ointment basis and its solubility in organic solvents rarely leads to complications in the determination of active ingredients. Greenwood and Guppy (1974) have examined a number of preparations of benzoic acid and of salicylic acid by direct titration with $0.1N$ NaOH using phenolphthalein as indicator. The results were well within the range normally allowed for quality control. It is usually the ingredients themselves, when present in admixture, that give rise to problems e.g. the determination of ammoniated mercury in Ammoniated Mercury and Coal Tar Ointment B.P.C. is similar to that for the substance except that a mixture of organic solvent is added to dissolve the paraffin basis. A different method, however, is prescribed when salicyclic acid is also present in the ointment.

Ammoniated Mercury, Coal Tar, and Salicylic Acid Ointment. *Determination of* NH_2HgCl *and* $C_7H_6O_3$

The determination of the compounds is based upon selective extraction of each, Ammoniated Mercury with acetic acid from a solvent ether/light petroleum suspension, and Salicylic Acid with sodium carbonate solution from a mixture of the ointment and light petroleum. Each substance is determined by characteristic reactions.

Ammoniated Mercury

$$NH_2HgCl \xrightarrow[\text{Na}_2\text{S}_2\text{O}_3]{\text{NaOH}} NH_3$$
$$NH_2HgCl \equiv NH_3 \equiv H$$
$$0.02521 \text{ g} \equiv 1 \text{ ml } 0.1N$$

This reaction is used in the determination of N in organic compounds by the Kjeldahl method (HgO catalyst) to ensure complete recovery of NH_3 by decomposing organo-mercury compounds (Chapter 5).

Salicylic Acid

Conversion to tribromophenol (Chapter 8).

Ammoniated Mercury. Critically examine the ointment and, if satisfactorily uniform in appearance and frèe from aggregates, weigh accurately about 10 g of ointment into a 100 ml beaker. Add a mixture of solvent ether and light petroleum b.p. 60–80° (1:1, 40 ml) and mix thoroughly whilst warming to disperse the ointment. Transfer to a separator (Note 1) and complete the quantitative transfer by rinsing the beaker with more of the solvent mixture (20, 20 ml). Extract the ammoniated mercury with warm acetic acid (4 × 25 ml) (Note 2) and transfer the extracts to the flask of an ammonia distillation apparatus. Cool, add carefully to form a layer below the acid extract sodium hydroxide solution (50 per cent, 50 ml) and sodium thiosulphate (2 g). Connect to the ammonia distillation, mix (Note 3) and distil the liberated ammonia into $0.1N$ H_2SO_4 (20 ml). Titrate the excess acid with $0.1N$ NaOH using methyl red as indicator.

Salicylic Acid. Disperse the ointment (about 2 g accurately weighed) in light petroleum b.p. 40–60° as described for ammoniated mercury using about 50–60 ml of solvent in all. Extract the salicylic acid with sodium carbonate solution (0.5 per cent, 4 × 10 ml) and pass each extract through the same previously moistened small filter paper into a 500 ml glass stoppered flask (iodine flask). Add phenolphthalein indicator and then hydrochloric acid until effervescence ceases (Note 4) and swirl the contents to remove as much CO_2 as possible. Add $0.5N$ NaOH until the solution is just alkaline to phenolphthalein followed by $0.1N$ Bromine (50 ml), Water (50 ml) and hydrochloric acid (5 ml). Complete the assay as described for phenol (p. 202).

Note 1. Use a small glass funnel and glass rod to assist in this transfer as ethereal solutions tend to trickle down the outside of the beaker in such transfers.

Note 2. Ether boils at 35° and the utmost care must be taken at this stage to avoid loss of contents due to excessive pressure. This must be carefully and constantly released.

Note 3. The contents *must* be mixed before heat is applied otherwise disastrous results will ensue.

Note 4. Care should be exercised at this point as the addition of too much acid too quickly may cause loss of solution in the vigorous effervesence.

Cognate Determinations

Salicylic Acid and Sulphur Ointment. The sulphur is determined by conversion to thiosulphate by refluxing with sodium sulphite solution and titration with $0.1N$ iodine. Interference by excess sulphite is removed by addition of formaldehyde, and the oily basis by filtration of the reaction mixture.

Haemorrhoid Ointment. *Determination of Ephedrine Hydrochloride, Lignocaine Hydrochloride and Allantoin*

This example is chosen to illustrate how the solubility properties of the ingredients of an ointment can be used to full advantage to determine each, free from interference by the other. The preparation can be made to the following formulation:

Ephedrine Hydrochloride		0.25 per cent
Lignocaine Hydrochloride		0.5 per cent
Allantoin		0.5 per cent
Ointment Basis (Paraffin-type)	to	100 per cent

The determination of ephedrine by normal procedures for synthetic organic bases (Chapter 13) is complicated by marked losses due to water-solubility of the base during the washing of the bulked ether extracts. This property, however, can be turned to advantage in separating ephedrine from lignocaine.

Allantoin is soluble in acid, neutral and alkaline aqueous media and, although showing end-absorption only in acid, peak absorption at about 225 nm appears in alkaline solution.

Isolation of ingredients. Weigh the ointment (about 2 g accurately weighed) into a small beaker and disperse thoroughly in ether (25 ml). Transfer to a separator (Note 1) (100 ml) exercising the normal precautions (page 336) and complete the transfer with ether (20, 20 ml) and *0.1N* H_2SO_4 (10 ml). Shake thoroughly, run off the acid layer into a separator (50 ml) containing ether (10 ml). Shake, allow to separate, and run off the acid layer into a third separator (50 ml). Complete the extraction of allantoin and salts in the same manner with Water (10, 10 ml).

Isolation of bases. Make the bulked aqueous extracts alkaline with sodium hydroxide solution (20 per cent, 2 ml) and extract the bases with ether (5×20 ml). Bulk the ether extracts and wash once with saturated sodium chloride solution (3 ml). Add the washing to the reserved alkaline aqueous liquid for the determination of allantoin.

Ephedrine. Extract the ether layer with sodium bicarbonate solution (1 per cent, 8×10 ml) washing each extract with ether (10 ml). Make up to 100 ml with more bicarbonate solution, mix well and record the absorption curve of the solution using 1 per cent sodium bicarbonate as blank in 4 cm cells (Note 2). On the same chart record the absorption curve of a 0.01 per cent solution of ephedrine hydrochloride in 1 per cent sodium bicarbonate.

Calculate the percentage of ephedrine hydrochloride in the ointment.

Lignocaine. Extract the ethereal layer with *0.1N* H_2SO_4 (15, 10, 10 ml), washing each extract with the same ether (10 ml) as was used in the determination of ephedrine. Transfer to a 100 ml volumetric flask and make up to volume with more *0.1N* H_2SO_4. Record the absorption curve in 4 cm cells using a standard lignocaine hydrochloric solution (0.01 per cent in *0.1N* H_2SO_4) for comparison with *0.1N* H_2SO_4 as blank.

Calculate the percentage of lignocaine hydrochloride in the ointment.

Allantoin. Dilute the reserved alkaline aqueous solution to 100 ml in a volumetric flask with Water. Dilute the solution (5 ml) to 50 ml with *0.1N* NaOH and also 5 ml to 50 ml with *0.1N* H_2SO_4. Record the absorption curve for each solution using respective blanks of *0.1N* NaOH and *0.1N* H_2SO_4 in cells over the region 220–300 nm.

Calculate the percentage of allantoin by means of the ΔE value at 225 nm and the ΔE value of a standard allantoin solution (0.001 per cent) in *0.1N* NaOH and *0.1N* H_2SO_4 respectively.

Note 1. In all experiments involving precision ultraviolet absorption measurements it is advisable to use as little grease as possible on separator taps. Ideally, water alone should be used as there is only slight sticking of taps at worst.

Note 2. Ephedrine has zero absorbance at 285 nm and the pen should be set at zero at this wavelength to take account of any slight haziness that might occur in the solution over a 4 cm path length. The trace itself should show no evidence of irrelevant absorption except at about 230 nm and below. The advantage of recording the absorption of the pure sample on the same chart is the confirmation obtainable that this is so. A scale-expansion accessory is invaluable in this determination.

SUPPOSITORIES

The bases for suppositories may be either fatty, e.g. theobroma oil and hydrogenated vegetable oil, or water-soluble, e.g. macrogols. Standards imposed on suppositories include uniformity of both internal and external *Appearance* as a control on the uniform distribution of medicament; *Uniformity of Weight* in which twenty suppositories are weighed singly and no suppository should deviate by more than 5 per cent of the average weight except that two may deviate by not more than 7.5 per cent; and a *Disintegration Test* under carefully controlled conditions whereby, unless otherwise specified, disintegration should be complete within thirty minutes.

Except for Glycerol Suppositories which are dissolved directly in water, representative samples of suppositories for assay are obtained either by cutting into small pieces and weighing the required quantity of these, or by melting together, allowing to cool, with stirring until set and weighing the required quantity of the mass. The latter method is adopted in the British Pharmaceutical Codex where fatty bases are used in the preparations.

In spite of the variation in bases they rarely complicate the determination of active ingredients. Thus, non-aqueous titration is adopted for Bisacodyl Suppositories (perchloric acid) and for Phenylbutazone Suppositories (tetrabutylammonium hydroxide), whilst Indomethecin Suppositories are examined by ultraviolet absorption.

Glycerol Suppositories. *Determination of the percentage of* $C_3H_8O_3$

The determination depends upon the reaction expressed by the following equation:

$$\begin{array}{l} CH_2OH \\ | \\ CHOH \\ | \\ CH_2OH \end{array} + 2NaIO_4 \longrightarrow 2H.CHO + H.COOH + 2NaIO_3 + H_2O$$

$$H.COOH + NaOH \rightarrow H.COONa + H_2O$$
$$\therefore \quad C_3H_8O_3 \equiv H.COOH \equiv H \equiv 1000 \text{ ml } N$$
$$\therefore \quad 92.10 \text{ g } C_3H_8O_3 \equiv 1000 \text{ ml } N$$
$$\therefore \quad 0.00921 \text{ g } C_3H_8O_3 \equiv 1 \text{ ml } 0.1N \text{ NaOH}$$

The excess sodium metaperiodate is removed by the addition of propylene glycol when the following reaction occurs:

$$\begin{array}{l} CH_2OH \\ | \\ CHOH \\ | \\ CH_3 \end{array} + NaIO_4 \longrightarrow H.CHO + CH_3CHO + NaIO_3 + H_2O$$

It is essential to convert the periodate to iodate because it consumes sodium hydroxide with concomitant fading of the indicator and gradual appearance of a purple colour. No such effect is observed with iodate and a sharp colour change occurs at the end-point at about pH 7.

Method. Dissolve a sufficient number of suppositories to give about 8 g of glycerol and dissolve in Water. Dilute the solution to 250 ml in a volumetric flask. Dilute the solution (5 ml) with Water (about 150 ml), add 5 drops of bromocresol purple indicator solution and neutralise any acid by addition of 0.1N NaOH to the blue colour of the indicator. Add sodium metaperiodate (1.6 g) and allow to stand for 15 minutes. Add propylene glycol (3 ml), allow to stand for 5 minutes and titrate the formic acid with 0.1N NaOH to the same blue colour.

Suppositories incorporating a number of ingredients offer good opportunities of applying separation techniques and illustrating difficulties introduced by the formulation.

Locan Suppositories. *Determination of Bismuth Subnitrate, Zinc Oxide, Amethocaine Hydrochloride and Amylocaine Hydrochloride*

The suppositories are stated to be of the following composition:

Bismuth Subnitrate	60 mg
Zinc Oxide	120 mg
Amethocaine Hydrochloride	15 mg
Amylocaine Hydrochloride	5 mg

The nature of the base is not stated but from its property of dissolving in water and from its infrared absorption curve it appears to be of the macrogol type.

Properties used in the determination. The suppositories are completely soluble in a mixture of hydrochloric acid and water but the end-point obtained for bismuth by edetate titration under normal conditions (page 240) is poor, compare Bismuth Subgallate Suppositories, Compound B.P.C. Bismuth salts in acid solution give a yellow colour with thiourea and this reaction is used in the AOAC method for bismuth in drugs. It has been criticised by Plank (1972) who recommends a polarographic method as being more accurate and precise. This would appear to be the ideal method as zinc ions are also readily determined by polarography. However, macrogols interfere with the reduction of zinc ions and a reduction step for zinc at about -1.1 V does not appear. That for bismuth at -0.1 V remains unaffected and offers an alternative means of determining the ion.

Zinc salts are readily separated from bismuth salts as the latter form insoluble oxysalts in dilute acid solutions.

The synthetic organic bases are readily separated from inorganic salts in the conventional manner (Chapter 13) and show dissimilar ultraviolet absorption spectra suitable for assay purposes.

Zinc Oxide. Weigh 5 suppositories and cut into small pieces. To a quantity of the mass, accurately weighed, equivalent to about one suppository, add Water (20 ml), warm gently to dissolve the base and disperse insoluble ingredients. Add 0.1N HCl (40 ml) and continue warming for one minute. Cool and filter off the undissolved bismuth salt through a porosity 3 sintered glass crucible using warm Water to wash the residue. Cool, add ammonia buffer solution (page 244) and titrate the zinc solution with 0.05N disodium edetate using solochrome black as indicator.

$$0.004068 \text{ g ZnO} \equiv 1 \text{ ml } 0.05N \text{ disodium edetate}$$

Bismuth Subnitrate. Dissolve a quantity of the suppository mass accurately weighed, equivalent to about one suppository in Water (20 ml) and hydrochloric acid (20 ml) with gentle warming. Cool, dilute to 500 ml with Water and mix well (Note 1).

Pipette the solution (10 ml) into a 50 ml volumetric flask, add dilute hydrochloric acid (5 ml) and thiourea solution (10 per cent, 5 ml). Mix and dilute to 50 ml with Water. Measure the extinction at 395 nm using a blank of reagents. Calculate the amount of bismuth subnitrate by reference to a calibration prepared from aliquots of a solution of bismuth subnitrate (70 mg) dissolved in hydrochloric acid (40 ml) and diluted to 500 ml with Water.

Amethocaine and Amylocaine Hydrochlorides. Prepare a solution of the weighed suppository mass as directed under Zinc Oxide above. Add sodium chloride to give a strong solution and ammonia to make alkaline. Extract the organic bases with ether (60, 50, 50) (Note 2) and bulk the ether extracts. Wash the ether extracts with saturated sodium chloride solution (2 × 5 ml) and extract the washed ether with *0.1N* HCl (10.0 ml) followed by Water (2 × 25 ml). Transfer the acid and aqueous washings to a 100 ml volumetric flask and add *0.1N* NaOH (11.0 ml) (Note 3). Dilute to 100 ml with Water, mix well and dilute 10 ml of the solution to 100 ml with Water.

Record the absorption curve of the solution over the region 360–210 nm and use the method for two components with dissimilar spectra, (Part 2) or use the following data remembering that, strictly, constants should be obtained for the particular instrument available.

$$E_{1cm}^{1\%} \text{ Amylocaine Hydrochloride at 308 nm} = 0$$
$$E_{1cm}^{1\%} \text{ Amethocaine Hydrochloride at 308 nm} = 674$$
$$E_{1cm}^{1\%} \text{ Amylocaine Hydrochloride at 235 nm} = 472$$
$$E_{1cm}^{1\%} \text{ Amethocaine Hydrochloride at 235 nm} = 190$$

Identification. Prepare an ether extract of a suppository as described for *Amethocaine* and *Amylocaine Hydrochloride.* Evaporate the ether to small bulk (1–2 ml) and examine along with standards by thin-layer chromatography on Silica gel G plates using methanol: strong ammonia solution (100 : 1.5) as solvent. Iodine vapour readily detects the bases.

Note 1. If insufficient hydrochloric acid is used at this stage bismuth oxychloride gradually precipitates after dilution with water. This is particularly evident with the standard as the suppository base probably helps to maintain a clear solution.

Note 2. The sodium chloride assists separation into two phases as do the large volumes of ether.

Note 3. Maximum ultraviolet absorption of amethocaine occurs in alkaline solution and the quantities of acid and alkali are accurately measured for this purpose. The final solution must not be too alkaline as hydrolysis of amethocaine takes place readily.

TABLETS

Weight

Variation in the weight of individual tablets within a given batch is such that wider limits of tolerance are prescribed than is usual for the pure active ingredient from which the tablets are prepared. Since the individual tablets may vary considerably in weight (page 15) the assay is so adjusted that a reasonably average sample is taken for the determination. This is done by weighing a batch of at least twenty tablets, and calculating the *average weight* for each tablet. The final calculation of the content of active ingredient per tablet is based upon this average figure and therefore is itself only an average weight per tablet. In spite of the variation amongst individual tablets the tablet granules from which the tablets are prepared should be of reasonably constant composition (but see below), so that determination of the active ingredient content of the powdered tablets is an accurate measure of the amount of active principle present. For this reason it is essential to express the result of any such determination in two ways:

(a) the percentage of active ingredient in the powdered tablet, and

(b) the average weight of active ingredient per tablet.

However, if the active ingredient is present in small amount, for example 0.05–0.25 mg in a tablet of weight 50–100 mg, the use of (a) and (b) for expressing results may not be sufficient for complete control of the tablets. It is necessary to determine the medicament in individual tablets to confirm that a reasonably uniform distribution has been achieved in manufacture. The limits are wider than those for large-dose tablets and the requirement for Digoxin Tablets (0.25 mg) is as follows:

'the content of each tablet is between 80 and 120 per cent of the average except that for one tablet the content may be between 75 and 125 per cent of the average.'

Disintegration

The function of the test is to ensure that tablets break up completely in water at 36–38° in not more than 15 minutes for most uncoated tablets and not more than one hour for sugar-coated and film-coated tablets.

Apparatus

The Pharmacopoeia specifies certain dimensions for a relatively simple system. It consists of a glass or plastic tube 80–100 mm long, with internal and external diameters of 28 and 30 to 31 mm respectively. Support for tablets is provided by a disc of rustproof wire gauze of size corresponding to No. 1.70 sieve fitted neatly to the tube to form a type of basket.

Also required are a glass cylinder of about 45 mm diameter and flat base, a thermostatically controlled water-bath and a system by which the basket can be raised and lowered in the glass cylinder thirty times per minute for a complete up and down movement.

Method. Place water in the cylinder to a depth of not less than 15 cm and raise the temperature to 36–38°. Check that when the basket moves up and down, the gauze just breaks the surface of the water at the highest position and that the rim of the basket remains just above the surface of the water at the lowest position.

Place five tablets in the basket and set the basket in motion. Note the time and proceed with the test until no particle, except fragments of coating, remains above the gauze which would not readily pass through it. Note the time at this point which, unless otherwise stated in the monograph for the tablet, should be not more than fifteen minutes in the case of uncoated tablets.

If the tablets fail the test it may be repeated using a guided disc in the basket above the tablets. The guided disc (Fig. 35) of weight 2.0 ± 0.1 g and carefully controlled dimensions fits neatly into the basket without too much play.

It is possible, because of the nature of the medicament or presence of excipients, that the tablets may aggregate and fail the test. In such an event the tablets may be tested individually as described above initially without, and if necessary, with the guided disc.

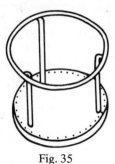

Fig. 35

The apparatus is also used in the Disintegration Test for Enteric-coated Tablets but the water is replaced with 225 to 275 ml of 0.06N hydrochloric acid to simulate the acid conditions of the stomach. After three hours (one hour for Erythromycin Tablets) during which time particles of tablet coating only pass through the gauze the tablets are washed rapidly with water and the acid solution is replaced with solution of pH 6.8. The test is continued for one hour by which time disintegration, as defined above, should be complete. The guided disc may be used if necessary in a repeat test with five tablets.

Dissolution

The problem that branded and non-branded products of the same chemical entity may not be clinically equivalent is well recognised. Dissolution tests are more valid indicators of differences between brands and of bioavailability

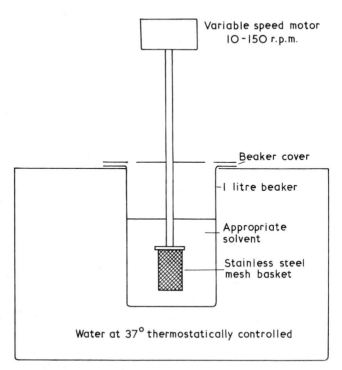

Fig. 36. Simple apparatus for tablet dissolution studies (not to scale)

than are disintegration tests. A convenient apparatus is shown diagrammatically in Fig. 36, but it must be emphasised that many modifications are possible. They are designed to take account of variables that make meaningful results difficult to obtain.

Excipients. Tablet excipients may have an adverse effect on the absorption of a drug as compared with that for a solution or dispersion of the drug.

Dissolution Medium. The medium is generally chosen to suit the appropriate conditions, for example 0.6 per cent hydrochloric acid for those drugs absorbed from the stomach and buffers of varying pH up to about pH 8 for those tablets designed to liberate the medicament over several hours or for enteric-coated tablets. Enzymes, e.g. pepsin for acid, or trypsin for alkaline media, are not normally added and the omission materially assists spectrophotometric end-methods of analysis.

The volume of dissolution medium is also important, perhaps more so for drugs of poor solubility in water or acid media. It is essential that sufficient medium is used such that the drug, even if completely dissolved, is present to no more than about 30–40 per cent of its saturation solubility. Usually 600 or 900 ml of medium are used.

Shape of Beaker. The results obtained with flat-bottomed beakers tend to differ from those obtaining with round-bottomed beakers. An explanation may lie in solvent flow past the tablet in the rotating basket as the following variables are also known to be significant in dissolution studies.

Stirring. The *position* of the basket in the beaker and the *rate of rotation* should be fixed to obtain comparable results on different occasions. Normally a speed of 120 r.p.m. is used for any one laboratory, therefore conditions must be carefully standardised even to the extent of deciding on whether or not the basket should be rotating as it is lowered into the dissolution medium. Clearly, solvent flow around the tablet in the basket is complex and modifications suggested to improve upon this position are continuous-flow dissolution apparatus as described by Tingstad *et al.* and a rotating filter-stationary basket system (Shah *et al.*). In filtration systems used for continuous-flow studies it is essential to avoid clogging of the filter even to the extent of providing a by-pass so that solution is passed through the filter and spectrophotometer cell at fixed intervals instead of continuously.

Temperature. To simulate body conditions a temperature of 37° is used.

Tetracycline Hydrochloride Tablets. *Comparison of dissolution behaviour of different brands of tablets*

Method. Place 0.06N HCl (500 ml) in a beaker (Fig. 36) and allow to stabilise at 37°. Place the tablet to be studied in the wire mesh basket, set the motor in motion at 25 r.p.m. and lower into the dissolution medium, noting the time ($t = 0$). Attach a simple filtration system to a 5 ml pipette (Note 1) and take samples at $t = 5, 10, 20, 30, 60$ and 90 minutes. Maintain the volume constant by adding 5 ml of 0.06N HCl after each removal of sample.

Determine the absorbance of each sample, making dilutions when necessary, over the region 250–400 nm (Note 2) and calculate the amount of tetracycline in solution from a previously constructed calibration curve. Plot the amount of drug released as a percentage of stated tablet content. Carry out 3 determinations and repeat for different brands (Note 3). Construct a Table of the results in the following form

Brand name	Manufacturer	Lot No.	Per cent drug released at (min) 10 30 60			Time for 50 per cent release
			(i) (ii) (iii)			(i) (ii) (iii)

Note 1. A small tube containing glass wool fitted to the end of the pipette should remove insoluble tablet basis or coating.

Note 2. The general shape of the absorption curves should be the same as that of a standard curve to show that tablet excipients are not interfering in any way with the tetracycline.

Note 3. As a comparison of different brands is being undertaken it is essential to obtain some idea of the reproducibility of the method and apparatus. The significance of between-brands variation could not otherwise be assessed.

Slow Lithium Carbonate Tablets. *Determination of Solution Rate*

In the treatment of some mental diseases it is essential that careful control of plasma concentrations of lithium in the range 0.6–1.6 millimoles per litre be maintained. Slow lithium tablets are designed to release lithium ions over a period of several hours thus avoiding high concentrations which put the patient at risk. Even so, monitoring of lithium levels in plasma is carried out. The apparatus for solution rate is that described for the Disintegration Test for Tablets except that an extinction thimble is used to hold the tablets.

Method. Place 3 tablets in the thimble and, using a suitable known volume of 0.6 per cent v/v of hydrochloric acid instead of Water, operate the apparatus for 2 hours at 37°. Reserve the acid extract (A).

Replace the acid with a buffer pH 6.0 and continue to extract for 1 hour. Reserve the extract (B).

Replace the buffer pH 6.0 with one of pH 6.8 and continue to extract for a further 5 hours. Reserve the extract (C).

Examine each extract by flame photometry (Part 2) for the lithium content.

The tablets comply if the following results are obtained

Extract A	30 per cent		
Extract A + B	30 per cent	and	50 per cent
Extract A + B + C	70 per cent	and	95 per cent of Li_2CO_3 available.

Cognate Determinations

Slow Potassium Chloride Tablets. The object of this preparation is to avoid high local concentrations of potassium chloride in contact with the lining of the intestine and so prevent irritation.

Slow Fe Tablets. Use the apparatus described under Tetracycline Hydrochloride Tablets and measure 1 ml samples for development of a colour with

o-phenanthroline. As the experiment proceeds the concentration of ferrous ions increases so much that dilution of samples becomes necessary. This is one reason why o-phenanthroline is not added to the dissolution medium; the other reason is that with tablets containing the equivalent of 50 mg of Fe, more than 0.5 g of o-phenanthroline would be required for each determination which becomes quite expensive.

Determination of Medicaments by Solvent Extraction

The presence of lubricants and substances to aid disintegration of compressed tablets may necessitate a preliminary extraction and separation but this is not always so and many examples of direct determination of the active ingredient by the method of the parent drug have already been given. Strong adsorption of active materials by constituents of formulations may sometimes occur, for example Gupta and Euler, obtained recoveries of less than 12 per cent of L-hyoscyamine from an antacid preparation containing magnesium trisilicate. Extraction procedures under these conditions, and those where mixtures of compounds of similar physical and chemical characteristics occur, become complicated, and each product requires individual treatment.

Extraction with Chloroform

Codeine Phosphates. *Determination of the weight of* $C_{18}H_{21}O_3N,H_3PO_4,1\frac{1}{2}H_2O$ *per tablet of average weight*

Method. Accurately weigh 20 tablets, and calculate the average weight per tablet. Handle the tablets carefully and as little as possible at this stage to minimise losses by mechanical damage. Powder the tablets and weigh accurately a quantity of powder equivalent to approximately 0.3 g of codeine phosphate. Dissolve as completely as possible in 20 ml of $0.5N$ H_2SO_4 and filter directly into a separator. This removes starch and other acid-insoluble matter such as stearic acid. Wash the residue with $0.5N$ H_2SO_4 until complete extraction of the alkaloid is effected. Make alkaline with ammonia to liberate codeine base and extract with successive portions of 30, 30, 10 and 10 ml of chloroform. Wash each successive chloroform extract with the same 10 ml of Water. Evaporate the chloroform. Add 5 ml of 95 per cent ethanol (neutralised to methyl red) and evaporate to dryness. Dissolve the residue in 1 ml of neutralised ethanol, add excess $0.5N$ HCl, and back-titrate the excess with $0.1N$ NaOH to methyl red.

Cognate Determinations

Diethylcarbamazine Tablets. The combined chloroform extracts are extracted with standard sulphuric acid solution and Water, and the excess acid back-titrated with standard sodium hydroxide solution.

Hyoscine Tablets

Methylprylone Tablets. Methylprylone is isolated by continuous extraction with chloroform, the chloroform evaporated and the residue titrated potentiometrically with $0.1M$ alkaline potassium ferricyanide.

The following tablets are extracted with chloroform and determined by non-aqueous titration:

Chlorpropamide Tablets
Dextromethorphan Tablets
Phendindamine Tablets
Pethidine Tablets

The following tablets are extracted with chloroform and determined gravimetrically:

Barbitone Sodium Tablets. Extraction with chloroform-ethanol-ether.

Ethisterone Tablets. The powdered tablets are extracted with light petroleum in a chromatography tube to remove non-steroid fatty material. The light petroleum effluent is rejected and the material extracted with chloroform and weighed.

The following tablets are extracted with chloroform and determined spectrophotometrically:

Betamethasone Tablets
Bisacodyl Tablets
Cortisone Tablets
Fludrocortisone Tablets
Methylprednisolone Tablets
Methyltestosterone Tablets

The following tablets are extracted with chloroform and determined colorimetrically:

Betamethasone Sodium Phosphate Tablets
Dexamethasone Tablets

Aspirin Phenacetin and Codeine Tablets. *Aspirin, Phenacetin and Codeine Phosphate are all determined*

Method (a) Acetylsalicylic Acid. Weigh and powder 20 tablets and transfer an accurately weighed sample (1.2 g) to a separator. Add N NaOH (20 ml) and shake for 2 minutes to dissolve the aspirin as its sodium salt. Extract the phenacetin and codeine by shaking gently (to prevent emulsification) with successive portions of chloroform. Extraction is complete when 2 ml of the chloroform fails to yield a residue on evaporation. Wash each chloroform extract with the same 10 ml of Water, and reserve the combined chloroform extracts for determination of phenacetin and codeine.

Combine the alkaline solution and the aqueous wash liquors in a flask, add sodium hydroxide solution and warm (water-bath) to remove dissolved chloroform. Reflux (beware frothing) for 10 minutes to hydrolyse the aspirin to sodium salicylate. Cool, neutralise with dilute hydrochloric acid and adjust to 500 ml. Transfer 50 ml of the solution to a glass-stoppered bottle, add 50 ml of *0.1N* bromine and concentrated hydrochloric acid (5 ml). Shake repeatedly during 15 minutes and set aside for 15 minutes. Add aqueous potassium iodide solution, shake thoroughly to replace excess of bromine by iodine. Wash the stopper and back-titrate the excess of iodine with $0.1N$ $Na_2S_2O_3$.

$$\therefore \quad 180.2 \text{ g } C_9H_8O_4 \equiv 3Br_2 \equiv 60\ 000 \text{ ml } 0.1N$$
$$\therefore \quad 0.003003 \text{ g } C_9H_8O_4 \equiv 1 \text{ ml } 0.1N$$

(b) *Phenacetin.* The first chloroform solutions, obtained as above by extraction of the alkaline solution contain both phenacetin and codeine. Shake out with three successive portions of $0.1N$ H_2SO_4 to remove the codeine. Wash the mixed acid solutions with two 10 ml portions of chloroform and wash each chloroform solution with the same 10 ml of Water. Combine the chloroform solutions, filter into a tared flask and evaporate to dryness. Weigh the dry residue of phenacetin.

(c) *Codeine Phosphate.* Mix the acid extracts and aqueous wash liquors, make alkaline with ammonia to precipitate codeine and extract with successive portions of chloroform (20 ml). Wash each chloroform solution with Water, filter and evaporate to dryness. Proceed as described under Codeine Phosphate Tablets (p. 345) titrating with $0.02N$ solutions, because of the small quantity of alkaloid present.

Further control is exercised over the tablets by limit tests for 4-chloroacetanilide which may be present in the phenacetin (TLC) and for salicylic acid which arises from possible hydrolysis of the aspirin on storage.

Amylobarbitone Sodium Tablets. *Determination of the weight of* $C_{11}H_{17}N_2NaO_3$ *per tablet of average weight*

Method. Dissolve the sample of powdered tablets (equivalent to approximately 0.3 g of Amylobarbitone Sodium) as completely as possible in dilute sodium hydroxide solution. Saturate the solution with sodium chloride and acidify with hydrochloric acid. The latter liberates the amylobarbitone from its sodium salt and the sodium chloride ensures that it is displaced from solution in water. Extract with ether in the usual way, dry the residual amylobarbitone and determine gravimetrically.

Cognate Determinations

Chloroquine Phosphate Tablets
Chloroquine Sulphate Tablets
Emetine and Bismuth Iodide Tablets
Levorphanol Tablets
Methadone Tablets
Pentobarbitone Tablets
Phenobarbitone Tablets
Phenobarbitone Sodium Tablets
Phenytoin Tablets
Quinalbarbitone Tablets
Triprolidine Tablets

Mepyramine Tablets. *Determination of the weight of* $C_{21}H_{27}O_5N_3$ per tablet of average weight

Method. Digest 20 tablets with Water (50 ml) and dilute hydrochloric acid (5 ml) until completely disintegrated and almost completely dissolved. Filter, wash the filter and make up to 200 ml with Water. Pipette a volume of this solution equivalent to 0.5 g of Mepyramine Maleate, make alkaline with sodium hydroxide, and extract with ether in the usual way. Wash the ether, evaporate to dryness and dissolve the residue in *0.1N* HCl. Back-titrate the excess acid with *0.1N* NaOH to methyl red.

Glutethimide. The ether-extracted material is decomposed by hydrolysis with ethanolic potassium hydroxide and the resulting potassium 4-phenyl-hexanoate determined by titration with *0.2N* HCl.

Methadone Tablets. *Determination of the weight of* $C_{21}H_{27}ON,HCl$ *per tablet of average weight*

Method. Dissolve the sample (equivalent to about 0.1 g Methadone Hydrochloride) in Water, make alkaline with sodium hydroxide solution and extract with successive portions (50, 25, 25 and 10 ml) of ether. Wash the combined ethereal solutions with two 10 ml portions of half-saturated brine, and re-extract the aqueous washings with ether. Shake the combined ethereal solutions with *0.02N* HCl (20 ml, pipette) and separate. Wash with two 10 ml portions of Water and combine the aqueous solutions. Warm to remove ether, cool and back-titrate the excess acid with *0.02N* NaOH to methyl red.

Cognate Determination

Chlorpheniramine Tablets. Ether-soluble material is removed in a preliminary extraction from *0.1N* sulphuric acid, before making alkaline with sodium hydroxide and extracting chlorpheniramine base with ether. The base is re-extracted with *0.5N* H_2SO_4 and determined spectrophotometrically.

References

Allen, J., Gartside, B. and Johnson, C. A., *J. Pharm. Pharmac.* (1962) **14**, 73T.

Ensor, P. P., *J. Assoc. Public Analysts* (1972) **10**, 56.

Greenwood, N. D., *Pharm. J.* (1972) **208**, 290.

Greenwood, N. D. and Guppy, I. W., *Pharm. J.* (1974) **212**, 30.

Gupta, V. Das and Euler, K. G., *J. Pharm. Sci.* (1972) **61**, 1458.

Kirk, W. F., *J. Pharm. Sci.* (1972) **61**, 262.

Lunn, A. K. and Morton, R. A. *Analyst* (1952) **77**, 718.

Plank, W. M. *J. Assoc. Official Anal. Chem.* (1972) **55**, 155.

Shah, A. C. and Ochs, J. F. *J. Pharm. Sci.* (1974) **63**, 110.

Somerville, T. G. and Gibson, M., *Pharm. J.* (1973) **211**, 128.

Tingstad, J. E. and Riegelman, S. *J. Pharm. Sci.* (1970) **59**, 692.

Index